METHODS IN MOLECULAR BIOLOGY™

Series Editor
John M. Walker
School of Life Sciences
University of Hertfordshire
Hatfield, Hertfordshire, AL10 9AB, UK

For other titles published in this series, go to
www.springer.com/series/7651

DNA Damage Detection *In Situ*, *Ex Vivo*, and *In Vivo*

Methods and Protocols

Edited by

Vladimir V. Didenko

*Department of Neurosurgery and Department of Molecular & Cellular Biology,
Baylor College of Medicine, Houston, Texas, USA*

💥 Humana Press

Editor
Vladimir V. Didenko, MD, Ph.D.
Department of Neurosurgery
and Department of Molecular & Cellular Biology
Baylor College of Medicine
Houston, Texas
USA
vdidenko@bcm.edu

ISSN 1064-3745   e-ISSN 1940-6029
ISBN 978-1-60327-408-1   e-ISBN 978-1-60327-409-8
DOI 10.1007/978-1-60327-409-8
Springer New York Dordrecht Heidelberg London

Library of Congress Control Number: 2010938360

© Springer Science+Business Media, LLC 2011
All rights reserved. This work may not be translated or copied in whole or in part without the written permission of the publisher (Humana Press, c/o Springer Science+Business Media, LLC, 233 Spring Street, New York, NY 10013, USA), except for brief excerpts in connection with reviews or scholarly analysis. Use in connection with any form of information storage and retrieval, electronic adaptation, computer software, or by similar or dissimilar methodology now known or hereafter developed is forbidden.
The use in this publication of trade names, trademarks, service marks, and similar terms, even if they are not identified as such, is not to be taken as an expression of opinion as to whether or not they are subject to proprietary rights.
While the advice and information in this book are believed to be true and accurate at the date of going to press, neither the authors nor the editors nor the publisher can accept any legal responsibility for any errors or omissions that may be made. The publisher makes no warranty, express or implied, with respect to the material contained herein.

Printed on acid-free paper

Humana Press is part of Springer Science+Business Media (www.springer.com)

Dedication

To my loving mother

Preface

The development of cellular sciences has now entered the stage which requires the evaluation of DNA damage at both the single-cell and the whole organism levels. New approaches are developed to satisfy this need, and the older established techniques were adapted to the task. Advances in organic chemistry, fluorescent microscopy, and materials science have created a whole new range of techniques and probe for imaging DNA damage in molecular and cell biology.

The volume presents all major assays used in molecular and cell biology for the labeling of DNA damage *in situ*, *ex vivo*, and *in vivo*. It brings together recently introduced techniques, as well as those established earlier, which detect and quantify DNA damage at the scales ranging from subcellular to the level of a whole, living organism.

Historically, many techniques which detect DNA damage were originally introduced to solve a utilitarian task of labeling apoptotic cells. These include methods designed to detect specific single- and double-stranded DNA breaks in tissue sections using terminal transferase (TUNEL assay), T4 DNA ligase (ISL assay), T7 DNA polymerase, and many others. These techniques' association with specific cellular process provides an additional bonus of their utilization. Therefore, their application for apoptosis detection is described in the volume. In such cases, detailed analysis from the DNA damage detection's point of view is also provided.

The book does not contain a description of MRI-based or other similar approaches, which use high-end medical diagnostic instrumentation. These belong to a different field due to their specialized nature.

Proper decision-making on what technique to chose requires clear understanding of the terms "*in vivo*, "*ex vivo*," and "*in situ*" in their application to the detection methods.

In vivo means "within the living" or "inside the living body." In application to detection techniques, it refers to measurements done in cells and tissues which remain in a whole, living organism.

Ex vivo translates as "out of the living" or "outside the body." In application of detection techniques, it refers to measurements done in cells or other materials taken from a living organism and performed outside the body. *Ex vivo* measurements are made under conditions impossible in the living organism. The term is used in many different contexts with different emphases. In a more narrow definition, not used here, *ex vivo* indicates a procedure in which cells are taken from a living organism for a treatment, and then put back into the body. Another related term – *in vitro* (i.e., "within a glass," "in a test tube"), is opposite *in vivo*, and indicates a work done in permanent cell cultures, when cells are grown and always remain in the artificial environment.

In situ means "on site" or "in place." The term is not in opposition to either *in vivo* or *ex vivo*. It just denotes the processes detected in their place of origin. In molecular and cell biology, this usually refers to undisrupted mounted cells or tissue sections. In that meaning, "*in situ*" is used as part of the terms "*in situ* PCR," "*in situ* transcription," "*in situ* hybridization," "*in situ* end labeling," and "*in situ* ligation." Sometimes, the "*in *situ*" term is applied at the subcellular level to cells disrupted in the process of analysis, for example, the detection of specific sequences in chromosomes in fluorescent *in situ* hybridization (FISH). Historically, the term was used primarily in methods dealing with nucleic acids. *In situ* methods are advantageous in the analysis of heterogeneous cellular populations. Their attractive features include single-cell detection level, potential to colocalize DNA damage and cellular proteins, the ability to use cellular morphology to verify cellular phenomena, and small sample size. The opposite of *in situ* assays are biochemical techniques, which detect their targets in bulk samples without reference to specific individual cells.

The book is divided into three parts. The first part deals with fixed tissue sections. It contains a complete set of enzymatic approaches to study DNA breaks and apoptosis. The second part describes the detection of DNA damage and apoptosis in cultured cells. It includes instrumental approaches, such as those which use flow- and image cytometry or electrophoresis of agarose-trapped cells. All of the techniques presented in these two parts are the *in situ* approaches. The third part describes methods developed either for *in vivo* detection, directly in a living organism, or for samples taken from the body, *ex vivo* – in blood, urine, and sperm. These are, somewhat, closer to diagnostic procedures. The presented assays often permit monitoring levels of DNA damage and can provide conclusions at the scale of the whole organism. They are either *in situ* assessments or biochemical bulk measurements.

The volume is self-sufficient and easy to understand the source of information needed to reproduce its protocols and interpret their results. Each chapter presents a single protocol or a group of techniques which are related either by their mechanisms or by the molecular targets they detect. This serves to better understand the concepts underlying the methods and the types of DNA damage they label. Such structure also makes it easier to select an approach which better suits the specific needs of a particular study.

The book is equally useful for newcomers to the field and for experienced researchers. A novice molecular biology scientist will use the volume as a guide for selecting and mastering the technique most suitable for his specific needs. To help with this task, the chapters provide detailed, simple-to-follow protocols and describe possible technical pitfalls and limitations.

The rapid growth of specialization makes it increasingly difficult even for the most experienced scientists to follow the never-ending stream of technical innovations which could be beneficial for their research. For the more experienced, the volume serves as a source of novel, recently developed methods and a resource of the new technical possibilities provided by familiar approaches.

It can also help scientists from other more distant fields, such as clinical scientists and nanobiotechnology specialists, who would like to familiarize themselves with research possibilities on DNA damage detection and explore the benefits of this unique technical arsenal.

Researchers in many fields, including molecular and cell biology, experimental and clinical pathology, toxicology, radiobiology, oncology, embryology, experimental pharmacology, drug design, and environmental science, benefit from the book.

I thank Candace Minchew for her highly efficient technical assistance. I am also grateful to Professor John Walker for his help in the review process.

Houston, TX   *Vladimir V. Didenko*

Contents

Preface .... *vii*
Contributors .... *xi*

PART I DETECTION IN TISSUE SECTIONS

1 *In Situ* Detection of Apoptosis by the TUNEL Assay:
   An Overview of Techniques 3
   Deryk T. Loo

2 Combination of TUNEL Assay with Immunohistochemistry for Simultaneous
   Detection of DNA Fragmentation and Oxidative Cell Damage 15
   Alexander E. Kalyuzhny

3 EM-ISEL: A Useful Tool to Visualize DNA Damage
   at the Ultrastructural Level 29
   Antonio Migheli

4 *In Situ* Labeling of DNA Breaks and Apoptosis by T7 DNA Polymerase ... 37
   Vladimir V. Didenko

5 *In Situ* Ligation: A Decade and a Half of Experience 49
   Peter J. Hornsby and Vladimir V. Didenko

6 *In Situ* Ligation Simplified: Using PCR Fragments for Detection
   of Double-Strand DNA Breaks in Tissue Sections 65
   Vladimir V. Didenko

7 5′OH DNA Breaks in Apoptosis and Their Labeling
   by Topoisomerase-Based Approach 77
   Vladimir V. Didenko

PART II DETECTION IN CELL CULTURES

8 Detection of DNA Strand Breaks in Apoptotic Cells by Flow- and
   Image-Cytometry 91
   Zbigniew Darzynkiewicz and Hong Zhao

9 Fluorochrome-Labeled Inhibitors of Caspases: Convenient *In Vitro*
   and *In Vivo* Markers of Apoptotic Cells for Cytometric Analysis 103
   *Zbigniew Darzynkiewicz, Piotr Pozarowski, Brian W. Lee,
   and Gary L. Johnson*

10 Combining Fluorescent *In Situ* Hybridization with the Comet Assay
   for Targeted Examination of DNA Damage and Repair 115
   Sergey Shaposhnikov, Preben D. Thomsen, and Andrew R. Collins

11 Simultaneous Labeling of Single- and Double-Strand DNA Breaks
   by DNA Breakage Detection-FISH (DBD-FISH) 133
   José Luis Fernández, Dioleyda Cajigal, and Jaime Gosálvez

12 Co-localization of DNA Repair Proteins with UV-Induced DNA Damage
    in Locally Irradiated Cells.. 149
    Jennifer Guerrero-Santoro, Arthur S. Levine, and Vesna Rapić-Otrin

PART III DETECTION IN LIVE TISSUES, BLOOD, URINE, SPERM

13 Ultrasound Imaging of Apoptosis: Spectroscopic Detection
    of DNA-Damage Effects at High and Low Frequencies 165
    Roxana M. Vlad, Michael C. Kolios, and Gregory J. Czarnota

14 Quantifying Etheno–DNA Adducts in Human Tissues, White Blood Cells,
    and Urine by Ultrasensitive ^{32}P-Postlabeling and Immunohistochemistry 189
    *Jagadeesan Nair, Urmila J. Nair, Xin Sun, Ying Wang, Khelifa Arab,
    and Helmut Bartsch*

15 ELISpot Assay as a Tool to Study Oxidative Stress in Peripheral
    Blood Mononuclear Cells.. 207
    Jodi Hagen, Jeffrey P. Houchins, and Alexander E. Kalyuzhny

16 Cytokinesis-Block Micronucleus Cytome Assay in Lymphocytes 217
    Philip Thomas and Michael Fenech

17 Buccal Micronucleus Cytome Assay 235
    Philip Thomas and Michael Fenech

18 γ-H2AX Detection in Peripheral Blood Lymphocytes, Splenocytes,
    Bone Marrow, Xenografts, and Skin 249
    *Christophe E. Redon, Asako J. Nakamura, Olivier Sordet,
    Jennifer S. Dickey, Ksenia Gouliaeva, Brian Tabb, Scott Lawrence,
    Robert J. Kinders, William M. Bonner, and Olga A. Sedelnikova*

19 Immunologic Detection of Benzo(a)pyrene–DNA Adducts................... 271
    Regina M. Santella and Yu-Jing Zhang

20 Non-invasive Assessment of Oxidatively Damaged DNA: Liquid
    Chromatography-Tandem Mass Spectrometry Analysis
    of Urinary 8-Oxo-7,8-Dihydro-2′-Deoxyguanosine 279
    *Vilas Mistry, Friederike Teichert, Jatinderpal K. Sandhu, Rajinder Singh,
    Mark D. Evans, Peter B. Farmer, and Marcus S. Cooke*

21 Assessing Sperm DNA Fragmentation with the Sperm Chromatin
    Dispersion Test.. 291
    *José Luis Fernández, Dioleyda Cajigal, Carmen López-Fernández,
    and Jaime Gosálvez*

Erratum E1

Index 303

Contributors

KHELIFA ARAB • *Division of Toxicology and Cancer Risk Factors and Division of Epigenomics and Cancer Risk Factor, German Cancer Research Center (DKFZ), Heidelberg, Germany*

HELMUT BARTSCH • *Division of Toxicology and Cancer Risk Factors, German Cancer Research Center (DKFZ), Heidelberg, Germany*

WILLIAM M. BONNER • *Laboratory of Molecular Pharmacology, Center for Cancer Research, National Cancer Institute, Bethesda, MD, USA*

DIOLEYDA CAJIGAL • *INIBIC-Complexo Hospitalario Universitario A Coruña (CHUAC), La Coruña, Spain*

ANDREW R. COLLINS • *Department of Nutrition, Faculty of Medicine, University of Oslo, Oslo, Norway*

MARCUS S. COOKE • *Department of Cancer Studies and Molecular Medicine, Department of Genetics, Radiation & Oxidative Stress Group, University of Leicester, Leicester, UK*

GREGORY J. CZARNOTA • *Department of Medical Biophysics, Department of Radiation Oncology, Faculty of Medicine, University of Toronto, Toronto, ON, Canada; Department of Radiation Oncology and Imaging Research, Sunnybrook Health Sciences Centre, Toronto, ON, Canada*

ZBIGNIEW DARZYNKIEWICZ • *Department of Pathology, New York Medical College, Brander Cancer Research Institute, Valhalla, NY, USA*

JENNIFER S. DICKEY • *Laboratory of Molecular Pharmacology, Center for Cancer Research, National Cancer Institute, Bethesda, MD, USA*

VLADIMIR V. DIDENKO • *Departments of Neurosurgery and Molecular & Cellular Biology, Baylor College of Medicine, and Michael E. DeBakey VA Medical Center, Houston, TX, USA*

MARK D. EVANS • *Department of Cancer Studies and Molecular Medicine, Radiation & Oxidative Stress Group, University of Leicester, Leicester, UK*

PETER B. FARMER • *Department of Cancer Studies and Molecular Medicine, Cancer Biomarkers and Prevention Group, University of Leicester, Leicester, UK*

MICHAEL FENECH • *CSIRO Human Nutrition, Adelaide, SA, Australia*

JOSE LUIS FERNANDEZ • *INIBIC-Complexo Hospitalario Universitario A Coruña (CHUAC), Centro Oncológico de Galicia, La Coruña, Spain*

JAIME GOSALVEZ • *Facultad de Biología, Universidad Autónoma de Madrid, Madrid, Spain*

KSENIA GOULIAEVA • *Laboratory of Molecular Pharmacology, Center for Cancer Research, National Cancer Institute, Bethesda, MD, USA*

JENNIFER GUERRERO-SANTORO • *Department of Microbiology and Molecular Genetics, Hillman Cancer Center, University of Pittsburgh School of Medicine, and University of Pittsburgh Cancer Institute, Pittsburgh, PA, USA*

JODI HAGEN • *R&D Systems, Inc, Minneapolis, MN, USA*

PETER J. HORNSBY • *Department of Physiology, Sam and Ann Barshop Center for Longevity and Aging Studies, University of Texas Health Science Center, San Antonio, TX, USA*

JEFFREY P. HOUCHINS • *R&D Systems, Inc, Minneapolis, MN, USA*

GARY L. JOHNSON • *Immunochemistry Technologies, Bloomington, MN, USA*

ALEXANDER E. KALYUZHNY • *R&D Systems, Inc, Minneapolis, MN, USA*

ROBERT J. KINDERS • *Pharmacodynamics Assay Development and Implementation Section (PADIS), Laboratory of Human Toxicology and Pharmacology, SAIC-Frederick, Inc., NCI-Frederick, Frederick, MD, USA*

MICHAEL C. KOLIOS • *Department of Medical Biophysics, University of Toronto, Toronto, ON, Canada; Department of Physics, Ryerson University, Toronto, ON, Canada*

SCOTT LAWRENCE • *Pathology and Histology Laboratory (PHL), SAIC-Frederick, Inc., NCI-Frederick, Frederick, MD, USA*

BRIAN W. LEE • *Immunochemistry Technologies, Bloomington, MN, USA*

ARTHUR S. LEVINE • *Department of Microbiology and Molecular Genetics, Hillman Cancer Center, University of Pittsburgh School of Medicine, and University of Pittsburgh Cancer Institute, Pittsburgh, PA, USA*

DERYK T. LOO • *MacroGenics, Inc, South San Francisco, CA, USA*

CARMEN LÓPEZ-FERNÁNDEZ • *Facultad de Biología, Universidad Autónoma de Madrid, Madrid, Spain*

ANTONIO MIGHELI • *Centro Regionale Diagnosi ed Osservazione delle Malattie Prioniche DOMP-ASL TO2, Turin, Italy*

VILAS MISTRY • *Department of Cancer Studies and Molecular Medicine, Radiation & Oxidative Stress Group, University of Leicester, Leicester, UK*

JAGADEESAN NAIR • *Division of Toxicology and Cancer Risk Factors, German Cancer Research Center (DKFZ), Heidelberg, Germany*

URMILA J. NAIR • *Division of Toxicology and Cancer Risk Factors, German Cancer Research Center (DKFZ), Heidelberg, Germany*

ASAKO J. NAKAMURA • *Laboratory of Molecular Pharmacology, Center for Cancer Research, National Cancer Institute, Bethesda, MD, USA*

PIOTR POZAROWSKI • *Brander Cancer Research Institute, New York Medical College, Valhalla, NY, USA; Department of Clinical Immunology, Medical University, Lublin, Poland*

VESNA RAPIĆ-OTRIN • *Department of Microbiology and Molecular Genetics, Hillman Cancer Center, University of Pittsburgh School of Medicine, and University of Pittsburgh Cancer Institute, Pittsburgh, PA, USA*

CHRISTOPHE E. REDON • *Laboratory of Molecular Pharmacology, Center for Cancer Research, National Cancer Institute, Bethesda, MD, USA*

JATINDERPAL. K. SANDHU • *Department of Cancer Studies and Molecular Medicine, Cancer Biomarkers and Prevention Group, University of Leicester, Leicester, UK*

REGINA M. SANTELLA • *Department of Environmental Health Sciences, Mailman School of Public Health, Columbia University, New York, NY, USA*

OLGA A. SEDELNIKOVA • *Laboratory of Molecular Pharmacology, Center for Cancer Research, National Cancer Institute, Bethesda, MD, USA*

SERGEY SHAPOSHNIKOV • *Department of Nutrition, Faculty of Medicine, University of Oslo, Oslo, Norway*

RAJINDER SINGH • *Department of Cancer Studies and Molecular Medicine, Cancer Biomarkers and Prevention Group, University of Leicester, Leicester, UK*

OLIVIER SORDET • *Laboratory of Molecular Pharmacology, Center for Cancer Research, National Cancer Institute, Bethesda, MD, USA*

XIN SUN • *National Center for Chronic and Non-communicable Disease Control and Prevention, Chinese Center for Disease Control and Prevention, Beijing, China*

BRIAN TABB • *Pathology and Histology Laboratory (PHL), SAIC-Frederick, Inc., NCI-Frederick, FrederickMD, USA*

FRIEDERIKE TEICHERT • *Department of Cancer Studies and Molecular Medicine, Cancer Biomarkers and Prevention Group, University of Leicester, Leicester, UK*

PHILIP THOMAS • *CSIRO Human Nutrition, Adelaide, SA, Australia*

PREBEN D. THOMSEN • *Faculty of Life Science, University of Copenhagen, Frederiksberg, Denmark*

ROXANA M. VLAD • *Department of Medical Biophysics, Radiation Medicine Program, Princess Margaret Hospital, University of Toronto, Toronto, ON, Canada; Department of Radiation Oncology and Imaging Research, Sunnybrook Health Sciences Centre, Toronto, ON, Canada*

YING WANG • *Tongji Hospital, Tongji Medical College, Huazhong Science & Technology University, Wuhan, China*

YU-JING ZHANG • *Department of Environmental Health Sciences, Mailman School of Public Health, Columbia University, New York, NY, USA*

HONG ZHAO • *Department of Pathology, New York Medical College, Brander Cancer Research Institute, Valhalla, NY, USA*

Part I

Detection in Tissue Sections

Chapter 1

In Situ Detection of Apoptosis by the TUNEL Assay: An Overview of Techniques

Deryk T. Loo

Abstract

Apoptosis, or programmed cell death, plays an important role in normal development and homeostasis of adult tissues. Apoptosis has also been linked to many disease states, including cancer. One of the biochemical hallmarks of apoptosis is the generation of free 3′-hydroxyl termini on DNA via cleavage of chromatin into single and multiple oligonuleosome-length fragments. The TdT-mediated dUTP-biotin nick end labeling (TUNEL) assay exploits this biochemical hallmark by labeling the exposed termini of DNA, thereby enabling visualization of nuclei containing fragmented DNA. This review outlines the general method for *in situ* TUNEL staining of cultured cells and tissue sections, and highlights recent improvements in the technique and limitations of the assay.

Key words: Apoptosis, Necrosis, TUNEL, DNA fragmentation

1. Introduction

Apoptosis, or programmed cell death, is a naturally occurring process that plays a critical role in the development and tissue homeostasis of adult tissues (1, 2). Apoptosis also plays a role in many disease states, including cancer, and components in apoptotic pathways are being pursued as targets for therapy (3–6). Therefore, the ability to identify and quantify apoptosis is critical for advancing these areas of investigation. One of the hallmarks of apoptosis is internucleosomal cleavage of genomic DNA into small fragments – typified by the laddering seen when the DNA is fractionated on agarose gels (7–9). In the late 1992 and early 1993, two laboratories described the development of the TdT-mediated dUTP-biotin nick end labeling (TUNEL) assay, which exploits this biochemical hallmark and enables visualization of nuclei containing fragmented DNA (10, 11). TUNEL staining utilizes the ability of the enzyme terminal deoxynucleotidyl transferase (TdT) to incorporate labeled dUTP onto the free 3′-hydroxyl termini of fragmented genomic DNA. Although TUNEL staining has been adopted as the method of choice for detecting apoptosis *in situ*, it is important to recognize that TUNEL staining is not limited to the detection of apoptotic cells (12, 13). Because TUNEL staining is nonspecific in the sense that the assay will label all free 3′-hydroxyl termini, irrespective of the molecular mechanisms that led to the development of these termini, TUNEL staining will also detect non-apoptotic cells – including necrotic degenerating cells (14, 15), cells undergoing DNA repair (16), cells damaged by mechanical forces (17, 18), and even cells undergoing active gene transcription (19). Therefore, TUNEL staining should be considered generally as a method for the detection of DNA damage (DNA fragmentation or others), and when used in conjunction with secondary apoptosis-specific assays, more specifically as a method for identifying apoptotic cells.

The goal of this chapter is to provide the reader with a general overview of the technique and step-by-step protocols for *in situ* TUNEL staining of fragmented nuclear DNA in both cultured cells and tissue sections. As noted above, the TUNEL assay was developed more than 15 years ago, and although there have been improvements reported for specific systems, the general assay method has changed very little. Thus, the protocols outlined in this chapter will have much in common with other published TUNEL methods. Two areas where there have been significant developments over the past several years are (1) the development of improved chemical labels for detection and (2) the accumulation of a richer pool of data examining the potential artifactual TUNEL staining of non-apoptotic cells under certain circumstances. Discussion of these two areas of TUNEL assay development is expanded upon in the Notes section at the end of this review.

2. Materials

2.1. Cultured Cells

1. Phosphate-buffered saline (PBS), pH 7.4.
2. 2% Buffered formaldehyde: dilute high-quality formaldehyde (v/v) in PBS prior to use.
3. 70% Ethanol.
4. TdT equilibrium buffer: 2.5 mM Tris–HCl (pH 6.6), 0.2 M potassium cacodylate, 2.5 mM $CoCl_2$, 0.25 mg/mL bovine serum albumin (BSA). Aliquots may be stored at −20°C for several months.

5. TdT reaction buffer: TdT equilibrium buffer containing 0.5 U/μL of TdT enzyme and 40 pmol/μL of biotinylated-dUTP (Roche Diagnostics Corp., Indianapolis, IN). Prepare fresh from stock solutions prior to use.

6. TdT staining buffer: 4× saline–sodium citrate (0.6 M NaCl, 60 mM sodium citrate), 2.5 μg/mL fluorescein isothiocyanate-conjugated avidin (Amersham Biosciences; Piscataway, NJ), 0.1% Triton X-100, and 1% BSA. Prepare fresh from stock solutions prior to use.

7. Hoechst 33342 counterstain: 2 μg/mL in PBS (Invitrogen/Molecular Probes; Eugene, OR). Stock solution may be stored at 4°C in the dark for several weeks.

8. Vectashield antifade mounting medium (Vector Laboratories; Burlingame, CA).

2.2. Tissue Sections

1. PBS, pH 7.4.
2. 4% Buffered formaldehyde: dilute high-quality formaldehyde (v/v) in PBS prior to use.
3. 20 μg/mL proteinase K (Roche Diagnostics Corp.). Stock solution may be stored at −20°C for several months.
4. 95, 90, 80, and 70% Ethanol in Coplin jars.
5. 2% Hydrogen peroxide. Prepare fresh from hydrogen peroxide reagent stock prior to use.
6. 2% BSA solution: 2% BSA (w/v) dissolved in PBS and passed through a 0.45-μm filter. Sterile stock solution may be stored at 4°C for several weeks.
7. 2× SSC buffer: 300 mM NaCl, 30 mM sodium citrate. Stock solution may be stored at room temperature for several months.
8. TdT equilibrium buffer: 2.5 mM Tris–HCl (pH 6.6), 0.2 M potassium cacodylate, 2.5 mM $CoCl_2$, 0.25 mg/mL BSA. Aliquots may be stored at −20°C for several months.
9. TdT reaction buffer: TdT equilibrium buffer containing 0.5 U/μL of TdT enzyme and 40 pmol/μL of biotinylated-dUTP (Roche Diagnostics Corp.). Prepare fresh from stock solutions prior to use.
10. Vectastain ABC-peroxidase stock solution (Vector Laboratories).
11. 3,3′-Diaminobenzidine (DAB) staining solution (Vector Laboratories).
12. TdT staining buffer: 4× saline–sodium citrate (0.6 M NaCl, 60 mM sodium citrate), 2.5 μg/mL fluorescein isothiocyanate-conjugated avidin (Amersham Biosciences), 0.1% Triton X-100, and 1% BSA. Prepare fresh from stock solutions prior to use.
13. Hematoxylin counterstain (Sigma-Aldrich; St. Louis, MO).

14. Hoechst 33342 counterstain: 2 µg/mL in PBS (Invitrogen/Molecular Probes). Stock solution may be stored at 4°C in the dark for several weeks.
15. Aqua-Poly/Mount mounting medium (Polysciences, Inc., Warrington, PA).

3. Methods (see Notes 1–4)

A flowchart of the general protocol for TUNEL staining of cells and tissues is shown in Fig. 1. Cells or tissues are fixed with formaldehyde and then permeabilized with ethanol to allow penetration of the TUNEL reaction reagents into the cell nucleus. Following fixation and washing, incorporation of biotinylated-dUTP onto the 3′ ends of fragmented DNA is carried out in the reaction containing TdT enzyme. Depending on the specific needs of the investigator and/or available equipment, the incorporated biotinylated-dUTP may be visualized by (1) fluorescence microscopy or flow cytometry following staining with fluorescent-tagged avidin (see Note 5) or (2) light microscopy following staining with horseradish peroxidase-conjugated avidin–biotin complex in conjunction with a colorimetric substrate. Additionally, TUNEL may be combined with immunohistochemical staining in order to simultaneously label expressed antigens on cells within the TUNEL-stained sample (see Note 6).

3.1. Cultured Cells

3.1.1. Suspension Cells

1. Collect cells by centrifugation ($\sim 160 \times g$), wash with PBS, and resuspend cells at a concentration of $1–2 \times 10^7$/mL in PBS. Transfer 100 µL of cell suspension to a V-bottomed 96-well plate.
2. Fix cells by addition of 100 µL of 2% formaldehyde in PBS, pH 7.4 (see Note 7). Incubate on ice for 15 min.
3. Collect cells by centrifugation, wash once with 200 µL of PBS, and then postfix/permeabilize with 200 µL of 70% ice-cold ethanol. The cells may be stored in 70% ethanol at –20°C for several days.
4. Collect the cells by centrifugation and wash twice with 200 µL PBS.
5. Resuspend the cells ($1–5 \times 10^5$) in 50 µL of TdT equilibration buffer. Incubate the cell suspension at 37°C for 10 min with occasional gentle mixing.
6. Resuspend the cells in 50 µL of TdT reaction buffer. Incubate the cell suspension at 37°C for 30 min with occasional gentle mixing.
7. Collect the cells by centrifugation and wash with 200 µL PBS.

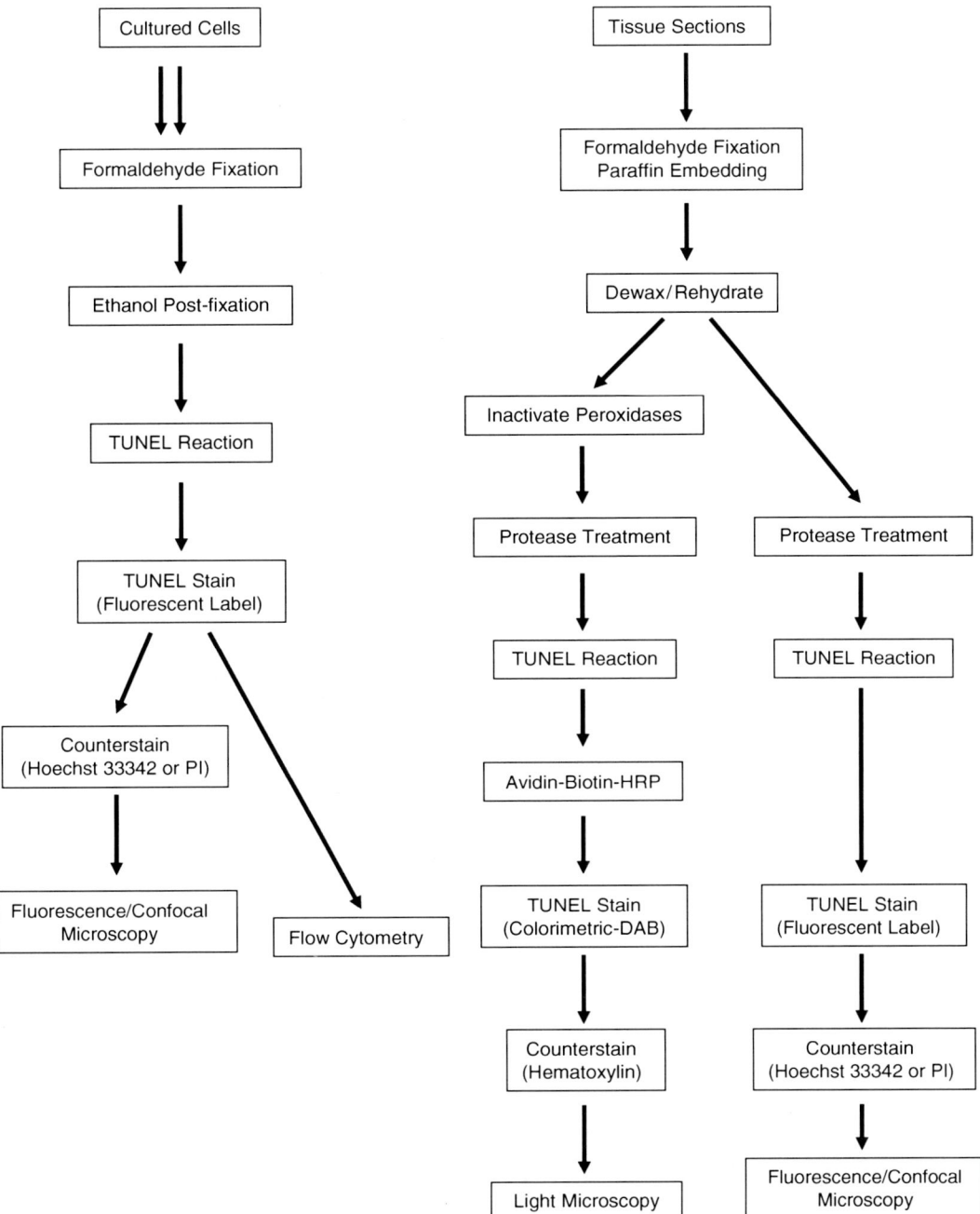

Fig. 1. General flow chart outlining the TUNEL assay protocol steps described in this chapter for staining cultured cells and tissue sections. Reprinted with Permission from Humana Press, (29).

8. Resuspend the cells in 100 µL of TdT staining buffer. Incubate the cell suspension at room temperature for 30 min in the dark.
9. Collect the cells by centrifugation, wash twice with 200 µL PBS, and then resuspend in PBS at $2–8 \times 10^6$/mL. For fluorescence

microscopy, attach coverslips using Vectashield antifade mounting medium.

10. Examine the cells by fluorescence microscopy, confocal microscopy, or flow cytometry.

3.1.2. Cytospin Preparation of Suspension Cells

TUNEL staining and subsequent fluorescent microscopic or confocal examination of suspension cells may be conveniently carried out on cells attached to glass slides. The cell suspension protocol may be easily modified to accommodate cytospin samples, as described below.

3.1.3. Adherent Cells

1. Collect cells by centrifugation, wash with PBS, and then collect on glass slides pretreated with aqueous 0.01% poly-l-lysine using a cytospin device. Routinely $1–5 \times 10^5$ cells are collected on a single slide.
2. Fix the cells by covering with a puddle of 1% formaldehyde in PBS for 15 min.
3. Rinse the slides with PBS, then transfer to a Coplin jar containing ice-cold 70% ethanol for 1 h. The slides may be stored overnight in 70% ethanol at 4°C.
4. Rinse the slides with PBS and pipet 25–50 µL of TdT buffer onto the slides, enough to cover the cells. Incubate the slides in a humidified chamber for 30 min at 37°C. In order to conserve reagents, a reduced volume of TdT buffer may be applied and then carefully covered with a glass coverslip during the incubation. Take care to avoid trapping air bubbles which may lead to staining artifacts.
5. Rinse the slides with PBS, then pipet 25–50 µL of TdT staining buffer onto the slides. Incubate for 30 min at room temperature in the dark.
6. Rinse the slides with PBS, air dry, and attach coverslips using Vectashield antifade mounting medium.
7. Examine the cells by fluorescence or confocal microscopy.

3.2. Tissue Sections

3.2.1. Colorimetric Staining for Light Microscopic Examination

1. Fix tissue samples in 4% formaldehyde in PBS for 24 h and embed in paraffin. Adhere 4–6-µm paraffin sections to glass slides pretreated with 0.01% aqueous solution of poly-l-lysine.
2. Deparaffinize sections by heating the slides for 30 min at 60°C (or 10 min at 70°C), followed by two 5-min incubations in a xylene bath at room temperature in Coplin jars. Rehydrate the tissue samples by transferring the slides through a graded ethanol series: 2×3 min 90% ethanol, 1×3 min 80% ethanol, 1×3 min 70% ethanol, and 1×3 min double-distilled water (DDW).

3. Carefully blot away excess water and pipet 20 μg/mL of proteinase K solution to cover the sections. Incubate for 15 min at room temperature.
4. Following proteinase K treatment, wash the slides 3×5 min with DDW.
5. Inactivate endogenous peroxidases by covering the sections with 2% hydrogen peroxide for 5 min at room temperature. Wash the slides 3×5 min with DDW.
6. Carefully blot away excess water, then cover the sections with TdT reaction buffer. Incubate the slides in a humidified chamber for 30 min at 37°C. In order to conserve reagents, a reduced volume of TdT buffer may be applied and then carefully covered with a glass coverslip during the incubation. Take care to avoid trapping air bubbles which may lead to staining artifacts.
7. Stop the reaction by incubating the slides 2×10 min in 2×SSC.
8. Rinse the slides in PBS, then block nonspecific binding by covering the tissue sections with 2% BSA solution for 30–60 min at room temperature.
9. Wash the slides 2×5 min in PBS, then incubate in Vectastain ABC-peroxidase solution for 1 h at 37°C.
10. Wash the slides 2×5 min in PBS, then stain with DAB staining solution at room temperature. Monitor color development until desired level of staining is achieved (typically 10–60 min). Stop the reaction by incubating the slides in DDW.
11. Lightly counter-stain the tissue sections with hematoxylin.
12. Cover the tissue sections with coverslips using Aqua-Poly/Mount mounting medium.
13. Observe the sections under light microscopy.

3.2.2. Fluorescent Staining

1. Follow steps 1–9 outlined above for the colorimetric staining of tissue sections (Subheading 3.2.1), omitting the hydrogen peroxide inactivation step.
2. Wash the slides 2×5 min in PBS, then cover the tissue sections with TdT staining buffer. Incubate the slides at room temperature for 30 min in the dark.
3. Wash the slides 2×5 min in PBS.
4. Lightly counterstain the sections with hematoxylin, Hoechst 33342, or other appropriate counterstain (see Note 8).
5. Wash the slides with PBS, air dry, and attach coverslips using Vectashield antifade mounting medium.
6. Examine the tissue sections by fluorescence or confocal microscopy.

4. Notes

1. The protocols outlined in this chapter represent a general method, along with several variations, that has been used successfully for TUNEL staining of a broad variety of cultured cells and tissues. However, specific cell types and tissues may require modification or optimization of the staining conditions to obtain successful results. Two of the steps that often require optimization are formaldehyde fixation and proteinase K treatment. A lack of observed TUNEL staining in the positive control sample may result from over-fixation (20). This result may be remedied by using shorter fixation times and/or reducing the formaldehyde concentration. Additionally, over-treatment with proteinase K may cause loss of TUNEL staining in the positive control sample. Positive staining in the negative control sample may also result from inadequate inactivation of endogenous peroxidase activity (colorimetric detection), or nonspecific binding of the fluorescent-conjugated avidin reagent (fluorescent detection). In the case of fluorescent detection, increasing the BSA concentration to 2–5% in the TdT staining buffer and/or reducing the fluorescent-conjugated avidin concentration may reduce the background fluorescence. It is important to optimize the staining conditions empirically with the control samples prior to examining and interpreting data from experimental samples.

2. While this chapter provides the investigator with a step-by-step method for TUNEL staining that requires a modest amount of reagents, ready-to-use TUNEL staining kits are commercially available. Listed below are kits from four manufacturers that are supplied in a variety of formats to allow flexibility with the method of visualization.

    (a) *In Situ* Cell Death Detection Kits – Roche Diagonostics Corp., Indianapolis, IN. Available with directly conjugated fluorescein-dUTP or tetramethylrhodamine-dUTP for use with flow cytometry or fluorescence microscopy. The kits are also available with anti-fluorescein antibody phosphatase conjugates for use with light microscopy.

    (b) APO-BrdU TUNEL Assay Kit – Invitrogen/Molecular Probes, Inc., Eugene, OR. Available with 5-Bromo-dUTP (BrdUTP) and anti-BrdU antibody conjugated to Alexa Fluor 488 for use with flow cytometry or fluorescence microscopy. The Alexa Fluor 488 dye is brighter and more stable than commonly used fluorescein conjugates. An anti-BrdU antibody phosphatase conjugate could also be used in conjunction with light microscopy. The kit contains propidium iodide to determine total cellular DNA content.

(c) ApoTarget APO-BRDU kit – Biosource International, Inc., Camarillo, CA. Available with BrdUTP and anti-BrdU antibody conjugated to a fluorescein dye. The kit contains propidium iodide to determine total cellular DNA content.

(d) ApopTag kit – Millipore Corporation, Billerica, MA. Available in an indirect format utilizing digoxigenin-dUTP with an anti-digoxigenin antibody conjugated to either a rhodamine fluorochrome or peroxidase enzyme, and in a direct format utilizing fluorescein-dUTP. A TRITC-conjugated anti-BrdU antibody-based kit optimized for Guava instruments is also available from this vendor.

3. An increasing body of literature identifying artifactual false-positive TUNEL staining highlights the importance of confirming apoptotic cell death via additional methods (12, 13, 21). Artifacts have been attributed to physical treatment – proteinase K digestion has been reported to expose preexisting free DNA ends in fixed non-apoptotic cells for TdT labeling (22), fixation and tissue handling techniques can alter the outcome of TUNEL staining (13, 18), and histological sectioning through nuclei can cause DNA strand breaks in non-apoptotic cells and lead to TUNEL reactivity (17, 18). Artifacts have also been linked to normal physiological processes: In addition to the commonly observed TUNEL reactivity of nuclei from degrading necrotic cells (14), TUNEL reactivity has also been observed in cells undergoing active gene transcription (19) and cells undergoing the process of DNA repair (16). Thus, it is important to confirm results from the TUNEL assay with other detection methods (nuclear morphology, internucleosomal DNA laddering, Annexin V staining, and caspase activation) in order to be confident that the TUNEL reactivity is identifying events.

4. It is critical to include control samples with each staining experiment to facilitate interpretation of the staining results. As a positive control, treat cells and tissues with DNase I (1 μg/mL in 30 mM Tris–HCl (pH 7.2), 140 mM potassium cacodylate, 4 mM $MgCl_2$, and 0.1 mM DTT) for 10 min at room temperature. Following DNase I treatment, wash the samples 3×2 min in DDW, then proceed with TUNEL staining. As a negative control, omit the TdT enzyme from the TdT reaction buffer.

5. In addition to the commonly used fluorochromes described in this review, a new generation of fluorochromes in a range of emission wavelengths, which provide high sensitivity and photostability, is available in TUNEL assay-ready formats. The Alexa fluor™ series of fluorochromes are available conjugated to an anti-BrdU antibody or streptavidin (Invitrogen, Carlsbad, CA). The Cy™ series of fluorochromes are available conjugated to streptavidin or dUTP (Amersham, Piscataway, NJ).

6. Immunohistochemical staining of cultured cells or tissue sections for cell surface or intracellular antigens may be performed simultaneously with TUNEL using colorimetric, fluorescent, or a combination of colorimetric and fluorescent detection systems (23–27). Additionally, TUNEL may be combined with nuclear dyes, including propidium iodide and 4′,6-diamidino-2-phenylindole, in order to correlate apoptosis with cell cycle parameters (28). Multi-stained cells and tissue sections can be analyzed by microscopy and, in case of suspension cell cultures, by flow cytometry.

7. Formaldehyde-fixed cell culture samples have been successfully stored in 70% ethanol at −20°C for several weeks prior to TUNEL staining. The suitability of prolonged storage should be determined empirically for the individual culture system employed.

8. Hoechst 33342 binds to DNA and serves as a nuclear stain. Combining Hoechst 33342 staining with TUNEL staining allows one to compare TUNEL-positive nuclei with surrounding normal nuclei and observe changes in nuclear size and morphology. Hoechst 33342 also serves as a counterstain, enabling visualization of anatomical structures in both TUNEL-positive and TUNEL-negative cells in cultured cells and tissues (20).

References

1. Kerr, J.F.R., Wyllie, A.H., and Currie, A.R. (1972) Apoptosis: A basic biological phenomenon with wide-ranging implications in tissue kinetics *Br J Cancer* **26**, 239–57.

2. Wyllie, A.H. (1980) Cell death: The significance of apoptosis, in *Int. Rev. Cytol*, (Bourne, G.H., Danielli, F.J., and Jeon, K.W., eds.) Academic Press, New York, NY, vol. **68**, pp. 251–306.

3. Bremer, E., van Dam, G., Kroesen, B.J., de Leij, L., and Helfrich, W. (2006) Targeted induction of apoptosis for cancer therapy: Current progress and prospects *Trends Mol Med* **12**, 382–93.

4. Ashkenazi, A. (2008) Targeting the extrinsic apoptosis pathway in cancer *Cytokine Growth Factor Rev* **19**, 325–31.

5. Plummer, R., Attard, G., Pacey, S., Li, L., Razak, A., Perrett, R., Barrett, M., Judson, I., Kaye, S., Fox, N.L., Halpern, W., Corey, A., Calvert, H., and de Bono, J. (2007) Phase I and pharmacokinetic study of lexatumumab in patients with advanced cancers *Clin Cancer Res* **13**, 6187–94.

6. Hotte, S.J., Hirte, H.W., Chen, E.X., Siu, L.L., Le, L.H., Corey, A., Iacobucci, A., MacLean, M., Lo, L., Fox., N.L., and Oza, A.M. (2008) A Phase I study of mapatumumab (fully human monoclonal antibody to TRAIL-R1) in patients with advanced solid malignancies *Clin Cancer Res* **14**, 3450–55.

7. Arends, M.J., Morris, R.G., and Wyllie, A.H. (1990) Apoptosis: the role of the endonuclease *Am J Pathol* **136**, 593–608.

8. Bortner, C.D., Oldenburg, N.B.E., and Cidlowski, J.A. (1995) The role of DNA fragmentation in apoptosis *Trends Cell Biol* **5**, 21–6.

9. Loo, D.T. and Rillema, J.R. (1998) Measurement of cell death, in *Methods in Cell Biol* (Mather, J.P. and Barnes, D., eds.), Academic Press, San Diego, CA, vol. 57, pp. 251–64.

10. Gavrieli, Y., Sherman, Y., and Ben-Sannon, S.A. (1992) Identification of programmed cell death in situ via specific labeling of nuclear DNA fragmentation *J Cell Biol* **119**, 493–501.

11. Gorczyca, W., Gong, J., and Darzynkiewicz, Z. (1993) Detection of DNA strand breaks in individual apoptotic cells by the in situ terminal deoxynucleotidyl transferase and nick translation assays *Cancer Res* **53**, 1945–51.
12. Watanabe, M., Hitomi, M., van der Wee, K., Rothenberg, F., Fisher, S.A., Zucker, R., Svoboda, K.K.H., Goldsmith, E.C., Heiskanen, K.M., and Nieminen, A.-L. (2002) The pros and cons of apoptosis assays for use in the study of cells, tissues, and organs *Microsc Microanal* **8**, 375–91.
13. Huerta, S., Goulet, E.J., Huerta-Yepez, S, and Livingston, E.H. (2007) Screening and detection of apoptosis *J Surg Res* **139**, 143–56.
14. Ansari, B., Coates, P.J., Greenstein, B.D., and Hall, P.A. (1993) In situ end-labeling detects DNA strand breaks in apoptosis and other physiological and pathological states *J Pathol* **170**, 1–8.
15. Didenko, V.V. and Hornsby, P.J. (1996) Presence of double-strand breaks with single-base 3' overhangs in cells undergoing apoptosis but not necrosis *J Cell Biol* **135**, 1369–76.
16. Kanoh, M., Takemura, G., Misao, J., Hayakawa, Y., Aoyama, T., Nishigaki, K., Noda, T., Fujiwara, T., Fukuda, K., Minatoguchi, S., and Fujiwara, H. (1999) Significance of myocytes with positive DNA in situ nick end-labeling (TUNEL) in hearts with dilated cardiomyopathy: not apoptosis but DNA repair *Circulation* **99**, 2757–64.
17. Sloop, G.D., Roa, J.C., Delgado, A.G., Balart, J.T., Hines III, M.O., and Hill, J.M. (1999) Histologic sectioning produces TUNEL reactivity: a potential cause of false-positive staining *Arch Pathol Lab Med* **123**, 529–32.
18. Jerome, K.R., Vallan, C., and Jaggi, R. (2000) The TUNEL assay in the diagnosis of graft-versus-host disease: caveats for interpretation *Pathology* **32**, 186–90.
19. Kockx, M.M., Muhring, J., Knaapen, M.W., and de Meyer, G.R. (1998) RNA synthesis and splicing interferes with DNA in situ end labeling techniques used to detect apoptosis *Am J Pathol* **152**, 885–88.
20. Whiteside, G., Cougnon, N., Hunt, S.P., and Munglani, R. (1998) An improved method for detection of apoptosis in tissue sections and cell culture, using the TUNEL technique combined with Hoechst stain *Brain Res Prot* **2**, 160–4.
21. Hughes, S.E. (2003) Detection of apoptosis using in situ markers for DNA strand breaks in the failing human heart. Fact or epiphenomenon? *J Pathol* **201**, 181–86.
22. Gal, I., Varga, T., Szilagyi, I., Balazs, M., Schlammadinger, J., and Szabo Jr., G. (2000) Protease-elicited TUNEL positivity of non-apoptotic fixed cells *J Histochem Cytochem* **48**, 963–69.
23. Kishimoto, H., Surh, C.D., and Sprent, J. (1995) Upregulation of surface markers on dying thymocytes *J Exp Med* **181**, 649–55.
24. Christina, M., Angelika, H.L., Bernd, P., and Martina, P. (2006) Simultaneous detection of a cell surface antigen and apoptosis by microwave-sensitized TUNEL assay on paraffin sections *J Immunol Methods* **316**, 163–6.
25. Furukawa, H., Oshima, K., Tung, T., Cui, G., Laks, H., and Sen, L. (2008) Overexpressed exogenous IL-4 and IL-10 paradoxically regulate allogeneic T-cell and cardiac myocytes apoptosis through FAS/FASL pathway *Transplantation* **85**, 437–46.
26. Kuge, Y., Kume, N., Ishino, S., Takai, N., Ogawa, Y., Mukai, T., Minami, M., Shiomi, M., and Saji, H. (2008) Prominent lectin-like oxidized low density lipoprotein (LDL) receptor-1 (LOX-1) expression in atherosclerotic lesions is associated with tissue factor expression and apoptosis in hypercholesterolemic rabbits *Biol Pharm Bull* **31**, 1475–82.
27. Hewitson, T.D., Bisucci, T., and Darby, I.A. (2006) Histochemical localization of apoptosis with in situ labeling of fragmented DNA *Methods Mol Biol* **326**, 227–34.
28. Darzynkiewicz, Z., Galkowski, D., and Zhao, H. (2008) Analysis of apoptosis by cytometry using TUNEL assay *Methods* **44**, 250–4.
29. Loo, D. T. 2002. TUNEL assay. An overview of techniques, in: In Situ *Detection of DNA Damage, Methods in Molecular Biol.*, Humana Press, ed. V. Didenko, *Vol.* **203**:*pp. 21–30, Figure 1.*

Chapter 2

Combination of TUNEL Assay with Immunohistochemistry for Simultaneous Detection of DNA Fragmentation and Oxidative Cell Damage

Alexander E. Kalyuzhny

Abstract

Oxidative cell damage causes disruption of DNA via formation of 8-hydroxy-2′-deoxyguanosine and can trigger apoptotic cell death. The cells damaged by oxidative stress can either become apoptotic, or recover. Therefore, it is helpful to employ a parallel assay that would confirm whether cells experiencing oxidative damage undergo apoptosis. Our paper describes the technique that combines immunohistochemical detection of 8-hydroxy-2′-deoxyguanosine with the TUNEL assay. This permits simultaneous detection of oxidative damage and apoptosis at a single-cell level. We have developed simple and reliable protocols which can be used with cultured cells and slide-mounted tissue sections. These techniques can be employed in research dealing with high-throughput drug screening, toxicology, and cancer.

Key words: Oxidative stress, 8-hydroxy-2′-deoxyguanosine, Immunohistochemistry, TUNEL assay, Apoptosis, Cultured cells, Tissue sections, Double-labeling

1. Introduction

The TUNEL assay is a very powerful technique for the detection of DNA fragmentation in cells. Based on terminal deoxynucleotidyl transferase (TdT)-mediated nick end-labeling, TUNEL allows for *in situ* detection of DNA breaks in both apoptotic and necrotic cells (1, 2). That's why the TUNEL assay is widely used to study cellular mechanisms underlying embryonic development and morphogenic redistribution of cells (3), aging (2, 4), tumorigenesis (5, 6), and neurodegenerative processes (7). Although TUNEL detects DNA fragmentation, which appears to represent the end-point of apoptotic DNA degradation, it cannot detect the causes of such degradation which can be, for example, an oxidative stress induced by reactive oxygen species (ROS) and oxygen radicals. Oxidative damage has been implicated in many neurodegenerative disorders including Alzheimer's disease, amyotrophic lateral sclerosis, Huntington's disease, and Parkinson's disease (8–11). Oxidative stress leads to the formation of 8-hydroxy-2′-deoxyguanosine (8-OHdG) which, in turn, causes disruptive DNA modifications. Immunohistochemical detection of 8-OHdG appears to be a valuable technique, allowing for identification of possible oxidative DNA damage induced by ROS. However, measuring the extent of oxidative damage alone has a limited value because it does not answer the question whether it triggers apoptotic DNA damage or not.

Since apoptotic DNA fragmentation is preceded by oxidative DNA damage, the sequential application of each technique allows quantification of cells during the early phase (8-OHdG-positive) and late phase (TUNEL-positive) of apoptosis. Because oxidative damage may not result in cell death, measuring the number of TUNEL- and 8-OHdG-positive cells helps to determine a threshold of cellular sensitivity to oxidative stimuli at which DNA oxidation results in DNA damage.

Protocols that combine the TUNEL assay and immunohistochemical labeling for 8-OHdG can be used with cultured cells and with sections. Briefly, after doing immunohistochemistry for 8-OHdG labeling, the same specimens are subjected to the TUNEL assay. This technique is easy to perform and it can be customized for a large variety of cells and tissues.

2. Materials

1. Dissociated (primary culture) rat dorsal root ganglia (DRG) neurons from 3-day-old Sprague-Dawley rat pups.
2. DRG dissociation solution: 10 mg/mL of collagenase/dispase (Roche Molecular Biochemicals, Indianapolis, IN) in Hank's Balanced Salt Solution (HBSS; Gibco BRL, Grand Island, NY).
3. Culture media for DRG neurons (all reagents except NGF are from Sigma–Aldrich, St. Louis, MO): Ham's medium (F-12; Gibco BRL, Grand Island, NY) supplemented with 5% heat-inactivated horse serum and 5% fetal bovine serum, 50 ng/mL nerve growth factor (R&D Systems, Minneapolis, MN), 4.4 mM glucose, 2 mM l-glutamine, penicillin (50 µg/mL) and streptomycin (50 µg/mL).
4. Chemical fixation of cultured DRG neurons: 4% formaldehyde fixative.

Paraformaldehyde powder is toxic and it is required to wear mask and gloves and use the chemical hood when working with this chemical.

(a) Prepare solution A (phosphate buffered saline, PBS): Fill 1 L beaker with 900 mL of distilled water and dissolve 0.23 g of NaH_2PO_4 (anhydrous), 1.15 g Na_2HPO_4 (anhydrous), and 9 g NaCl. Adjust pH to 7.4 using 1 M NaOH and/or 1 M HCl;

(b) Prepare solution B (8% formaldehyde): Slowly dissolve 8 g of paraformaldehyde powder in 100 mL of deionized water using heating stir plate. After temperature reaches 58°C turn the heat off and add one to two drops of 1 M NaOH to clear formaldehyde solution. Continue stirring for another 20–30 min, cool it down and filter it using filter paper such as Whatman #1;

(c) Monitor the temperature of formaldehyde solution using a thermometer to avoid heating this solution above 58°C. Formaldehyde solution heated above 58°C is not useful as a fixative: it should be discarded and new batch has to be made;

(d) Four percent formaldehyde fixative: is prepared by mixing one part of Solution A with one part of Solution B. For solution stored at 4°C its expiration date is about 4 weeks.

5. DRG culture dishes and other accessories: Multiwell (4 or 8 wells) chamber slides and sterile 35 mm Petri dishes, 70% alcohol, Pasteur pipettes, scissors, forceps, dissecting microscope, and sterile hood for dissection and dissociation of DRG neurons. 37°C/CO_2 humidified incubator to culture dissociated DRG neurons.

6. Reagent to induce oxidative stress: prepare 0.5 mM solution of H_2O_2 by adding 0.5 µL of 30% H_2O_2 to 8.8 mL of Hank's Balanced Salt Solution (HBSS; Gibco BRL, Grand Island, NY). Prepare 5 µM solution of H_2O_2 by adding 100 µL of 1 mM H_2O_2 solution to 9.9 mL of HBSS. Store at 4°C. Calculate the volume of each reagent needed for stimulation cells in two multiwell (4-well or 8-well) chamber slides.

7. Dilution buffer: PBS containing 1% bovine serum albumin, 1% normal donkey serum, 0.3% Triton X-100 (v/v), and 0.01% sodium azide.

8. Anti-8-OHdG (primary) antibodies for immunocytochemistry: Prepare working solution of 5 µg/mL of mouse monoclonal anti-8-OHdG antibodies (Cat # 12501, clone 15A3; QED Bioscience Inc., San Diego, CA) in a dilution buffer. This primary antibody solution will be used to incubate both dissociated DRG neurons and tissue sections mounted onto

histological slides. This solution can be stored at 4°C for up to 1 month and for longer storage it is recommended to make small volume aliquots and store them at −20°C for up to 1 year and avoid their repeated freeze-thaw cycles.

9. Fluorescent detection (secondary) antibodies for 8-OHdG visualization: Donkey anti-mouse conjugated with NL-493 (Cat # NL009; R&D Systems, Inc., Minneapolis, MN). Dilute secondary antibodies 1:100 with dilution buffer and use it as a working solution (can be stored at 4°C for up to 1 month).

10. Fluorescent detection of TUNEL reaction: Prepare 1:100 working solution of Streptavidin conjugated with NL-557 (Cat # NL999; R&D Systems, inc., Minneapolis, MN) with dilution buffer (can be stored at 4°C for up to 1 month).

11. Mounting medium for fluorescent labels: i-BRITE Plus mounting medium (Cat # SF40000-1; Neuromics, Edina, MN). This medium minimizes loss of fluorescence by fluorescent probes due to photobleaching during examination under the fluorescence microscope.

12. TUNEL assay: TACS 2 TdT DAB kit (Cat # 4810-30-K; Trevigen, Gaithersburg, MD; http://www.trevigen.com). This kit is also available for ordering through VWR International, LLC (http://www.vwr.com).

13. Chromogenic system for 8-OHdG detection: HRP-AEC Cell and Tissue Staining kit (CTS 003; R&D Systems, Inc., Minneapolis, MN);

14. Chromogenic system for TUNEL detection: HRP-DAB Cell and Tissue Staining kit (CTS 002; CTS 005 or CTS 008; R&D Systems, Inc., Minneapolis, MN) combined with DAB enhancer (CTS010; R&D Systems, Inc., Minneapolis, MN);

15. Mounting medium for chromogenic labels: Aqueous mounting medium (CTS011; R&D Systems, Inc., Minneapolis, MN).

16. Waterproof pen to label slides: SHUR/MARK pen

17. PAP pen: to draw a hydrophobic line surrounding tissue section mounted onto histological slide (Cat # Z377821, Sigma-Aldrich, St Louis, MO). This hydrophobic barrier prevents leakage of incubation reagents from the slide.

18. Slide incubation chamber: "SlideShow" tray with transparent cover (Cat # 6844-30CL; Newcomer Supply of Middleton WI, http://www.newcomersupply.com)

19. Microscopy: Bright field/fluorescence microscope (Provis; Olympus, Hauppauge, NY) equipped with microscope digital camera (DP71; Olympus Hauppauge, NY) and fluorescence filter sets (460–490 nm excitation/510–550 nm emission and 541–551 nm excitation/572–607 nm emission).

3. Methods

All procedures were performed at room temperature unless stated otherwise. If protocol calls for incubation at room temperature, reagents stored at 4°C should be adjusted to room temperature before they added to cell and tissue samples. Since 3-amino-9-ethylcarbazole (AEC) and 3,3′Diaminobenzidine (DAB) are potential carcinogens wear gloves to avoid contact of these reagents with skin. It is recommended that each staining experiment be performed in at least duplicate, in case some samples dry out during the incubation and cannot be used to complete the experiment.

3.1. Double-Labeling of Cultured DRG Neurons for 8-OHdG and TUNEL Reaction

1. Prepare suspension of rat neonatal DRG neurons in a sterile hood (see Note 1). Culture DRG neurons in multiwell chamber slides in the 37°C/CO_2 humidified incubator for 2–3 days and after that cells can be used in double-labeling experiments.
2. Transfer chamber slides with cells from the 37°C/CO_2 humidified incubator into the sterile hood. Wait for approximately 20 min to allow temperature of the culture medium in the chamber slide to decrease from 37°C to ambient.
3. Using a sterile pipette gently remove the culture medium from each well: position the tip of the pipette into the corner of the well to avoid disturbing cells that can cause their detachment. Save collected culture medium in a sterile tube for further use.
4. Add reagents to designated wells to induce oxidative stress. In one or two wells add plain HBSS: cells in these wells will serve as control groups (see Note 2). Place the chamber slide back into the 37°C/CO_2 humidified incubator and incubate for 30 min.
5. Transfer the chamber slide from the incubator into a sterile hood. Discard oxidative stress reagents from designated wells and rinse them three times with sterile HBSS. Discard HBSS and add culture medium that has been collected in step 3. Return chamber slides with cells into the 37°C/CO_2 humidified incubator and incubate for additional 18–24 h (see Note 3).
6. Transfer the chamber slide from the incubator to a lab bench, discard the culture medium and add 4% formaldehyde into each well. Fix cells for 10 min and then wash them 3 × 15 min with PBS.
7. Remove the upper chamber from the slide using tools provided by the slide chamber vendor.
8. Place slides horizontally into humidity chamber, add primary anti-8-OHdG antibodies and incubate 3 h. Alternatively, cells

may be incubated overnight at 4°C. Adding too much primary antibodies may cause their leakage from the slide causing samples to dry.

9. Wash slide 3×15 min in PBS, place it horizontally into a humid chamber, add secondary anti-mouse NL-493 conjugated antibodies and incubate for 1 h.

10. Wash slide 3×15 min in PBS and start TUNEL assay.

11. Rinse slide with DNase-free water and permeabilize cells for 30 min with CytoPore reagent provided with TUNEL kit.

12. Rinse slide with DNase-free water and then transfer it into a Coplin jar containing TdT labeling buffer: follow dilution recommendations provided with TUNEL kit. Incubate for 5 min.

13. Prepare TdT labeling mixture. Calculate the volume of TdT mixture: 150 µl is required to cover the entire area of the slide with cells. Combine TdT labeling components (TdT dNTP mix, 50× Mn^{2+}, TdT enzyme and TdT labeling buffer) as recommended in the TUNEL kit insert (see Note 4). Apply TdT labeling mixture and cover it with 22×60 mm coverslip. Do not press the coverslip too hard to avoid the leakage of the labeling mixture. Place slide horizontally into a humid chamber and incubate at 37°C for 1 h.

14. Transfer slide with coverslip into a Coplin jar containing TdT Stop Buffer: follow dilution recommendations provided in TUNEL kit protocol. Keep slide in stop buffer for 2 min and then pull it slowly out of the buffer to remove coverslip by washing it off. If coverslip remains sticking to the slide, lift it gently with fine forceps. Transfer uncovered slides with tissue section into another Coplin jar filled with Stop Buffer and incubate for 5 min.

15. Wash slide in PBS for 5 min.

16. Place slide horizontally into a humid chamber, apply three to five drops of Streptavidin-NL-557 solution and incubate for 1 h.

17. Stop the reaction by rinsing slides with PBS and then wash them 3×10 min with PBS in a Coplin jar.

18. Mount under coverslips using mounting medium for fluorescent labels and examine the staining under the fluorescence microscope (see Note 5).

3.2. Double-Labeling for 8-OHdG and TUNEL Reaction on Paraffin-Embedded Tissue Sections

Double-labeling immunofluorescence appears to be more convenient than the procedure utilizing chromogenic labels: it is faster and results in a clear distinction between color labels (see Note 6). However, the presence of autofluorescent pigment, lipofuscin, in human tissues can obscure fluorescent labels used for labeling tissue targets. For example, high density of lipofuscin in

Alzheimer's brain leaves no other option but to employ chromogenic rather than fluorescent labels for immunohistochemical experiments. We utilized Alzheimer's brain tissue sections to develop a two-color chromogenic detection protocol that can be used on the same tissue section. This technique allows obtaining a good contrast and accurate separation of color labels and can be used for simultaneous immunohistochemical detection of DNA fragmentation (see Note 7) and oxidative damage to RNA and DNA on the same tissue section.

1. Mark slides with tissue sections using waterproof SHUR/MARK pen.

2. Outline the tissue section with Pep-pen making at least a 2 mm gap between the outer edge of the tissue section and the inner edge of the Pep-pen line. Allow Pep-pen drawing to dry for at least 5 min.

3. Place slide into a Coplin jar and deparaffinize paraffin-embedded tissue sections by treating them sequentially with the following reagents:
    (a) Xylene – two baths, 10 min each;
    (b) 100% ethanol – two baths, 10 min each;
    (c) 95% ethanol – two baths, 10 min each;
    (d) 70% ethanol – two baths, 10 min each.

4. Wash slides 2×5 min in PBS.

5. Wash slides in DNase-free water for 5 min.

6. Treat tissue section with Proteinase K solution (see Notes 8 and 9). It will be necessary to calculate the working volume of Proteinase K solution assuming that 50 is required to cover 2×1 cm^2 tissue area. Dilute Proteinase K concentrate with DNase-free water using the ratio recommended in the TUNEL kit data sheet. Apply Proteinase K solution onto tissue section and cover them gently with the coverslip of appropriate size. Do not press coverslip too hard to avoid the leakage of Proteinase K solution. Place this slide horizontally into a humid chamber and incubate for 30 min.

7. Transfer the slide with coverslips into a Coplin jar containing DNase-free water and keep it there for 2 min. Pull the slide slowly from water to facilitate removal of the coverslip. Transfer uncovered slides with tissues into another Coplin jar with fresh portion of DNase-free water.

8. Remove slide from the Coplin jar, shake excess water, place it horizontally and add three to five drops of H_2O_2 blocking reagent from Cell and Tissue Staining kit. Incubate 10 min in a humid chamber.

9. Rinse slide in DNase-free water for 5 min.

10. Transfer slide into a Coplin jar containing TdT labeling buffer prepared according to dilution recommendations specified in the TUNEL kit protocol. Incubate for 5 min.
11. Prepare TdT labeling mixture. Calculate the volume of TdT mixture assuming that 50 µL is required to cover 2×1 cm^2 of tissue area. Combine TdT labeling components (TdT dNTP mix, 50× Mn^{2+}, TdT enzyme and TdT labeling buffer) as recommended in the TUNEL kit protocol. Apply 50 µL of TdT labeling mixture onto tissue section and cover it gently with coverslip of appropriate size. Do not press the coverslip too hard to avoid the leakage of the labeling mixture. Place slide horizontally into a humid chamber and incubate it for 1 h at 37°C.
12. Transfer slide with coverslip into a Coplin jar containing TdT stop buffer: follow up dilution recommendations provided in TUNEL kit protocol. Keep slide in a stop buffer for 2 min and then pull it slowly from the buffer to facilitate the removal of the coverslip. Transfer uncovered slide into another Coplin jar with fresh portion of stop buffer and incubate for 5 min.
13. Wash slide in PBS for 5 min.
14. Place slide horizontally into a humid chamber and apply three to five drops of Streptavidin-HRP solution from Cell and Tissue staining kit. Incubate 30 min.
15. Wash slide 2×5 min in PBS.
16. Place slide horizontally onto a microscope stage and apply mixture of DAB (from Cell and Tissue Staining kit, CTS002; R&D Systems, Inc., Minneapolis, MN) with DAB enhancer to cover entire tissue section. Monitor the development of dark-blue color under the microscope using 4× or 10× power lens. Three to ten minutes is needed for developing strong blue-black labeling of nuclei in apoptotic and/or necrotic cells.
17. Discard DAB mixture from the slide.
18. Wash slide in PBS for 20 min.
19. Place slide horizontally and add three to five drops of H_2O_2 blocking reagent from Cell and Tissue Staining kit. Incubate 10 min in a humid chamber.
20. Rinse slides in PBS and perform avidin-biotin blocking procedure using reagents from Cell and Tissue Staining kit. Place slide horizontally into a humid chamber and add three to five drops of avidin blocking solution and incubate for 15 min. Rinse slide in PBS, place it horizontally and add three to five drops of biotin blocking solution and incubate 15 min.
21. Wash slide in PBS for 20 min.
22. Place slide into a humid chamber and apply primary anti-8-OHdG antibodies. 80–150 µL of antibody solution is needed to cover the tissue section on the slide (see Notes 10 and 11).

23. Incubate with primary anti-8-OHdG antibodies 16–24 h at 4°C.
24. Wash slide 3 × 15 min in PBS.
25. Repeat step 14.
26. Repeat step 15.
27. Place slide horizontally onto the microscope stage and apply AEC solution from Cell and Tissue Staining kit (CTS003; R&D Systems, Inc., Minneapolis, MN) to cover entire tissue section. Monitor the development of red color under the microscope using 4× or 10× lens. Strong red color usually develops in 1–3 min.
28. Repeat step 18.
29. Cover tissue sections with aqueous mounting medium and let them dry.
30. Use bright-field microscope equipped with digital color camera to collect images of tissue section double-labeled for TUNEL and 8-OHdG.

4. Notes

1. Isolation of DRGs, their cell dissociation, and culturing described in this chapter includes the following steps. Decapitate 3-day-old Sprague-Dawley rats (use 70% alcohol to sterilize surgical instruments (scissors and forceps) and decapitated bodies). Under dissection microscope dissect DRGs from the thoracic and lumbar segments of the spinal cord and incubate them in collagenase/dispase digestion solution for 30 min in the 37°C/CO_2 incubator. Dissociation of cells can be accomplished by triturating DRGs through large, medium, and small diameter flame constricted pipettes: 50–60 times/per pipette. Digestion is terminated by adding 10 mL of sterile HBSS into cell suspension and centrifuging it at 500 × g for 10 min. Supernatant is discarded and the pellet is resuspended in the same volume of HBSS followed by centrifugation. Supernatant is discarded and the resulting pellet is resuspended in 5 mL of F-12 culture medium. This cell suspension is added into wells of sterile multiwell chamber slides for further manipulations.
2. It is necessary to include different types of controls to be sure that TUNEL signal is confined to true apoptotic or/and/ necrotic cells rather than was generated nonspecifically in normal cells. There are three types of controls addressing the

specificity issue: (a) *reagent-specific*, (b) *target-specific* and (c) *positive tissue control*.

(a) *Reagent-specific* control answers the question whether components of the TdT mixture interact with each other in a specific manner. This type of control includes incubating cell and tissue specimens with TdT mixture that does not contain terminal deoxynucleotidyl transferase (TdT enzyme). It is expected that that cell nuclei will not be labeled. Some postmortem tissues stored for a long time in formalin may show staining caused by nonspecific tissue binding of either streptavidin-HRP or DAB chromogen. However, the cytoplasmic pattern of such tissue staining can be easily recognized from specific TUNEL reaction which is always confined to cell nuclei.

(b) *Target-specific* control helps to determine whether TUNEL assay detects only apoptotic/necrotic cells. This is done by comparing labeling in normal tissues versus tissues undergoing apoptosis or necrosis. To employ a target-specific control, it is recommended to include normal cells and tissues with specimens undergoing apoptosis/necrosis. Although normal specimens may contain cells with TUNEL-positive nuclei (due to naturally occurring apoptosis), their number is expected to be much lower in comparison with apoptotic/necrotic specimens.

(c) *Positive tissue* control is done to prove that TUNEL assay is both reagent- and target-specific. This type of control includes inducing apoptosis/necrosis in normal cells and tissues by treating them with nuclease. Nuclease treatment causes DNA breaks and results in increased number of cells with TUNEL-positive nuclei.

3. Cells and tissue sections should not dry during the incubation and washing steps: specimens found dry should be excluded from experiment because this may result in nonspecific background. Because partial drying may be overlooked by an operator fatigued by processing a large number of specimens, hence it is recommended to interpret staining of tissue margins (which is more prone to partial drying) with caution.

4. The TUNEL kit manufactured by Trevigen includes Streptavidin-HRP and DAB chromogen reagents which can be used to obtain strong single-color brown-labeled nuclei. It appears that using similar reagents from R&D Systems' Cell and Tissue Staining kit and DAB chromogen, combined with R&D Systems' DAB enhancer producing dark-blue color, gives a better color separation when combined with AEC (red color) chromogen. Alternatively, similar reagents can be purchased separately from other vendors. However, using

staining kits saves a lot of time because they don't require any time-consuming optimization.

5. When manipulating digital images, brightness and contrast should be adjusted on "control" and "experimental" samples using identical parameters to avoid misrepresentation of staining results.

6. The double-staining protocol presented in this chapter can be used for other tasks including embryonic development, cancer research, aging and neurodegenerative disorders, and toxicological studies. For example, this technique can be employed for qualitative and quantitative pollution monitoring using cultured cells as a pollution probe.

7. Apoptosis and/necrosis determined by TUNEL needs to be also confirmed by morphological criteria such as cell membrane blebbing, shrinkage of cells, and the formation apoptotic bodies.

8. Proteinase K treatment may cause tissue damage recognized as detached (floating) tissue fragments and/or holes of irregular size. In this case, it is recommended to reduce the duration of enzymatic treatment from 30 min to 10–20 min. Cultured cells are more sensitive to the damaging effects of Proteinase K and therefore enzymatic treatment on cytological samples is not recommended. Alternatively, cells may be permeabilized by incubating them in 0.1–0.3% Triton X-100/PBS solution for 20 min.

9. To enhance the signal intensity of 8-OHdG labeling, cells and tissues may be treated with Proteinase K (10 µg/mL in PBS) from 10 to 30 min in a humid chamber at 37°C, but see Note 8.

10. The specificity of labeling for 8-OHdG is determined by two factors: (a) specificity of 8-OHdG antibodies and (b) specificity of immunohistochemical reagents. In addition, the specificity of staining may be evaluated using (c) negative control.

    (a) Evaluating the specificity of anti-8-OHdG antibodies is needed to confirm that immunohistochemical labeling resulted from the interaction of antibodies with their cognate (8-OHdG) target, rather than caused by their cross-reactivity with irrelevant antigens. This can be accomplished using a so-called absorption control. Briefly, mix anti- 8-OHdG antibodies (taken in working dilution, i.e. 5 µg/mL) with 10 µg/mL purified 8-OHdG antigen (Sigma-Aldrich, St Louis, MO). Mix well and incubate either 5 h at room temperature or overnight at 4°C. Since these antibodies recognize RNA-derived 8-OHdG as well, it is also helpful to do absorption control using 8-OHdG reactant. It is expected that intensity of labeling will decrease when using absorption control mixtures since soluble antigens occupy binding sites on anti-8-OHdG antibodies and thus reduce capacity to interact with intracellular 8-OHdG targets.

(b) Nonspecific signal can sometimes be present due to sticking of immunohistochemical reagents to fixed cells and sections. The simplest way to test this would be the incubation of specimens by omitting anti-8-OHdG antibody from the working solution: lack of labeling proves specificity of immunohistochemical reagents. However, if nonspecific labeling is observed, additional steps are required to block it. For example, nonspecific background staining may be caused by free aldehyde groups present in tissues that are fixed with formaldehyde or glutaraldehyde solutions: free aldehyde groups are capable of reacting with secondary antibodies "cross-linking" them to the tissue. To neutralize free aldehyde groups, specimens can be treated with 0.5 mg/mL of sodium borohydrate ($NaBH_4$) for 10–20 min at room temperature. Alternatively, free aldehyde groups can be blocked by incubating samples in 10% normal horse or donkey serum for 5–30 min at room temperature. The blocking of free aldehyde groups is done before applying primary antibodies.

(c) A negative control can be a specimen that is known to lack the antigen of interest. For example, the antigen can be intentionally destroyed. Since anti-8-OHdG antibodies target DNA and/or RNA, their degradation is expected to result in reduced or abolished cell and tissue labeling. To do such a control, DNA and RNA may be degraded by pretreatment of specimen with either 10 U/µL of DNase I or 10 U/µL S1 DNase or 5 µg/µL of RNase from 30 min to 1 h in a humid chamber at 37°C. Alternatively, specimens can be treated with all three enzymes combined together which appear to degrade DNA and RNA targets more efficiently than using individual enzymes.

11. The advantage of using anti-8-OHdG antibodies produced by QED Bioscience is that this antibody interacts with RNA localized in the cytoplasm and therefore immunohistochemical labeling for 8-OHG does not obscure labeling for TUNEL, which is confined to cell nuclei.

References

1. Muppidi, J., Porter, M., and Siegel, R. M. (2004) Measurement of apoptosis and other forms of cell death. *Curr. Protoc. Immunol. Chapter*, Unit 3.17.
2. Kerr, W. G. (2008) Analysis of apoptosis in hematopoietic stem cells by flow cytometry. *Methods Mol. Biol.* **430**, 87–99.
3. Alarcon, V. B., and Marikawa, Y. (2004) Molecular study of mouse peri-implantation development using the in vitro culture of aggregated inner cell mass. *Mol. Reprod. Dev.* 67, 83–90.
4. Goncalves, J. S., Sasso-Cerri, E., and Cerri, P. S. (2008) Cell death and quantitative reduction of

rests of Malassez according to age. *J. Periodontal Res.* **43**, 478–481.
5. Harada, K., Ferdous, T., Itashiki, Y., Takii, M., Mano, T., Mori, Y., and Ueyama, Y. (2009) Effects of cepharanthine alone and in combination with fluoropyrimidine anticancer agent, S-1, on tumor growth of human oral squamous cell carcinoma xenografts in nude mice. *Anticancer Res.* **29**, 1263–1270.
6. Kang, Y., Zhang, X., Jiang, W., Wu, C., Chen, C., Zheng, Y., Gu, J., and Xu, C. (2009) Tumor-directed gene therapy in mice using a composite nonviral gene delivery system consisting of the piggyBac transposon and polyethylenimine. *BMC Cancer* **9**, 126.
7. Anderson, A. J., Stoltzner, S., Lai, F., Su, J., and Nixon, R. A. (2000) Morphological and biochemical assessment of DNA damage and apoptosis in Down syndrome and Alzheimer disease, and effect of postmortem tissue archival on TUNEL. *Neurobiol. Aging* **21**, 511–524.
8. Lee, S. Y., Moon, Y., Hee Choi, D., Jin Choi, H., and Hwang, O. (2007) Particular vulnerability of rat mesencephalic dopaminergic neurons to tetrahydrobiopterin: Relevance to Parkinson's disease. *Neurobiol. Dis.* **25**, 112–120.
9. Kadota, T., Shingo, T., Yasuhara, T., Tajiri, N., Kondo, A., Morimoto, T., Yuan, W. J., Wang, F., Baba, T., Tokunaga, K., Miyoshi, Y., and Date, I. (2009) Continuous intraventricular infusion of erythropoietin exerts neuroprotective/rescue effects upon Parkinson's disease model of rats with enhanced neurogenesis. *Brain Res.* **1254**, 120–127.
10. Ohta, Y., Kamiya, T., Nagai, M., Nagata, T., Morimoto, N., Miyazaki, K., Murakami, T., Kurata, T., Takehisa, Y., Ikeda, Y., Asoh, S., Ohta, S., and Abe, K. (2008) Therapeutic benefits of intrathecal protein therapy in a mouse model of amyotrophic lateral sclerosis. *J. Neurosci. Res.* **86**, 3028–3037.
11. Park, J. E., Lee, S. T., Im, W. S., Chu, K., and Kim, M. (2008) Galantamine reduces striatal degeneration in 3-nitropropionic acid model of Huntington's disease. *Neurosci. Lett.* **448**, 143–147.

Chapter 3

EM-ISEL: A Useful Tool to Visualize DNA Damage at the Ultrastructural Level

Antonio Migheli

Abstract

A method for the localization of DNA strand breaks at the ultrastructural level is presented. The technique involves the use of terminal deoxynucleotidyl transferase and labeled dUTP. Incorporation of labeled nucleotides is visualized through colloidal gold labeling. Cells undergoing apoptotic or necrotic cell death, as well as cells showing death-unrelated DNA damage, can be easily distinguished. The technique uses tissues routinely processed for electron microscopy. It has been successfully applied to study DNA damage and apoptosis in different pathologic conditions. The feasibility of this technique for retrospective studies on archival material is emphasized.

Key words: DNA damage, *In situ* end-labeling, Electron microscopy, Apoptosis, Necrosis

1. Introduction

DNA damage has been recognized as a fundamental player in human diseases (1). Single- and double-strand breaks arise as a consequence of either attacks from exogenous and endogenous toxic agents, or genetic and acquired defects in DNA repair mechanisms. Genomic instability and transcriptional infidelity are at the basis of cancer, immune dysfunction, radiosensitivity, and degenerative diseases. The central nervous system (CNS) is particularly prone to DNA damage, especially in the form of single-strand breaks, because of the burden of genotoxic reactive oxygen species that are continuously formed. DNA damage may alter the expression of vulnerable genes involved both in neuronal survival and in highly specialized functions such as learning and memory. Several evidences point out the role of extensive DNA damage and/or defective DNA repair in brain aging, ischemia, developmental defects, and neurodegeneration (2, 3).

DNA degradation is a critical step in cell death. Both apoptosis and necrosis are the consequence of extensive and irreversible DNA damage. Apoptosis is a major, although not unique, mechanism of programmed cell death (PCD) (4, 5). Morphologically, it is characterized by cell shrinkage, chromatin condensation and margination, nuclear pyknosis, and late fragmentation into apoptotic bodies, while cell membranes and organelles are preserved. All of these changes are due to cleavage of various cytoplasmic and nuclear substrates through caspase-dependent and independent pathways (6, 7). During apoptosis, DNA is degraded into oligonucleosomal fragments, multiples of 180–200 base pairs, through activation of several nucleases (8).

DNA damage also occurs in necrosis. At variance with apoptosis, necrosis is not a programmed event, and is characterized morphologically by early swelling, disintegration of membranes and organelles and absence of chromatin condensation, and is due to activation of non-caspase proteases (9). Necrotic DNA degradation may occur both in a random and in an oligonucleosomal fashion. 3'-OH and 5'-OH ends of DNA strand breaks are generated in both types of death, although with a different ratio (10). However, double-strand breaks with single-base 3' overhangs appear to be specific for apoptosis but not for necrosis (11).

A major advance in the study of DNA damage in single cells has been the development of techniques for the *in situ* end-labeling (ISEL) of fragmented DNA, through the application of labeled nucleotides and incorporation of enzymes such as DNA polymerase and terminal deoxynucleotidyl transferase (TdT) (12–14). In the TUNEL assay (12) that employs TdT, only 3'-OH ends are labeled; on the contrary, both 3'-OH and 5'-OH ends are labeled with the DNA polymerase *in situ* nick translation (ISNT) assay (13, 14). In spite of this theoretical advantage of ISNT, comparative *in vitro* (15) and *in vivo* (16) studies have shown a greater labeling capacity for TUNEL, which has been mostly used thereafter.

Notably, both apoptotic and necrotic cell death will be labeled by ISEL techniques (17–20). Furthermore, ISEL may also label DNA damage in cells that are not actually dying (21–24), or it may detect DNA damage that is due to artifactual conditions rather than to pathologic events (25–27). Discriminating between these various conditions is often difficult at the light microscopy level. All these limitations have been obviated by modifying the ISEL technique in order to visualize the DNA damage at the electron microscopy (EM) level (28, 29). The EM-ISEL assay has a number of potential applications: it can be used as a genotoxicity assay (23, 28); most notably, however, it can be used as a cell death assay, since morphologic changes of apoptosis or necrosis in labeled cells are easily demonstrated at the ultrastructural level. Finally, since the technique has been developed on tissues routinely processed for EM, retrospective studies on archival material can be easily performed. Following its original description and application in neurodegenerative diseases characterized by apoptotic neuronal death (29, 30), the assay has been successfully applied by other investigators to study DNA damage and apoptosis in diverse conditions such as human development (31, 32), heart and vascular diseases (33–35), myopathies (36, 37), and renal (38), blood (39), and skin (40) diseases.

2. Materials

2.1. Tissues

Tissue blocks are prepared according to routine EM procedures, i.e., fixation in 2.5% glutaraldehyde for 1–3 h, postfixation in 1–2% osmium for 1–2 h, and embedding in an epoxy resin (see Notes 1–3). After selecting the area of interest on toluidine blue-stained semithin sections, thin sections are cut with an ultramicrotome, collected on Formvar-coated nickel grids, and stored until use.

2.2. Reagents

All solutions are prepared using double distilled water (DDW).

1. TdT (Roche Applied Science, Indianapolis, IN), 50 U/µL solution, is stored at –20°C.

2. Digoxigenin-11-dUTP (Roche Applied Science, Indianapolis, IN), 1 mM solution, is stored at –20°C (see Note 4).

3. Anti-digoxigenin goat immunoglobulins coupled with 10 nm colloidal gold (Electron Microscopy Sciences, Hatfield, PA) are stored at 4°C (see Note 4).

4. TdT buffer: 25 mM Tris-HCl, 200 mM potassium cacodylate, 2.5 mM cobalt chloride, pH 6.6. This buffer is prepared using two 2× stock solutions: (**a**) 50 mM Tris-HCl, 0.4 M potassium cacodylate, pH 6.6; (**b**) 5 mM cobalt chloride in DDW. After autoclaving, the two stock solutions can be stored at room temperature. Prepare the TdT buffer by mixing equal volumes of stock solutions **a** and **b** just before use.

5. 2× Saline sodium citrate (SSC) solution: 300 mM sodium chloride, 30 mM sodium citrate, pH 7.0. Prepare a 20× stock solution: 3 M sodium chloride, 0.3 M sodium citrate, pH 7.0. After autoclaving, the stock solution can be kept at room temperature. Make a 2× solution with DDW just before use.

6. Tris-buffered saline (TBS). Prepare a 10× stock solution: 6.055 g Tris base, 40.91 g NaCl in 500 ml DDW, pH 7.4. After autoclaving, the stock solution can be kept at room temperature. Make a 1× solution with DDW just before use.

7. TBS-bovine serum albumin (BSA) buffer. Dissolve 100 mg of BSA, fraction V, in 10 mL of TBS, pH 8.2. Store at 4°C.
8. 2% Uranyl acetate in DDW. Prepare fresh before use and keep in the dark.

2.3. Other Requirements

1. Anti-static tweezers are needed to handle nickel grids.
2. Incubation chamber: to create a humid chamber, a convenient way is to use a glass Petri dish. On the bottom, place a wet round filter paper and apply a square sheet of Parafilm™ over the filter. Prepare two such chambers.
3. Oven: set at 37°C and used for the TdT reaction.

3. Methods

1. Collect silver–gold thin sections on Formvar-coated nickel grids and store until use.
2. Place each grid over a 30-µL drop of TdT buffer inside the first humid chamber.
3. While the grids are floating, prepare the labeling solution in an Eppendorf tube (keep all tubes on ice, as TdT can be easily degraded) as follows: 1 U of TdT, 1 nmole of digoxigenin-11-dUTP in 100 µL of TdT buffer.
4. Divide the labeling solution into 20 µL drops in the first humid chamber and place the chamber in the oven set at 37°C for 10 min.
5. Quickly transfer the grids from the drops of TdT buffer to the drops of the labeling solution and keep the chamber inside the oven at 37°C for 10 min (see Notes 5 and 6).
6. While the grids are being incubated in the oven, prepare two series of 50 µL drops of 2× SSC and one series of 50 µL drops of TBS in a second humid chamber, and keep it at room temperature.
7. At the end of the incubation in the labeling solution, remove the grids from the first chamber and rinse them on the 50-µL drops of 2× SSC in the second chamber (two changes, 5 min each).
8. Place each grid on the 50-µL drop of TBS for 5 min.
9. While the grids are being rinsed in TBS, prepare a 1/30 solution of colloidal gold-coupled anti-digoxigenin goat immunoglobulins in TBS-BSA, pH 8.2, and divide it into 50-µL drops in the second humid chamber.
10. Incubate the grids in the anti-digoxigenin antiserum overnight at 4°C.

11. The following day, rinse the grids extensively in several drops of TBS and DDW.
12. Stain the grids with 2% uranyl acetate for 20 min and observe in a transmission electron microscope (see Note 7).

4. Notes

1. Best results are obtained with tissues that have been routinely processed for EM. We have found that fixation in glutaraldehyde-osmium gives more reproducible results compared to that in glutaraldehyde alone, or when paraformaldehyde is used instead of glutaraldehyde. Tissues can be fixed both by immersion and by perfusion (e.g., laboratory animals). Sections of biopsy material retrieved from paraffin blocks and re-embedded in epoxy resin are also suitable (unpublished observations).
2. Two types of epoxy resins have been studied, i.e., Epon 812 and Araldite, without any appreciable difference.
3. As an alternative to epoxy resins, acrylic resins such as LR White and LR Gold (30, 41) and Lowicryl (32, 35, 40) can be used. However, the morphologic detail is greatly decreased. A potential advantage of acrylic over epoxy resins is that with the former, EM-ISEL can be combined with immunoelectron microscopy (IEM) (40). Likewise, ultrathin cryosections have been recently used for colocalizing EM-ISEL and IEM (38).
4. Fluorescein-12-dUTP labeled nucleotides (Roche Applied Science) and 10-nm gold-conjugated anti-fluorescein immunoglobulins (EMS) can be used instead of digoxigenin-11-dUTP and anti-digoxigenin antibodies, with an equivalent intensity of reaction (unpublished observations). Other variants include (a) biotin-dUTP labeled nucleotides followed by gold-conjugated streptavidin (32); (b) ultra-small 1-nm gold-conjugated anti-digoxigenin antibodies (39) or ultra-small 1-nm gold-conjugated streptavidin (32), followed by silver enhancement.
5. No etching of the resin is needed prior to staining. In fact, etching with oxidizing agents (10% H_2O_2 or 10% $NaIO_4$) resulted in weak to absent staining, suggesting that the 3'-OH ends of DNA breaks are altered by the oxidizing agents and are no longer recognized by TdT (30).
6. The concentration of TdT and the length of the incubation step in the labeling solution indicated in the protocol are those that have given the best results in most cases.

Higher TdT concentrations, or longer incubation times, generally result in a progressive increase in the background staining of nuclei of normal cells. The reason for the latter finding is unclear. DNA nicks might be produced by the cutting procedure itself and be revealed under favorable staining conditions. Alternatively, the labeling might refer to sites of active gene transcription that are located in the condensed chromatin and have been demonstrated using a TdT-based approach (42). The latter finding, however, usually requires DNase I pretreatment of sections. However, since differences in the labeling intensity may occur with different tissues, it is strongly suggested that a series of grids be initially prepared, varying both TdT concentration and the incubation time in the labeling mix.

7. Quantitative analysis of gold labeling as a measure of free 3'-OH DNA end density can be performed using various programs (e.g., NIH Image Program) (10).

Acknowledgment

The financial support of Regione Piemonte-Ricerca Sanitaria Finalizzata is gratefully acknowledged.

References

1. McKinnon, P.J. and Caldecott, K.W. (2007) DNA strand break repair and human genetic disease. *Annu. Rev. Genom. Human Genet.* **8**, 37–55.
2. Katyal, S. and McKinnon, P.J. (2008) DNA strand breaks, neurodegeneration and aging in the brain. *Mech. Ageing Dev.* **129**, 483–491.
3. Martin, L.J. (2008) DNA damage and repair: relevance to mechanisms of neurodegeneration. *J. Neuropathol. Exp. Neurol.* **67**, 377–387.
4. Kerr, J.F.R., Wyllie, A.H. and Currie, A.R. (1972) Apoptosis: a basic biological phenomenon with wide-ranging implications in tissue kinetics. *Br. J. Cancer* **26**, 239–257.
5. Jaattela, M. (2004) Multiple cell death pathways as regulators of tumour initiation and progression. *Oncogene* **23**, 2746–2756.
6. Kumar, S. (2007) Caspase function in programmed cell death. *Cell Death Differ.* **14**, 32–43.
7. Turk, B. and Stoka, V. (2007) Protease signalling in cell death: caspases versus cysteine cathepsins. *FEBS Lett.* **581**, 2761–2767.
8. Parrish, J.Z. and Xue, D. (2006) Cuts can kill: the roles of apoptotic nucleases in cell death and animal development. *Chromosoma* **115**, 89–97.
9. Golstein, P. and Kroemer, G. (2007) Cell death by necrosis: towards a molecular definition. *Trends Biochem. Sci.* **32**, 37–43.
10. Hayashi, R., Ito, Y., Matsumoto, K., Fujino, Y. and Otsuki, Y. (1998) Quantitative differentiation of both free 3'-OH and 5'-OH DNA ends between heat-induced apoptosis and necrosis. *J. Histochem. Cytochem.* **46**, 1051–1059.
11. Didenko, V.V. and Hornsby, P.J. (1996) Presence of double-strand breaks with single-base 3' overhangs in cells undergoing apoptosis but not necrosis. *J. Cell Biol.* **135**, 1369–1376.
12. Gavrieli, Y., Sherman, Y. and Ben-Sasson, S.A. (1992) Identification of programmed cell death in situ via specific labeling of nuclear DNA fragmentation. *J. Cell Biol.* **119**, 493–501.
13. Gold, R., Schmied, M., Rothe, G., Zischler, H., Breitschopf, H., Wekerle, H. and

Lassmann, H. (1993) Detection of DNA fragmentation in apoptosis: application of an in situ nick translation to cell culture systems and tissue sections. *J. Histochem. Cytochem.* **41**, 1023–1030.

14. Wijsman, J.H., Jonker, R.R., Keijzer, R., Van de Velde, C.J.H., Cornelisse, C.J. and Van Dierendonck, J.H. (1993) A new method to detect apoptosis in paraffin sections: in situ end-labeling of fragmented DNA. *J. Histochem. Cytochem.* **41**, 7–12.

15. Gorczyca, W., Gong, J. and Darzynkiewicz, Z. (1993) Detection of DNA strand breaks in individual apoptotic cells by the in situ terminal deoxynucleotidyl transferase and nick translation assays. *Cancer Res.* **53**, 1945–1951

16. Migheli, A., Cavalla, P., Marino, S. and Schiffer, D. (1994) A study of apoptosis in normal and pathologic nervous tissue after in situ end-labeling of fragmented DNA. *J. Neuropathol. Exp. Neurol.* **53**, 606–616.

17. Grasl-Kraupp, B., Ruttkay-Nedecky, B., Koudelka, H., Bukowska, K., Bursch, W. and Schulte-Hermann, R. (1995) In situ detection of fragmented DNA (TUNEL assay) fails to discriminate among apoptosis, necrosis and autolytic cell death: a cautionary note. *Hepatology* **21**, 1465–1468.

18. van Lookeren Campagne, M., Lucassen, P.J., Vermeulen, J.P. and Balasz, R. (1995) NMDA and kainate induce internucleosomal DNA cleavage associated with both apoptotic and necrotic cell death in the neonatal rat brain. *Eur. J. Neurosci.* **7**, 1627–1640.

19. Mundle, S., Gao, X.Z., Khan, S., Gregory, S.A., Preisler, H.D. and Raza, A. (1995) Two in situ labeling techniques reveal different patterns of DNA fragmentation during spontaneous apoptosis in vivo and induced apoptosis in vitro. *Anticancer Res.* **15**, 1895–1904.

20. Gold, R., Schmied, M., Giegerich, G., Breitschopf, H., Hartung, H.P., Toyka, K.V. and Lassmann, H. (1994) Differentiation between cellular apoptosis and necrosis by the combined use of in situ tailing and nick translation techniques. *Lab. Invest.* **71**, 219–225.

21. Lopes, S., Jurisicova, A., Sun, J.G. and Casper, R.F. (1998) Reactive oxygen species: potential cause for DNA fragmentation in human spermatozoa. *Hum. Reprod.* **13**, 896–900.

22. Coates, P.J., Save, V., Ansari, B. and Hall, P.A. (1995) Demonstration of DNA damage/repair in individual cells using in situ end labelling: association of p53 with sites of DNA damage. *J. Pathol.* **176**, 19–26.

23. Assad, M., Lemieux, N. and Rivard, C.H. (1997) Immunogold electron microscopy in situ end-labeling (EM-ISEL): assay for biomaterial DNA damage detection. *Biomed. Mater. Eng.* **7**, 391–400.

24. Kisby, G.E., Kabel, H., Hugon, J. and Spencer, P. (1999) Damage and repair of nerve cell DNA in toxic stress. *Drug Metab. Rev.* **31**, 589–618.

25. Tateyama, H., Tada, T., Hattori, H., Murase, T., Li, W.X. and Eimoto, T. (1998) Effects of prefixation and fixation times on apoptosis detection by in situ end-labeling of fragmented DNA. *Arch. Pathol. Lab. Med.* **122**, 252–255.

26. Schallock, K., Schulz-Schaeffer, W.J., Giese, A. and Kretzschmar, H.A. (1997) Postmortem delay and temperature conditions affect the in situ end-labeling (ISEL) assay in brain tissue of mice. *Clin. Neuropathol.* **16**, 133–136.,

27. Labat-Moleur, F., Guillermet, C., Lorimier, P., Robert, C., Lantuejoul, S., Brambilla, E. and Negoescu, A. (1998) TUNEL apoptotic cell detection in tissue sections: critical evaluation and improvement. *J. Histochem. Cytochem.* **46**, 327–334.

28. Migheli, A., Piva, R., Wei, J., Attanasio, A., Casolino, S., Dlouhy, S.R., Bayer, S.A. and Ghetti, B. (1997) Diverse cell death pathways result from a single missense mutation in weaver mouse. *Am. J. Pathol.* **151**, 1629–1638.

29. Migheli, A., Attanasio, A. and Schiffer, D. (1995) Ultrastructural detection of DNA strand breaks in apoptotic neural cells by in situ end-labelling techniques. *J. Pathol.* **176**, 27–35.

30. Depault, F., Cojocaru, M., Fortin, F., Chakrabarti, S. and Lemieux, N. (2006) Genotoxic effects of chromium (VI) and cadmium (II) in human blood lymphocytes using the electron microscopy in situ end-labeling (EM-ISEL) assay. *Toxicol. In Vitro* **20**, 513–518.

31. Kumagai, K., Otsuki, Y., Ito, Y., Shibata, M.A., Abe, H. and Ueki, M. (2001) Apoptosis in the normal human amnion at term, independent of Bcl-2 regulation and onset of labour. *Mol. Hum. Reprod.* **7**, 681–689.

32. Al-Lamki, R.S., Skepper, J.N., Loke, Y.W., King, A. and Burton, G.J. (1998) Apoptosis in the early human placental bed and its discrimination from necrosis using the in-situ DNA ligation technique. *Hum. Reprod.* **13**, 3511–3519.

33. Kanoh, M., Takemura, G., Misao, J., Hayakawa, Y., Aoyama, T., Nishigaki, K., Noda, T., Fujiwara, T., Fukuda, K., Minatoguchi, S. and Fujiwara, H. (1999) Significance of myocytes with positive DNA

in situ nick end-labeling (TUNEL) in hearts with dilated cardiomyopathy-Not apoptosis but DNA repair. *Circulation* **99**, 2757–2764.

34. Ohno, M., Takemura, G., Ohno, A., Misao, J., Hayakawa, Y., Minatoguchi, S., Fujiwara, T. and Fujiwara, H. (1998) "Apoptotic" myocytes in infarct area in rabbit hearts may be oncotic myocytes with DNA fragmentation. Analysis by immunogold electron microscopy combined with in situ nick end-labeling. *Circulation* **98**, 1422–1430.

35. Hegyi, L., Skepper, J.N., Cary, N.R.B. and Mitchinson, M.J. (1996) Foam cell apoptosis and the development of the lipid core of human atherosclerosis. *J. Pathol.* **180**, 423–429.

36. Ikezoe, K., Nakagawa, M., Yan, C.Z., Kira, J., Goto, Y. and Nonaka, I. (2002) Apoptosis is suspended in muscle of mitochondrial encephalomyopathies. *Acta Neuropathol.* **103**, 531–540.

37. Ikezoe, K., Nakagawa, M., Osoegawa, M., Kira, J. and Nonaka, I. (2004) Ultrastructural detection of DNA fragmentation in myonuclei of fatal reducing body myopathy. *Acta Neuropathol.* **107**, 439–442.

38. Kalaaji, M., Fenton, K.A., Mortensen, E.S., Olsen, R., Sturfelt, G., Alm, P. and Rekvig, O.P. (2007) Glomerular apoptotic nucleosomes are central target structures for nephritogenic antibodies in human SLE nephritis. *Kidney Int.* **71**, 664–672.

39. Bunting, R.W. and Selig, M.K. (2002) Localization of DNA in ultrascopic nuclear appendages of polymorphonuclear white blood cells from patients with low serum B-12. *J. Histochem. Cytochem.* **50**, 1381–1388.

40. Ishida-Yamamoto, A., Yamauchi, T., Tanaka, H., Nakane, H., Takahashi, H. and Iizuka, H. (1999) Electron microscopic in situ DNA nick end-labeling in combination with immuno-electron microscopy. *J. Histochem. Cytochem.* **47**, 711–717.

41. Goping, G., Wood, K.A., Sei, Y. and Pollard, H.B. (1999) Detection of fragmented DNA in apoptotic cells embedded in LR White: a combined histochemical (LM) and ultrastructural (EM) study. *J. Histochem. Cytochem.* **47**, 561–568.

42. Thiry, M. (1991) In situ nick translation at the electron microscopic level: a tool for studying the location of DNAse I-sensitive regions within the cell. *J. Histochem. Cytochem.* **39**, 871–874.

Chapter 4

In Situ Labeling of DNA Breaks and Apoptosis by T7 DNA Polymerase

Vladimir V. Didenko

Abstract

The native T7 DNA polymerase is a fast and highly processive enzyme that can be used for *in situ* detection of apoptosis and various types of DNA breaks. The technique is quick and simple, and was shown to label earlier stages of apoptosis compared to the terminal transferase technique. The *in situ* labeling applications of T7 DNA polymerase are presented and summarized from the DNA damage detection standpoint. The detailed protocols are provided together with the discussion of their advantages and limitations.

Key words: DNA breaks, Apoptosis detection, T7 DNA polymerase, DNA damage, *In situ* labeling

1. Introduction

The technique using native T7 polymerase to identify apoptotic cells is one of several similar methods for the detection of DNA breaks *in situ*. The other enzymes employed in this type of labeling include terminal deoxyribonucleotidyl transferase (TdT) and DNA polymerases, such as *Escherichia coli* DNA polymerase I (Pol I) and Klenow fragment of DNA polymerase I.

Technically, all these approaches are comparable and, when applied to fresh-frozen or fixed cells and tissue sections, use 3′ termini at a variety of DNA breaks as priming points and synthesize new nucleic acid strands incorporating fluorescent or conventional tags. The methods are subdivided into three categories based on the mechanism of labeling:

1. TdT-mediated dUTP nick-end labeling (TUNEL) (1, 2), also sometimes called *in situ* nick-end labeling (ISNEL) (3), is based on the activity of TdT which, unlike Pol I or Klenow

fragment, is template independent and catalyzes long single-stranded DNA tails at any free 3' OH DNA ends.

2. *In situ* nick translation (ISNT) uses Pol I and the nuclease DNase I to label actively transcribing DNA regions; if DNase I is omitted, the Pol I labels the pre-existing single-stranded DNA breaks (4, 5).

3. *In situ* (DNA) end labeling (ISEL) employs Klenow fragment or T7 DNA polymerase to label various protruding 5' DNA ends via fill-in reaction (3, 4, 6, 7).

Although the major application of these polymerization-based methods is in apoptosis detection, they have also been successfully used to identify the specifics of DNA damage *in situ* (4, 8, 9).

The rationale for employing *in situ* labeling of DNA breaks for apoptosis detection is the assumption that DNA fragmentation is a characteristic apoptotic marker. In spite of the ever-increasing number of exceptions (10–13), orderly and extensive DNA fragmentation is considered a hallmark feature of apoptosis (1, 14). When combined with morphological verification and immunohistochemistry, it can be used as a sensitive and specific indicator of this process.

T7 DNA polymerase-based *in situ* labeling is faster, cheaper, and simpler than most other *in situ* approaches. The added advantage is in its ability to identify apoptotic cells at an earlier stage compared to TUNEL, the other popular technique (3).

Native T7 DNA polymerase is a heterodimeric 92-kDa enzyme. It was originally isolated from *E. coli* infected by T7 bacteriophage. It is composed of a polymerase proper- the T7 gene 5 protein, encoded by T7 bacteriophage, and a "booster" protein – thioredoxin, encoded by the *E. coli* bacterium. Both parts of the complex are essential for the efficient polymerization reaction. Without thioredoxin, the gene 5 protein is a nonprocessive DNA polymerase. It cannot continuously incorporate nucleotides without dissociating from the primer template. Thioredoxin significantly increases its affinity for the template, so that the DNA synthesis becomes processive for thousands of nucleotides (15, 16). By contrast, the Klenow fragment incorporates about 15 bases before dissociating, T4 DNA polymerase – about 10 bases, and T5 DNA polymerase – about 180 bases (17, 18). In fact T7 DNA polymerase is one of the most processive thermolabile DNA polymerases, with higher processivity than Pol I, Klenow fragment, or T4 DNA polymerase (15, 16). Importantly, T7 DNA polymerase efficiently incorporates not only normal dNTPs, but also biotinylated nucleotides (3).

The optimal polymerization templates for T7 DNA polymerase are 5' overhangs and single-stranded gaps in double-stranded DNA with available 3' OH DNA ends. The enzyme does not possess the 5′→3′ exonuclease activity, and consequently cannot catalyze strand displacement synthesis at nicks (phosphodiester breaks) in duplex DNA (19). In solution single-stranded DNA can serve as a polymerization template-primer for T7 DNA polymerase. However, in tissue sections, where DNA is fixed and immobile, this is unlikely to happen because the process starts only when a single DNA strand loops back and hybridizes to itself. Any unpaired nucleotides at the 3′ terminus are then removed by the 3′→5′ exonuclease activity until a base-paired terminus is reached and is used to prime the new strand synthesis (15, 19).

A significant feature of native T7 DNA polymerase, important for its labeling applications, is its strong 3′→5′ exonuclease (proofreading) activity. The activity degrades a DNA strand in the direction opposite to the direction of polymerization. The hydrolysis reaction initiates from terminal 3′ OH ends in both single- and double-stranded DNA. The double-strand exonuclease activity of T7 DNA polymerase is about six times stronger than its single-strand activity (20). However, in the presence of four dNTPs, the 3′→5′ exonuclease activity of T7 DNA polymerase is suppressed. dNTPs at 30 µM concentration completely inhibit 3′→5′ exonuclease activity for either single- or double-stranded DNA (21). This probably reflects the inaccessibility of the 3′ termini to the exonuclease because of the ongoing polymerization reaction at the termini (21, 22).

In the absence of nucleoside triphosphates, the enzyme initiates DNA hydrolysis at nicks and 3′OH termini in DNA (15). Starting from the available 3′OH ends, it will reduce 3′ overhangs and blunt-ended DNA breaks into long 5′ overhangs. It will expand nicks into gaps and will enlarge the single-stranded gaps in the partially double-stranded DNA. Excessive incubation can result in complete DNA degradation.

The high processivity, strong 3′→5′ exonuclease activity, and the ability to tolerate various tagged nucleoside triphosphates are the properties most important for the application of T7 DNA polymerase for *in situ* labeling of DNA damage. For additional key biochemical properties and constants of this enzyme, see Note 1.

What molecular targets are labeled by T7 DNA polymerase and justify its application for apoptosis detection?

Studies of apoptotic DNA fragmentation showed that the DNA breaks in it were either blunt-ended or possessed short, single nucleotide overhangs at 3′ or 5′ ends (23–26). The cutting properties of the major executioner apoptotic nucleases were found to be responsible for this effect, so that DNase I type enzymes produced blunt-ends or single-base 3′ overhangs (23, 24), whereas caspase-3-activated deoxyribonuclease (CAD) generated mainly single-base 5′ overhangs (25). From this standpoint, apoptosis labeling techniques that concentrate on the detection of these specific types of DNA brakes would be more specific. This was confirmed with the introduction of *in situ* ligation, which is highly specific for apoptosis and uses T4 DNA ligase to label selectively blunt ends or single-base 3′ overhangs (23, 27, 28). However, T7 DNA polymerase preferentially fills in 5′ overhangs and single-stranded gaps in double-stranded DNA, so that neither blunt ends nor 3′ overhangs are normally labeled. T7 DNA polymerase also cannot label nicks (single phosphodiester bond interruptions) in a DNA duplex because it is unable to displace the 5′-terminal strand at the nick.

Blunt ends and 3′ overhangs become the targets of polymerase-based labeling due to the utilization of replacement synthesis. The replacement synthesis procedure was initially developed for *in vitro* applications of T4 DNA polymerase and employed the strong 3′→5′ exonuclease activity of that enzyme (29). The approach was adapted for T7 DNA polymerase because it possesses an equally strong 3′→5′ exonuclease activity. To make use of the replacement synthesis, the labeling procedure includes the step where, prior to the labeling reaction, the tissue sections are pre-incubated with T7 DNA polymerase without adding dNTPs. This pretreatment significantly expands the spectrum of DNA breaks labeled by T7 DNA polymerase because its 3′→5′ exonuclease activity converts blunt-ended DNA breaks and 3′ overhangs into the detectable 5′ overhangs. It also expands nicks into single-stranded gaps and lengthens the pre-existing 5′ overhangs and gaps in DNA, making them more prominent. Although an overly long pre-incubation can extensively degrade DNA, short pre-incubation serves as a signal amplifier.

Two modifications of the *in situ* labeling technique using T7 DNA polymerase are described in this chapter. Both presented protocols efficiently label apoptotic cells and proceed through similar technical steps. The first technique was initially introduced by Wood et al. (6, 7). It employs non-fluorescent detection and labels apoptosis in unfixed, fresh-frozen sections and in cultured cells. The second protocol was developed by Tanaka and co-authors (3). It uses fluorescence labeling and is applicable to paraffin-embedded sections. This technique can also be performed in a double-staining format with immunohistochemistry for additional verification of cell death by apoptosis-related antibodies. The formalin-fixed paraffin-embedded sections generally provide significantly better morphology compared to frozen sections and permit individual morphologic evaluations of labeled cells.

The T7 DNA polymerase-based technique was shown to label cells in an earlier stage of apoptosis compared to TUNEL (3). In experiments, the TdT-based and T7 polymerase-based labeling of apoptotic rat brain neurons were compared with immunohistochemical detection of active caspase-3 in these cells. The earlier stage was indicated because the T7 DNA polymerase labeled more neurons with cytoplasmic localization of active caspase-3 compared with the TdT technique (40% vs. 13%). The T7-labeled cells also had morphologic signs of earlier apoptosis, such as significantly larger nuclei than those labeled by TdT (17.0 ± 2.3 μm^2 vs. 9.1 ± 0.8 μm^2) (3).

The reasons for this preference toward earlier apoptosis detection are not completely clear. The explanation provided in the original paper is not convincing. It stated that TdT preferentially labels single-stranded DNA, whereas T7 DNA polymerase exclusively labels DNA duplexes, It stated that TdT preferentially labels single-stranded DNA, whereas T7 DNA polymerase exclusively labels DNA duplexes, concentration of which supposedly decreases as apoptosis progresses (3). A more likely explanation could be in the ability of T7 DNA polymerase to expand the few nicks or single-stranded DNA breaks produced in earlier apoptosis into wide single-stranded gaps that are efficiently labeled by the enzyme.

Below we describe two protocols using T7 DNA polymerase for apoptosis detection in frozen cells and tissue sections and in formalin-fixed tissues. The descriptions are aimed at the beginner in this field without previous experience with the method.

2. Materials

1. Cryostat-cut tissue sections (20-μm thick) from fresh frozen tissue blocks, unfixed cultured cells, or 5–6-μm-thick sections cut from paraformaldehyde-fixed, paraffin-embedded tissue blocks.
2. Superfrost Plus microscope slides (Fisher Scientific, Pittsburgh, PA) for cryostat sections and unfixed cells.
3. ProbeOn™ Plus charged and precleaned slides (Fisher Scientific, Pittsburgh, PA) for the sections cut from paraformaldehyde-fixed, paraffin-embedded tissue blocks. Other slide brands can also be used if they retain tissue well.
4. 10× Phosphate-buffered saline (PBS). For 1 L: dissolve 2.0 g KH_2PO_4, 11.5 g Na_2HPO_4, 80 g NaCl, 2 g KCl in 800 mL distilled water, mix and adjust pH to 7.4 with HCl. Add water to bring the volume to 1 L.
5. 4% Formaldehyde. Dilute 37% formaldehyde (Sigma, St. Louis, MO) in PBS 1:9. Store at 4°C, protect from light. Avoid using after 3 weeks of storage.
6. 70%, 80%, 95%, and 100% ethanol.
7. Chloroform (Sigma, St. Louis, MO).
8. Unmodified T7 DNA polymerase 10 U/μL (Epicentre Biotechnologies Inc., WI) (see Note 2).

9. 10× T7 Reaction buffer (Epicentre Biotechnologies Inc., WI) – the reaction buffer for T7 DNA polymerase: 330 mM Tris–acetate, pH 7.8, 66 mM potassium acetate, 100 mM magnesium sulfate, 5 mM dithiothreitol.
10. 0.5 mM EDTA, pH 8.0, in PBS.
11. Nucleotides: 200 mM dGTP, dTTP, dCTP, 20 mM dATP and 40 mM biotin-14-dATP (Sigma, St. Louis, MO).
12. Streptavidin–Cy2 conjugate or streptavidin–fluorescein conjugate or streptavidin–Texas Red conjugate (Molecular Probes, Eugene, OR).
13. 0.1% BSA (USB Corporation, Cleveland, OH).
14. Biotin-blocking reagents, such as those provided in the Cell and Tissue Staining kit (R&D Systems, Minneapolis, MN).
15. Streptavidin–peroxidase conjugates (Rockland Immuno-chemicals, Inc., Gilbertsville, PA).
16. ImmPACT™ DAB Peroxidase Substrate (Vector Laboratories, Burlingame, CA). DAB is a potential carcinogen.
17. Permount mounting media (Fisher Scientific, Pittsburgh, PA).
18. Vectashield (Vector Laboratories, Burlingame, CA) antifading-counterstaining solution with DAPI (1 µg/mL). Store at 4°C in the dark. DAPI is a potential carcinogen.
19. Glass or plastic coverslips.
20. Fluorescent or conventional microscope with appropriate filters and objectives.

3. Methods

3.1. Protocol 1. Fresh-Frozen Cryostat Sections or Unfixed Cells

1. Air-dry unfixed cryostat sections, and then place them for 5 min in PBS. For cells, air-dry 100 µL of cells (approx. 1×10^6 cells/mL) on a glass microscope slide, then wash gently in PBS.
2. Place sections or cells in 4% formaldehyde for 10 min, then wash 2 × 5 min in PBS.
3. Dehydrate sections or cells through serial ethanol washes for 2 min in each 70%, 80%, 95%, and 100% ethanol.
4. Place in chloroform for 2 min.
5. Rehydrate by passing through graded ethanol concentrations: 95% Ethanol – 2 min; 70% Ethanol – 2 min; PBS – 3 × 5 min (see Note 3).
6. Apply T7 Reaction Buffer. Incubate for 10 min at room temperature (23°C) (see Note 4).

7. Replace buffer with 0.1 U/μL unmodified T7 DNA polymerase in T7 reactionbuffer. Incubate for 10 min at room temperature (23°C).
8. Replace with 0.1 U/μL unmodified T7 DNA polymerase, 0.2 mM each dCTP, dGTP, dTTP, 0.02 mM dATP, 0.04 mM biotin-14-dATP in T7 reaction buffer. Incubate for 10 min at room temperature (23°C) (see Note 5).
9. Aspirate the solution and stop the reaction by adding 0.5 mM of EDTA, pH 8.0, in PBS.
10. Wash sections or cells in three changes of PBS.
11. For color development, incubate the slide in 0.1% BSA for 30 min, then rinse twice in PBS.
12. Cover the specimen with 10 μg/mL of streptavidin-peroxidase in PBS for 30 min in the dark.
13. Wash the specimen in three changes of PBS.
14. Incubate the specimens with ImmPACT™ peroxidase DAB substrate solution for 5–15 min at room temperature. Wear gloves as DAB is a suspected carcinogen.
15. Stop the reaction by washing in three changes of distilled water (see Note 6).
16. For viewing, the specimen is dehydrated for 10 min each in 70%, 95%, and then 100% ethanol, and clarified in xylene for 10 min. Mount a coverslip using Permount. A positive apoptosis reaction is indicated by a dark purple color (see Note 7).

3.2. Protocol 2. Formaldehyde-Fixed, Paraffin-Embedded Tissues

1. Place the sections (5–6-μm-thick) cut from paraformaldehyde-fixed, paraffin-embedded tissue blocks onto ProbeOn™ Plus charged and precleaned slides (Fisher Scientific, Pittsburgh, PA). Other slide brands can also be used if they retain tissue well.
2. Dewax the sections in xylene for 15 min, transfer to a fresh xylene bath for additional 5 min.
3. Rehydrate by passing through graded ethanol concentrations: 95% ethanol – 2×5 min; 80% ethanol – 5 min; water – 2×5 min (see Note 8).
4. Equilibrate in T7 buffer for 10 min at room temperature (23°C) (see Note 4).
5. Replace buffer with 1 U/μL unmodified T7 DNA polymerase in T7 reaction buffer. Incubate for 10 min at room temperature (23°C) (see Note 9).
6. Apply the full reaction solution composed of 0.1 U/μL of T7 DNA polymerase, 200 mM dGTP, dTTP, dCTP, 20 mM dATP, and 40 mM biotin-14-dATP in T7 buffer. Incubate for 1 h at 37°C in a humidified chamber (see Note 10).

7. Aspirate the solution and stop the reaction by incubating for 10 min in 50 mM EDTA, pH 8.0, in PBS at room temperature.
8. Wash the sections 3 × 10 min in distilled water by gently immersing the slides in a coplin jar containing water at room temperature.
9. Dilute 2 µL of streptavidin–Cy2 conjugate in 1 mL of sodium bicarbonate buffer (see Note 11). Add 100 µL of this solution to the section. Incubate for 45 min at room temperature in a covered humidified chamber (see Note 12).
10. Wash 3 × 10 min in distilled water.
11. Add Vectashield with DAPI and a coverslip (see Note 7).

4. Notes

1. Here is a brief synopsis of the properties of native (unmodified) T7 DNA polymerase relevant to its usage for *in situ* labeling of DNA breaks and apoptosis.

   Temperature sensitivity. Optimal temperature 37°C; increasing the temperature to 42°C inhibits polymerase and exonuclease activities with half-time ~3 min (in the absence of Mg^{2+}) (21). Incubation at 75 C for 10 min completely inactivates the enzyme (in the presence of Mg^{2+}). Lowering the temperature below 37°C reduces polymerization rates and processivity. One week at room temperature (23°C) results in 30% loss of enzymatic activity (15). Enzyme is stable when stored at –20°C.

   pH sensitivity. Optimal pH 7.6–7.8 in potassium phosphate buffer. The enzyme is less active (retains ~40% activity) in Tris buffer in the same pH range (30).
   The enzyme is remarkably stable at alkaline pH values, whereas it is irreversibly inactivated at acidic pH (retains 83% activity after 2 h at pH 12, while it is only 4% active after 2 h at pH 2.2) (31).

   Sensitivity to ions. Mg^{2+} or Mn^{2+} ions are absolutely required for polymerization reaction. The Mg^{2+} optimum is 10 mM at 150 µM of each of the four nucleotides (31). Optimal concentration of Mn^{2+} is 0.1 mM (32). Substitution of Mn^{2+} for Mg^{2+} changes some enzyme properties such as fidelity (15).
   KCl at 20 mM increases enzyme activity by 20% followed by complete inhibition at 300 mM (31).

   Effects of chemicals. Sensitive to sulfhydryl inactivation and in order to remain active requires 2-mercaptoethanol or dithiothretol. When 2-mercaptoethanol is omitted from the incubation mixture, the reaction proceeds at only half the optimal rate (30). EDTA chelates Mg^{2+} and is a strong polymerase inhibitor.

Use of 30 µM dNTPs (but not rNTPs) inhibits 3′→5′ single-strand and double-strand exonuclease activities (21).

Nucleotide incorporation: Shows the comparatively high K_m value (Michaelis constant) for an equimolar mixture of the four dNTPs = 80 µM (31). The rate of synthesis is ~100 nucleotides per second and a single enzyme molecule is capable of synthesizing thousands of nucleotides from the same primer without dissociating (33).

Efficiently incorporates ddNTPs, dNTPαS, dITP, 7-deaza-dGTP, and biotin-14-dATP (3, 15). The efficiency of incorporation significantly increases when Mn^{2+} (2 mM) is substituted for Mg^{2+} (5 mM) (32). The native enzyme cannot incorporate 1,N^6 etheno-2′-deoxyadenosine 5′-triphosphate unlike modified enzyme without exonuclease activity (33).

Native T7 DNA polymerase cannot catalyze strand-displacement synthesis and stalls when synthesizing DNA at a nick (33). The unmodified native enzyme also does not polymerize nucleotides through hairpin structures and terminates the extension within three to four nucleotides of the base of the hairpin (33).

2. Unmodified, i.e., native T7 DNA polymerase is used in this labeling technique. In modified T7 DNA polymerase, such as Sequenase™ from USB Corporation, the 3′→5′ exonuclease activity of the original enzyme has been removed. Although such an enzyme will still label 5′ overhangs, its utility for apoptosis detection *in situ* has not been tested.

3. Alternately, after fixing, the cells or sections can be permeabilized by incubating 5 min in 0.1% Triton X-100, followed by three washes in PBS. To suppress endogenous peroxidase activity, treat the sections with 0.1% hydrogen peroxide for 20 min. To block endogenous biotin, incubate with biotin-blocking reagents, such as those provided in the Cell and Tissue Staining kit. Wash 2× in PBS.

4. Pre-incubation with T7 buffer ensures even saturation of the section prior to addition of the enzyme and dNTPs for uniform staining. Use 50–200 µL of solution, as needed, to cover the section or cells.

5. Direct fluorescent labeling is possible by using fluorescent dNTPs instead of biotin. To ensure the absence of nonspecific background staining due to interactions between fluorophores and charged groups in cellular proteins, the signal should be verified using control sections incubated with fluorescent dNTPs without the enzyme.

6. At this stage, if required, the specimen may be counterstained using Eosin, Methyl Green, or Hematoxylin.

7. A mock reaction is recommended as a regular control in order to rule out nonspecific background staining. In a mock reaction,

all labeling reaction components except the enzyme are present. A positive control with verified apoptosis is also recommended (such as dexamethazone-treated rat thymus, etc. (23)).

8. At this point, many protocols that use other DNA polymerases include a tissue permeabilization step accomplished by proteinase K treatment (4). This permits better access of polymerases to the sites of the breaks in formaldehyde-fixed sections. However, the proteinase K step is absent in all T7 DNA polymerase protocols, developed for either frozen or formalin-fixed sections (3, 6, 7, 34). In frozen sections, the proteinase K digestion can easily detach frozen sections from glass slides. Therefore, cells and sections are in this case permeabilized by incubation for 5 min in 0.1% Triton X-100, followed by three washes in PBS. However, in case of formalin-fixed sections, the absence of proteinase K treatment limits the amounts of detectable DNA breaks and could be one of the factors responsible for the detection of earlier apoptotic DNA breaks by this modification of the T7 DNA polymerase technique.

9. Note that in case of formalin-fixed sections, the amount of T7 DNA polymerase in the pre-incubation solution is ten times higher compared to that in the case of fresh-frozen sections.

10. Note the higher temperature (37°C) and the longer time of the labeling reaction (1 h) in case of formalin-fixed sections compared to that in the case of fresh-frozen sections and unfixed cells.

11. Other fluorescent conjugates can be used instead, such as streptavidin–fluorescein, streptavidin–Texas Red, etc. Some tissue sections have higher nonspecific background labeling with one particular type of fluorophore but not the other. In this case, using a different conjugate can significantly improve the signal/noise ratio.

12. Non-fluorescent signal sometimes produces less background compared with fluorescent labeling. It can be achieved by substituting streptavidin-peroxidase for streptavidin-fluorophores, with subsequent color development by using ImmPACT™ peroxidase DAB substrate solution as described in Subheading 3.1. In the color development step, incubate tissue sections with the DAB substrate working solution at room temperature until suitable staining develops. The optimal development time is determined experimentally and generally is in the 2–15 min range.

References

1. Walker P.R., Carson C., Leblanc J., and Sikorska M. (2002) Labeling DNA damage with terminal transferase: applicability, specificity and limitations, in *In Situ Detection of DNA Damage: Methods and Protocols* (Didenko, V.V. ed.) Humana, Totowa, NJ, pp. 3–19.

2. Loo D.T. (2002) TUNEL assay: an overview of techniques, in *In Situ Detection of DNA Damage: Methods and Protocols* (Didenko, V.V. ed.) Humana, Totowa, NJ, pp. 21–30.

3. Tanaka M., Momoi T., and Marunouchi T. (2000) In situ detection of activated caspase-3 in apoptotic granule neurons in the developing cerebellum in slice cultures and in vivo. *Dev. Brain Res.* **121**, 223–28.

4. Dierendonck J.H. (2002) DNA damage detection using DNA polymerase I or its Klenow fragment: applicability, specificity, limitations, in *In Situ Detection of DNA Damage: Methods and Protocols* (Didenko, V.V. ed.) Humana, Totowa, NJ, pp. 81–108.

5. Thiry M. (2002) In situ nick translation at the electron microscopic level, in *In Situ Detection of DNA Damage: Methods and Protocols* (Didenko, V.V. ed.) Humana, Totowa, NJ, pp.121–130.

6. Wood K.A., Dipasquale B., and Youle R.J. (1993) In situ labeling of granule cells for apoptosis-associated DNA fragmentation reveals different mechanisms of cell loss in developing cerebellum. *Neuron* **4**, 621–32.

7. Wood K.A. and Youle R.J. (1995) The Role of Free Radicals and ~53 in Neuron Apoptosis in vivo. *J. Neurosci.* **15**, 5851–57.

8. Didenko V.V., Ngo H., and Baskin D.S. (2003) Early necrotic DNA degradation: presence of blunt-ended DNA breaks, 3' and 5' overhangs in apoptosis, but only 5' overhangs in early necrosis. *Am. J. Pathol.* **162**, 1571–78.

9. Otsuki Y. and Ito Y. (2002) Quantitative differentiation of both free 3' OH and 5' OH DNA ends using terminal transferase-based labeling combined with transmission electron microscopy, in In Situ *Detection of DNA Damage: Methods and Protocols* (Didenko, V.V. ed.) Humana, Totowa, NJ, pp. 41–54.

10. De Felici M., Lobascio A.M., and Klinger F.G. (2008) Cell death in fetal oocytes: many players for multiple pathways. *Autophagy* **4**, 240–42.

11. Hirata H., Hibasami H., Yoshida T., Morita A., Ohkaya S., Matsumoto M., Sasaki H., and Uchida A. (1998) Differentiation and apoptosis without DNA fragmentation in cultured Schwann cells derived from wallerian-degenerated nerve. *Apoptosis* **3**, 353–60.

12. Métrailler-Ruchonnet I., Pagano A., Carnesecchi S., Ody C., Donati Y., and Barazzone Argiroffo C. (2007) Bcl-2 protects against hyperoxia-induced apoptosis through inhibition of the mitochondria-dependent pathway. *Free Radic Biol Med.* **42**, 1062–74.

13. Cohen G.M., Sun X.-M., Snowden R.T., Dinsdale D., and Skilleter D.N. (1992) Key morphological features of apoptosis may occur in the absence of internucleosomal DNA fragmentation. *Biochem. J.* **286**, 331–34.

14. Zhang J. and Xu M. (2002) Apoptotic DNA fragmentation and tissue homeostasis. *Trends Cell Biol.* **12**, 84–9.

15. Eun H.-M. (1996) *Enzymology Primer for Recombinant DNA Technology.* Academic Press, San Diego, CA. pp. 377–407.

16. Tabor S., Struhl K., Scharf S.J., and Gelfand D.H. (1997) DNA-dependent DNA polymerases, in *Current Protocols in Molecular Biology* (Ausubel F.M., Brent R., Kingston R.E., Moore D.D., Seidman J.G., Smith J.A., and Struhl, K. eds.), John Wiley & Sons, Hoboken, NJ, pp. 3.5.1–3.5.15.

17. Das S.K. and Fujimura R.K. (1979) Processiveness of DNA polymerases. A comparative study using a simple procedure. *J. Biol. Chem.* **254**, 1227–32.

18. Bambara R.A., Uyemura D., and Choi T. (1978) On the processive mechanism of *Escherichia coli* DNA polymerase I. *J. Biol. Chem* **253**, 413–23.

19. Lehman I.R. (1981) T-Phage DNA Polymerases, in *The Enzymes* **14** (Boyer P.D. ed.) Academic Press, New York p. 64.

20. Tabor S., Huber H.E., and Richardson C.C. (1987) Escherichia coli thioredoxin confers processivity on the DNA polymerase activity of the gene 5 protein of bacteriophage T7. *J. Biol. Chem.* **262**, 16212–223.

21. Adler S. and Modrich P. (1979) T7-induced DNA polymerase. Characterization of associated exonuclease activities and resolution into biologically active subunits. *J. Biol. Chem.* **254**, 11605–614.

22. Hori K., Mark D.F., and Richardson C.C. (1979) Deoxyribonucleic acid polymerase of bacteriophage T7. Purification and properties of the phage-encoded subunit, the gene 5 protein. *J. Biol. Chem.* **254**, 11591–97.

23. Didenko V.V. and Hornsby P.J. (1996) Presence of double-strand breaks with single-base 3' overhangs in cells undergoing apoptosis but not necrosis. *J. Cell Biol.* **135**, 1369–76.

24. Didenko V.V., Tunstead J.R., and Hornsby, P.J. (1998) Biotin-labeled hairpin oligonucleotides. Probes to detect double-strand breaks in DNA in apoptotic cells. *Am. J. Pathol.* **152**, 897–902.
25. Widlak P., Li P., Wang X., and Garrard W.T. (2000) Cleavage preferences of the apoptotic endonuclease DFF40 (caspase-activated DNase or nuclease) on naked DNA and chromatin substrates. *J Biol Chem.* **275**, 8226–32.
26. Staley K., Blaschke A.J., and Chun J. (1997) Apoptotic DNA fragmentation is detected by a semiquantitative ligation-mediated PCR of blunt DNA ends. *Cell Death Diff.* **4**, 66–75.
27. Didenko V.V. (2002) Detection of specific double-strand DNA breaks and apoptosis in situ using T4 DNA ligase, in *In Situ Detection of DNA Damage: Methods and Protocols* (Didenko V.V. ed.) Humana, Totowa, NJ, pp. 143–51.
28. Al-Lamki R.S., Skepper J.N., Loke Y.W., King A., and Burton G.J. (1998) Apoptosis in the early human placental bed and its discrimination from necrosis using the in-situ DNA ligation technique. *Hum Reprod.* **13**, 3511–9.
29. Challberg M.D. and Englund P.T. (1980) Specific labeling of 3' termini with T4 DNA polymerase. *Methods Enzymol.* **65**, 39–43.
30. Grippo P. and Richardson C.C. (1971) Deoxyribonucleic acid polymerase of bacteriophage T7. *J. Biol. Chem.* **246**, 6867–73.
31. Nordstrom B., Randahl H., Slaby I., and Holmgren A. (1981) Characterization of bacteriophage T7 DNA polymerase purified to homogeneity by antithioredoxin immunoadsorbent chromatography. *J. Biol. Chem.* **256**, 3112–17.
32. Tabor S. and Richardson C.C. (1989) Effect of manganese ions on the incorporation of dideoxynucleotides by bacteriophage T7 DNA polymerase and Escherichia coli DNA polymerase I. *Proc. Natl. Acad. Sci. U S A* **86**, 4076–80.
33. Tabor S. and Richardson C.C. (1989) Selective inactivation of the exonuclease activity of bacteriophage T7 DNA polymerase by in vitro mutagenesis. *J. Biol. Chem.* **264**, 6447–58.
34. Wood K.A. (1994) Unmodified T7 DNA Polymerase for the In Situ Detection of DNA Fragmentation Associated with Apoptosis. Epicentre Forum **1**, 1. (http://www.epibio.com/fl_2/fl_2t7.asp).

Chapter 5

In Situ Ligation: A Decade and a Half of Experience

Peter J. Hornsby and Vladimir V. Didenko

Abstract

The *in situ* ligation (ISL) methodology detects apoptotic cells by the presence of characteristic DNA double-strand breaks. A labeled double-stranded probe is ligated to the double-strand breaks *in situ* on tissue sections. Like the popular TUNEL assay, ISL detects cells in apoptosis based on the ongoing destruction of DNA by apoptotic nucleases. In comparison to TUNEL, it is more specific for apoptosis versus other causes of DNA damage, both repairable damage and necrosis. In the decade and a half since its introduction, ISL has been used in several hundred publications. Here we review the development of the method, its current status, and its uses and limitations.

Key words: ISL, ISOL, TUNEL, Apoptosis, *In situ* assays, Pathophysiology, Nucleases

1. Introduction

In the early 1990s, the authors were studying the role of cell cycle inhibitors such as p21WAF1/CIP1 in tissue damage and aging (1). As part of these studies, we needed an accurate measure for the incidence of apoptosis in the tissues we were studying. This is a frequent issue in biomedical science and pathological diagnosis: the need for an accurate measure of the number and location of apoptotic cells in fixed tissue. At the time, and continuing to the present day, the popular method for detecting apoptotic cells was the TUNEL (terminal deoxynucleotidyl transferase dUTP nick end labeling) technique (2). This labeling method depends on the ability of terminal deoxynucleotidyl transferase (TdT) to add nucleotides to breaks in DNA. During the late stages of apoptosis, double-strand breaks are produced when activated nucleases cleave DNA. Terminal transferase is used in this assay to add labeled nucleotides to the 3′ ends of cleaved DNA molecules, thus providing a sensitive assay for detecting apoptotic cells in tissues (see Note 1).

In our experiments, it soon became apparent that many cells that were TUNEL positive could not in fact be undergoing apoptosis. In particular, in studies of the adrenal cortex, we found that some treatments caused large numbers of cells to become TUNEL+ (1). Despite this, the gland did not disappear over the next few days – as would have been expected if such a large portion of the gland were comprised of cells that were actually in a terminal phase of cell death. Instead, the gland appeared to make a full recovery. This indicated that the strand breaks detected by the TUNEL assay were not indicative of apoptosis, but represented sites of temporary damage and potential repair (see Note 2).

We then set out to devise an assay to detect strand breaks in apoptotic cells that would be more specific than the TUNEL assay. We thought of adapting the concept of *in situ* detection of DNA strand breaks by using a process that did not depend on terminal transferase end labeling. The distinctive feature of apoptosis is the presence of double-strand breaks, which may have either blunt or staggered ends. We wanted to label double-strand breaks in such a way that the double-stranded nature of the DNA ends would become an essential part of the labeling process; single-stranded DNA ends would not be labeled. We developed the idea that double-strand breaks could be labeled by ligation of a double-stranded DNA tag. This method is in essence an *in situ* adaptation of ligation methods used commonly in molecular biology, both during subcloning procedures and in analytical procedures, such as ligation-mediated PCR (3) (see Note 3).

In our first experiments we used PCR to make digoxigenin-tagged double-stranded fragments that could be ligated to DNA double-strand breaks in apoptotic cells (4). These probes were incubated with deparaffinized or frozen sections of tissue in a mix of buffer, T4 DNA ligase, and ATP (required for the ligase reaction). The probes were allowed to become covalently attached to available sites on the section, and unattached probes were then washed away. The attached probes were detected by an antibody against digoxigenin or were directly observed by fluorescence microscopy. Practical tests of the method showed that it did indeed label apoptotic cells specifically. As a positive control we used rat thymus 24 h after administration of glucocorticoid, a model for apoptosis well-established by previous investigators (5). These experiments resulted in the introduction of a new assay for *in situ* detection of apoptosis.

2. An Assay with Many Names

The *in situ* ligation assay was designed to mark apoptotic cells via detection of two specific types of DNA damage. It selectively labels 5′ phosphorylated double-strand DNA breaks, which have either blunt ends or 3′ single base overhangs. Its basic components are the enzyme T4 DNA ligase and a DNA-based probe, which is ligated to the ends of cellular DNA breaks.

In the initial 1996 paper, where we described the technique using PCR fragments as probes, we did not give the assay a special name (4). This later resulted in an unusual consequence of the ligation assay having multiple different names.

Soon after the assay introduction, as we gained more experience with *in situ* ligation, we realized that the major practical problems were to make sufficient amounts of the PCR product and to purify it away from unincorporated labeled dNTPs, which could produce increased background on the section. We thought that making a probe chemically rather than enzymatically might solve these problems. To that end, we designed a double-stranded oligonucleotide that could be used to label double-strand breaks. The first generation of these hairpin probes had a stem-loop configuration resembling a tennis racket and became known as "looped hairpins" (6) (see Note 4). The design offered the advantages of structural uniformity and stability. The probes carried five biotins located in the loop area and were easy to prepare in large quantities. However, the loop area had some tendency to stick to sections and the *in situ* labeling procedure still required lengthy washing steps to remove the unligated probe. To address these issues, we designed a new oligonucleotide probe which became known as a "loopless hairpin" (7). In the new probe design, the reactive single-stranded loop was eliminated. In order to avoid steric hindrance problems and to create better conditions for the reaction between biotin and streptavidin in probe detection, the number of biotins was reduced from five to one. The design substantially reduced the cost of the probe and simplified the assay, transforming it into the *in situ* oligonucleotide ligation technique (ISOL), a convenient and robust modification of ISL methodology detecting apoptosis and DNA damage in tissue sections.

These developments came one after the other in short intervals, so all of the different probe designs were put into use in rapid succession and were employed concurrently. As a result, the ISL assay is known under many names depending on the probe design used by a particular research group. In general, these names follow the evolution of the ligation probes. So that when PCR probes are employed, the assay is presented as "*Taq* and *Pfu* labeling techniques" (8–10) or "*Taq* and *Pfu* polymerase *in situ* ligation assay" (11–16). The "looped" probes usage is acknowledged by such names as "hairpin probe assay" (17) or "HPP staining" (18–20). The arrival of the "loopless" probe and the popular commercial kit (ApopTag® Peroxidase *In Situ* Oligo Ligation kit from Millipore) which used it, resulted in the names of "*in situ* hairpin-1 ligation assay" (21) and "*in situ* oligonucleotide ligation" (ISOL) (22, 23). In addition to those, several more general or exotic names are simultaneously used, such as "*in situ* DNA ligase method" (24), "3'-overhang ligation" (25), "PCR *in situ* ligation assay" (26), and even "the Didenko and Hornsby ligation technique" (25).

We, as many others, prefer the most general title of "*in situ* ligation" (ISL) which is not influenced by variations in probe design (27–31).

3. Technological Evolution of *In Situ* Ligation Assay

The *in situ* ligation technique underwent several cycles of redevelopment since its inception in 1996. The majority of improvements concentrated in three areas: new ligation probes, expansion of detection targets, and increased sensitivity of detection.

3.1. Ligation Probes

We have already discussed the three different designs of *in situ* ligation probes. Although, with each new design our intent was to develop a probe which would surpass the earlier construct on all counts, yet all of these probe configurations appeared to have their own advantages and limitations. This probably explains why all of them are simultaneously employed by different groups.

The PCR-derived probes contain dozens of tags, which results in high sensitivity because every detected DNA break is labeled by multiple dyes. Another attractive feature of these probes is that they can be easily produced ad hoc when needed in any molecular biology laboratory. Their disadvantage is in the longer and more numerous washing steps because they tend to stick to tissue sections.

The advantage of the loopless hairpin probes, such as those used in the ApopTag ISOL kits, is the complete absence of non-specific background staining, as all unligated probes are easily washed away from the sections. Oligonucleotide probes also provide the opportunity to design any kind of double-stranded DNA ends, both for apoptotic cell labeling and for labeling any types of double-strand breaks in other biological materials. The hairpin configuration offers an efficient solution to the problem of forming double-stranded probes that are uncontaminated by single-stranded DNA. However, loopless hairpins only place a single tag at the end of each detected DNA break and, as a result, have the lowest sensitivity. This can be partially improved by using the postlabeling enzymatic amplification of signal (28).

The "looped" hairpins occupy a position intermediate between the PCR-derived and the "loopless" probes. They contain at least five tags but also possess a single-stranded region, which necessitates more vigorous washing of sections.

While the advances in ligation probes focused mainly on background reduction and more specific detection, the other direction in ISL evolution concentrated on increasing the variety of detected types of DNA damage and on co-labeling of other cellular markers.

3.2. Expansion of Targets for Detection

Per se the ISL assay exclusively detects only two DNA targets: it labels 5′ phosphorylated double-strand DNA breaks which have either blunt ends or 3′ single base overhangs. These particular types of breaks are important because they mark apoptotic cell death and can be used in its discrimination from necrosis. However, the spectrum of detectable biological targets and the utility of the assay can be expanded by combining ISL with other assays in dual- and triple-staining procedures. This advantage was explored in recent years by combining the ISL staining with other techniques, such as antibody immunohistochemistry, TUNEL, Klenow polymerase-based labeling, vaccinia topoisomerase-based labeling, and T4 DNA kinase section pretreatment. Such multilabeling procedures generate more information as compared with single-staining. They increase the variety of detected DNA breaks, show their relative distribution, and covisualize protein-based markers of cellular processes. The approach is also useful for validating results of individual techniques employed in experiments.

3.2.1. Colabeling with TUNEL

ISL-TUNEL colabeling was, perhaps, the earliest used combination. In terms of DNA breaks detection, the popular TUNEL assay labels both single- and double-strand breaks (32). The only requirement for labeling to work is the presence of free 3′OH groups. This assay is frequently employed to label apoptotic cells; however, it suffers from low specificity due to the presence of DNA breaks with 3′OH in many other processes (4, 33–37). The combination of TUNEL and ISL permits improved detection of apoptotic cells and better analysis of DNA damage present *in situ*. This costaining could not be used with PCR-derived probes because 3′OH groups at the unligated ends of the probes would be labeled by TUNEL. In addition, nicks and single-stranded breaks with 3′OH could be present in the long PCR fragments used as probes, thus creating additional priming sites for TUNEL. However, the problem of ISL-TUNEL co-labeling was resolved with the introduction of short hairpin oligo probes, permitting simultaneous visualization of single-strand and double-strand DNA breaks at the subcellular level (6, 28, 38).

3.2.2. Co-labeling with Klenow Polymerase

Although, on its own *in situ* ligation visualizes only double-strand DNA breaks that are either blunt-ended or possess 3′ overhangs, when modified, it also permits the selective detection of both 3′→5′ and 5′→3′ exonuclease activities *in situ*, via labeling of 3′ and 5′ DNA overhangs of all lengths. We developed this expanded approach for the purpose of comparative analysis of DNA damage in apoptosis and necrosis (39). The technique relies on Klenow enzyme pretreatment of cells before *in situ* ligation. If such pretreatment is performed in the presence of dNTPs it fills up all 3′ overhangs, converting them into blunt ends. However, when the pretreatment is done without added dNTPs, it reduces all 5′ overhangs to blunt ends, which can be subsequently detected by blunt end *in situ* ligation. Colabeling with Klenow polymerase has expanded the utility of ISL, permitting the detailed study of DNA damage *in situ*. Later this method was applied for analysis of free radical-induced DNA damage (29).

3.2.3. Dual Detection with T4 Polynucleotide Kinase Pretreatment

Based on the distribution of hydroxyl and phosphate groups at the ends of DNA breaks, two types of DNA breaks can potentially be present in apoptotic cells: $3'OH/5'PO_4$ breaks, generated by DNase I and DNase I-like nucleases; and $3'PO_4/5'OH$ DNA breaks with inverted distribution of these chemical groups, produced by DNase II and DNase II-like nucleases (40–42).

TUNEL and *in situ* ligation can visualize only $3'OH/5'PO_4$ breaks. These techniques cannot detect DNA fragmentation produced by nucleases generating $3'PO_4/5'$ breaks, because the TdT enzyme used in TUNEL does not react with $3'PO_4$ ends and T4 DNA ligase in the ligation assay is unable to attach the probe to 5′OH ends in DNA (43, 44).

Therefore, several protocols were developed in our laboratory for selective *in situ* labeling of double-strand DNA breaks with terminal 5′OH groups. The earliest of these procedures was a modification of the standard *in situ* ligation approach using a T4 kinase-based conversion of the breaks (45). The assay permitted fluorescent detection of 5′ hydroxyl-bearing double-strand breaks with blunt ends or short one to two base long 3′ overhangs.

The technique is based on the conversion of 5′ hydroxyls into 5′ phosphates with the help of the enzyme T4 polynucleotide kinase and their subsequent detection by *in situ* ligation. The procedure is performed in three stages. In the first stage, the unlabeled hairpin oligonucleotides are ligated to the section, blocking available 5′ phosphates, which may be present on the ends of DNA as a result of DNase I type nuclease activity. In the second stage, phosphate groups are added to the 5′OH ends by T4 polynucleotide kinase. In the third stage, an *in situ* ligation reaction is performed again using the hairpin probes.

The assay can be modified to simultaneously visualize both 5′ phosphates and 5′ hydroxyls using two different fluorophores. In this case, biotinylated hairpin probes (instead of unlabeled hairpins) are used in the first and third stages of the labeling reaction and visualized using different fluorophores. This double-detection assay (45), although highly useful, is time-consuming and we later substituted it with a much faster technology which uses vaccinia topoisomerase I (TOPO) (46, 47).

3.2.4. Dual-Detection with Vaccinia Topoisomerase I Assay

The vaccinia topoisomerase-based technique is a close relative of *in situ* ligation, and can be combined with ISL in a dual-detection procedure. First introduced in 2004 (46), combined staining using TOPO and ISL permits codetection of DNA breaks with 5'OH and 5' phosphate groups. In this regard, it is similar to the T4 kinase approach discussed above. However, the TOPO-ISL labeling is much faster and simpler. Its convenience resulted in a new dual-detection kit from Millipore (ApopTag® ISOL Dual Fluorescence Kit), combining these two techniques.

3.2.5. Codetection with Immunohistochemistry

The possibility of combining ISL with immunohistochemical detection was demonstrated in the first paper introducing the ligase technology (4). The paper showed that no false signal was generated in ISL even after prolonged heating of tissue sections during antigen retrieval, which is often needed for immunohistochemistry. The result was advantageous as compared to TUNEL, which produces intense artifactual staining in sections heated for antigen retrieval. Soon the dual detection approach combining ISL with antibody immunohistochemistry was tested by others and found useful (48). The codetection of DNA-based apoptotic markers and various protein-based targets has been reported in several publications (18, 20, 21, 49).

4. Advantages and Limitations of the ISL Methodology

It is clear that the ISL reaction gives researchers of apoptosis a valuable weapon in their arsenal to study this important cellular process. Probably, the first contribution of this technology was the clear realization of the imprecise and nonspecific nature of the widely-used TUNEL technique and other polymerase-based approaches, which in some situations provided overestimated numbers of apoptotic cells because they labeled other processes. This reinforced the necessity to use morphological and immunohistochemical verification of apoptosis when labeling it via detection of DNA breaks. Understandably, it resulted in a need to use the same strict verification standards in case of ISL too. Several studies addressed this issue and compared ISL with other apoptosis assays for specificity of apoptosis detection and its discrimination from necrosis.

Possibly, the first study which specifically evaluated ISL appeared in 1998 (48). It used ISL with PCR-derived probes, TUNEL, and immunohistochemical costaining. The study concluded that *in situ* DNA ligation was superior to TUNEL and, in association with immunohistochemistry, could readily distinguish apoptosis from necrosis, making it an attractive technique for discriminating these processes.

Since then, various ISL modifications and all three types of ligation probes were tested in many studies. For example, apoptosis detection by using ISL (PCR-derived probe), TUNEL, antibody-based labeling of single-stranded DNA and PARP-1 were compared (50). PARP-1 detection was chosen because this enzyme is a well-known target of the caspase protease activity associated with apoptosis. The study concluded that ISL and antibody-based single-stranded DNA detection, but not TUNEL, closely correlated with PARP-1 expression. Moreover, ISL and single-stranded DNA labeling were significantly more specific for apoptosis than TUNEL. The paper noted that, in contrast to T4 DNA ligase, the TUNEL assay stained cells marked by antibodies against proliferation marker Ki-67 or the splicing factor (SC-35), indicating false positivity. ISL did not stain these same cells (50). This confirmed the earlier work which first noted that cell proliferation, RNA synthesis, and splicing interfere with apoptosis detection by TUNEL (51).

In another work, which used the model of neonatal rat hypoxia-ischemia, the spatial and temporal activation of caspase-3 was correlated with three different markers of DNA damage (ISL, TUNEL and monoclonal antibody against single-stranded DNA) and with the loss of a neuronal marker, microtubule-associated protein 2 (52). The study demonstrated that ISL staining with "looped" 3' dA overhang hairpin probe (HPP) produced the best correlation with apoptotic caspase-3 activation. The same group later expanded their data by employing a slightly different set of approaches for apoptosis and DNA damage detection: Hoechst dye staining, TUNEL, and ISL with "looped" hairpin probes. In series of dual- and triple-staining experiments these techniques were combined with codetection of apoptosis-inducing factor (AIF), which triggers apoptosis in a caspase-independent manner (18).

Yet other research group compared TUNEL, ISEL (*in situ* end-labeling by Klenow polymerase), and ISL using "loopless" hairpins (ISOL). This group studied apoptosis of cardiomyocytes in explanted and transplanted hearts. It concluded that, in contrast to ISL, both TUNEL and ISEL had low specificity in this model. This drawback led to a high prevalence of false-positive results in myocardial studies and was exacerbated by the extreme care required in tissue processing for both these methods (22). While the study confirmed that the ISL assay detected apoptotic cells in a positive control specimen (involuting rat mammary gland), they found that the number of true apoptotic nuclei in all the hearts under study was extremely low. Indeed this is to be expected, based on the fact that apoptosis in mature organs is typically very low and also that the phase of apoptosis when DNA is actively undergoing cleavage by nucleases is very brief. Thus, one always should have a high degree of skepticism in accepting frequencies of apoptotic cells under normal and pathological conditions.

It is gratifying to see that many other groups, unaffiliated with the authors, have independently concluded that ISL is a more specific and sensitive approach for apoptosis detection, surpassing a variety of other *in situ* techniques. Although, the pros and cons of these different assays for apoptosis detection are now well-known (27, 28, 33–36, 53), nevertheless it would be true to say that ISL has not replaced TUNEL and other assays.

Quite the opposite, an assortment of multiple techniques is in current use for detecting apoptotic cells. Often this is done by combining several approaches in costaining. Such multiassay codetection compensates for the limitations of individual techniques. The most popular combinations include ISL performed in complex with TUNEL, active caspase-3 detection or other apoptosis-related antibodies, and electron microscopy. These multiangle assessments have now become standard in delineating contributions of apoptosis and necrosis in various systems.

For the sake of example, we can mention just some of the multi-technique approaches found in the literature. The combinations included ISL, TUNEL, nuclear morphology, and immunostaining for p53 and p21 (25); or ISL, TUNEL, H&E staining, and determination of DNA laddering by electrophoresis of labeled DNA (17); or the multitude of other groupings of techniques, all containing ISL (18, 26, 38, 39, 50, 52, 54).

In recent years, a new generation of methods aiming to detect apoptosis *in vivo* was introduced and tested (see other chapters in this volume). In an interesting application of the technique, ISL labeling was used as part of dual *in situ–in vivo* labeling of apoptosis. It was employed for verification of *in vivo* detection by (^{123}I) Annexin V. The degree of thymic apoptosis was codetermined in the same animals at 6 and 11 h after a single administration of dexamethasone by using two techniques: *in vivo* detection by radioactive Annexin V and *in situ* labeling by ISOL (55). *In situ* ligation demonstrated a 62- and 90-fold increase of the apoptotic index in thymic cortex at 6 and 11 h. Instead, the Annexin V-based signal fluctuated. It significantly decreased at 6 h and increased only 1.4-fold by the 11 h time-point. The study concluded that the specificity of the apoptotic signal provided by isotopic methods *in vivo* would always require confirmation by complementary *in vitro* techniques that verify the assessment of ongoing apoptosis accurately.

Are there limitations to the assay and reasons why it has not been used more often? In common with the TUNEL assay, ISL depends on the detection of DNA strand breaks. Like the TUNEL assay, it cannot be used as a sole method for apoptosis measurement. It is necessary to have other biological information to indicate that apoptosis is expected in the biological specimens being examined. Unfortunately, too often the TUNEL assay is used as proof of apoptosis, whether or not it is anticipated for other reasons. However, when apoptosis is well-established to occur in the tissue under the experimental circumstances being used, either the TUNEL or ISL assays will give equivalent results. For this reason, it is easy to understand why the familiar TUNEL assay has maintained its popularity.

Unlike the TUNEL assay, ISL is more specific for apoptosis versus other causes of DNA strand breaks, such as repairable DNA damage or necrosis. Thus, its particular usefulness is under experimental conditions where both apoptosis and other causes of DNA strand breaks are either simultaneously present or else both likely to occur. For example, in many pathophysiological circumstances of damage to organs, both apoptosis and other damage may be expected. The judicious use of the ISL methodology, in conjunction with other biochemical or cell biological assays, will enable the investigator to reliably determine the true extent of apoptosis present.

Interestingly, we noticed some "favorite" scientific fields where the popularity of ISL is much higher than in others. This refers to the fields of molecular cardiology and neuroscience. ISL was particularly attractive for these fields because they deal with terminally differentiated and highly specialized cells. The ligation technique is well-suited for apoptosis detection in heart and brain cells. In these organs, detection of apoptosis is most challenging and the existing methods, such as TUNEL, are often insufficient and do not provide clear answers about its extent or even existence.

The limitations of the assay stem from the fact that, strictly speaking, ISL is not an apoptosis or necrosis detection technology, but a technique labeling a specific subset of DNA breaks. On the strong side, the validity of these specific DNA breaks as apoptosis markers is well-established. In fact this type of DNA damage is considered more characteristic for apoptosis than any other DNA-based marker. However, as with any other biological marker, it is far from being absolute and always requires cautious interpretation.

5. Current Status and Perspectives of the ISL Methodology

In the decade and a half since the development of the original PCR-based assay, and its subsequent derivatives based on use of modified oligonucleotides, the assay has been used in a large number of publications. It is difficult to estimate the total number with great accuracy, because many articles have used the commercially available ISOL kits (now sold by Millipore Corporation) without citation of the original publications, precluding citation analysis as an accurate way of determining the popularity of the assay. However, based on search engine results, we can determine that the assay has been used in at least several hundred publications. The assay has been adopted by many groups as an alternative to the still hugely popular TUNEL assay. The reasons appear to be those that initially stimulated us to devise the ISL assay: the specificity of the assay for apoptosis versus various other forms of DNA damage.

However, ISL is more than just a sensitive assay for apoptosis detection. The distinctive feature of this assay is its unique selectivity for a single molecular target. Unlike other enzymatic *in situ* approaches, such as TUNEL or ISEL, ISL specifically detects only a single type of DNA breaks. Depending on the ligated probe end, it detects only blunt-ended or 3' staggered DNA breaks with terminal phosphates. Moreover, as we demonstrated, the ISL's range of detection targets can be expanded to include other characteristic types of DNA breaks. This makes the ligase-based approach a valuable tool for studying a variety of cellular events beyond apoptosis and necrosis. These future new fields of application could include studies in DNA recombination, reparable DNA damage, free radical biology, cell growth, and aging.

A necessary prerequisite for such a technique expansion is the increased sensitivity of DNA breaks detection. This is because much lower numbers of DNA strand breaks, up to a single break per cell, are expected in viable, non apoptotic cells which undergo other processes, such as V(D)J recombination or low-level irradiation (56). However, if needed, the intensity of the probe signal can be increased to levels approaching the detection of individual DNA breaks. This can be accomplished by using enzymatic amplification postligation, as we discussed (28). This will ultimately permit detection of solitary or very rare DNA breaks. The practical work in this direction has already started and, we hope, will bring about important new insights.

6. Notes

1. Because the substrate for apoptotic nucleases is DNA in chromatin, not naked DNA, specific double-strand break structures are predicted to occur. Specifically, double-strand breaks with single 3' overhangs are a signature for apoptotic DNA cleavage (4).
2. At the time of these experiments, there were already many indications that the TUNEL assay was not very specific for apoptosis. Other studies had suggested that necrotic cells stain positively in the TUNEL assay (33, 35).
3. As a historical note, we developed the ISL method not by using the ligation-mediated PCR example, but instead based

4. In the first type of oligonucleotide that we designed, we retained a single A 3' overhang as found in Taq polymerase-derived PCR products. The loop contained 5 deoxyuridine derivatives labeled with biotin. We also used a blunt-ended probe. Although the probe with a 3' overhang is probably more specific and was used most, its signal is also weaker as compared with the blunt-ended hairpin. In sum, both types of probes can detect apoptotic cells (4, 6, 39); the choice of which to use is best determined empirically in the biological system under investigation.

on a more general consideration of adapting common molecular biology techniques to *in situ* detection. However, independently a PCR-based method was developed for biochemical detection of double-strand breaks in apoptotic cell DNA (57).

References

1. Didenko, V.V., Wang, X., Yang, L., and Hornsby, P.J. (1999) DNA damage and p21WAF1/CIP1/SDI1 in experimental injury of the rat adrenal cortex and trauma-associated damage of the human adrenal cortex. *J. Pathol.* **189**, 119–126.
2. Gavrieli, Y., Sherman, Y., and Ben-Sasson, S.A. (1992) Identification of programmed cell death in situ via specific labeling of nuclear DNA fragmentation. *J. Cell Biol.* **119**, 493–501.
3. Steigerwald, S.D., Pfeifer, G.P., and Riggs, A.D. (1990) Ligation-mediated PCR improves the sensitivity of methylation analysis by restriction enzymes and detection of specific DNA strand breaks. *Nucleic Acids Res.* **18**, 1435–1439.
4. Didenko, V.V., and Hornsby, P.J. (1996) Presence of double-strand breaks with single-base 3' overhangs in cells undergoing apoptosis but not necrosis. *J. Cell Biol.* **135**, 1369–1376.
5. Wyllie, A.H. (1980) Glucocorticoid-induced thymocyte apoptosis is associated with endogenous endonuclease activation. *Nature* **284**, 555–556.
6. Didenko, V.V., Tunstead, J.R., and Hornsby, P.J. (1998) Biotin-labeled hairpin oligonucleotides. Probes to detect double-strand breaks in DNA in apoptotic cells. *Am. J. Pathol.* **152**, 897–902.
7. Didenko, V.V., Boudreaux, D.J., and Baskin, D.S. (1999) Substantial background reduction in ligase-based apoptosis detection using newly designed hairpin oligonucleotide probes. *Biotechniques* **27**, 1130–1132.
8. Fortuno, M.A., Gonzalez, A., Ravassa, S., Lopez, B., and Diez, J. (2003) Clinical implications of apoptosis in hypertensive heart disease. *Am. J. Heart Circ. Physiol.* **284**, H1495–H1506.
9. Gonzalez, A., Fortuno, M.A., Querejeta, R., Ravassa, S., Lopez, B., Lopez, N., and Diez, J. (2003) Cardiomyocyte apoptosis in hypertensive cardiomyopathy. *Cardiovasc. Res.* **59**, 549–562.
10. Kunapuli, S., Rosanio, S., and Schwarz, E.R. (2006) How do cardiomyocytes die? Apoptosis and Autophagic cell death in cardiac myocytes. *J. Card. Fail.* **12**, 381–391.
11. Koda, M., Takemura, G., Kanoh, M., et al. (2003) Myocytes positive for in situ markers for DNA breaks in human hearts which are hypertrophic, but neither failed nor dilated: a manifestation of cardiac hypertrophy rather than failure. *J. Pathol.* **199**, 229–236.
12. Hughes, S.E. (2003) Detection of apoptosis using in situ markers for DNA strand breaks in the failing human heart. Fact or epiphenomenon? *J. Pathol.* **201**, 181–186.
13. Okada, H., Takemura, G., Koda, M., Kanoh, M. Kawase, Y., Minatoguchi, S., and Fujiwara, H. (2005) Myocardial apoptotic index based on in situ DNA nick end-labeling of endomyocardial biopsies does not predict prognosis of dilated cardiomyopathy. *Chest* **128**, 1060–1062.
14. Jugdutt, B.I., and Idikio, H.A. (2005) Apoptosis and oncosis in acute coronary syndromes: assessment and implications. *Mol. Cell. Biochem.* **270**, 177–200.

15. Takemura, G., and Fujiwara, H. (2006) Morphological aspects of apoptosis in heart diseases. *J. Cell. Mol. Med.* **10**, 56–75.
16. Takemura, G., and Fujiwara, H. (2003) Doxorubicin-induced cardiomyopathy from the cardiotoxic mechanisms to management. *Prog. Cardiovasc. Dis.* **49**, 330–352.
17. Lukes, D.J., Tivesten, A., Wilton, J., Lundgren, A., Rakotonirainy, O., Kjellström, C., Isgaard, J., Karlsson-Parra, A., Soussi, B., and Olausson, M. (2003) Early onset of rejection in concordant hamster xeno hearts display signs of necrosis, but not apoptosis, correlating to the phosphocreatine concentration. *Transpl. Immunol.* **12**, 29–40.
18. Zhu, C., Qiu, L., Wang, X., Hallin, U., Candé, C., Kroemer, G., Hagberg, H., and Blomgren, K. (2003) Involvement of apoptosis-inducing factor in neuronal death after hypoxia-ischemia in the neonatal rat brain. *J. Neurochem.* **86**, 306–317.
19. Wang, X., Zhu, C., Qiu, L., Hagberg, H., Sandberg, M., and Blomgren, K. (2003) Activation of ERK1/2 after neonatal rat cerebral hypoxia-ischaemia. *J. Neurochem.* **86**, 351–62.
20. Plesnila, N., Zhu, C., Culmsee, C., Gröger, M., Moskowitz, M.A., and Blomgren, K. (2004) Nuclear translocation of apoptosis-inducing factor after focal cerebral ischemia. *J. Cereb. Blood Flow Metab.* **24**, 458–466.
21. Stein, A.B., Bolli, R., Guo, Y., Wang, O.L., Tan, W., Wu, W.J., Zhu, X., Zhu, Y., and Xuan, Y.T. (2007) The late phase of ischemic preconditioning induces a prosurvival genetic program that results in marked attenuation of apoptosis. *J. Mol. Cell. Cardiol.* **42**, 1075–1085.
22. Lesauskaite, V., Epistolato, M.C., Ivanoviene, L., and Tanganelli, P. (2004) Apoptosis of cardiomyocytes in explanted and transplanted hearts. Comparison of results from in situ TUNEL, ISEL, and ISL reactions. *Am. J. Clin. Pathol.* **121**, 108–116.
23. Sun, B., Huang, Q., Liu, S., Chen, M., Hawks, C.L., Wang, L., Zhang, C., and Hornsby, P.J. (2004) Progressive loss of malignant behavior in telomerase-negative tumorigenic adrenocortical cells and restoration of tumorigenicity by human telomerase reverse transcriptase. *Cancer Res.* **64**, 6144–6151.
24. Donath, S., Li, P., Willenbockel, C., Al-Saadi, N., Gross, V., Willnow, T., Bader, M., Martin, U., Bauersachs, J., Wollert, K.C., Dietz, R., and von Harsdorf, R. (2006) Apoptosis repressor with caspase recruitment domain is required for cardioprotection in response to biomechanical and ischemic stress. *Circulation.* **113**, 1203–1212.
25. Audo, I., Darjatmoko, S.R., Schlamp, C.L., Lokken, J.M., Lindstrom, M.J., Albert, D.M., and Nickells, R.W. (2003) Vitamin D analogues increase p53, p21, and apoptosis in a xenograft model of human retinoblastoma. *Invest. Ophthalmol. Vis. Sci.* **44**, 4192–4199.
26. Matsuoka, R., Ogawa, K., Yaoita, H., Naganuma, W., Maehara, K., and Maruyama, Y. (2002) Characteristics of death of neonatal rat cardiomyocytes following hypoxia or hypoxia-reoxygenation: the association of apoptosis and cell membrane disintegrity. *Heart Vessels* **16**, 241–248.
27. Hornsby, P.J., and Didenko, V.V. (2002) In situ DNA ligation as a method for labeling apoptotic cells in tissue sections: an overview, in *In Situ Detection of DNA Damage: Methods and Protocols* (Didenko, V.V. ed.), Humana, Totowa, NJ, pp.133–141.
28. Didenko, V.V. (2002) Detection of specific double-strand DNA breaks and apoptosis in situ using T4 DNA ligase, in *In Situ Detection of DNA Damage: Methods and Protocols* (Didenko, V.V. ed.), Humana, Totowa, NJ, pp.143–151.
29. Ribeiro, G.F., Côrte-Real, M., and Johansson, B. (2006) Characterization of DNA damage in yeast apoptosis induced by hydrogen peroxide, acetic acid, and hyperosmotic shock. *Mol. Biol. Cell.* **17**, 4584–4591.
30. Schoppet, M., Al-Fakhri, N., Franke, F.E., Katz, N., Barth, P.J., Maisch, B., Preissner, K.T., and Hofbauer, L.C. (2004) Localization of osteoprotegerin, tumor necrosis factor-related apoptosis-inducing ligand, and receptor activator of nuclear factor-kappaB ligand in Mönckeberg's sclerosis and atherosclerosis. *J. Clin. Endocrinol. Metab.* **89**, 4104–4112.
31. Frustaci, A., Chimenti, C., Pieroni, M., Salvatori, L., Morgante, E., Sale, P., Ferretti, E., Petrangeli, E., Gulino, A., and Russo, M.A. (2006) Cell death, proliferation and repair in human myocarditis responding to immunosuppressive therapy. *Mod. Pathol.* **19**, 755–765.
32. Walker, P.R., Carson, C., Leblanc, J., and Sikorska, M. (2002) Labeling DNA damage with terminal transferase: applicability, specificity and limitations, in *In Situ Detection of DNA Damage: Methods and Protocols* (Didenko, V.V. ed.), Humana, Totowa, NJ pp.3–19.
33. Charriaut-Marlangue, C., and Ben-Ari, Y. (1995) A cautionary note on the use of the TUNEL stain to determine apoptosis. *Neuroreport* **7**, 61–64.

34. Wolvekamp, M.C., Darby, I.A., and Fuller, P.J. (1998) Cautionary note on the use of end-labeling DNA fragments for detection of apoptosis. *Pathology* **30**, 267.
35. Grasl-Kraupp, B., Ruttkay-Nedecky, B., Koudelka, H., Bukowska, K., Bursch, W., and Schulte-Hermann, R. (1995) In situ detection of fragmented DNA (TUNEL assay) fails to discriminate among apoptosis, necrosis, and autolytic cell death: a cautionary note. *Hepatology* **21**, 1465.
36. Sloop, G.D., Roa, J.C., Delgado, A.G., Balart, J.T., Hines, M.O., and Hill, J.M. (1999) Histologic sectioning produces TUNEL reactivity. A potential cause of false-positive staining. *Arch. Pathol. Lab. Med.* **123**, 529.
37. Bassotti, G., Villanacci, V., Fisogni, S., Cadei, M., Galletti, A., Morelli, A., and Salerni, B. (2007) Comparison of three methods to assess enteric neuronal apoptosis in patients with slow transit constipation. *Apoptosis* **12**, 329–332.
38. Didenko, V.V., Ngo, H., Minchew, C.L., Boudreaux, D.J., Widmayer, M.A, and Baskin, D.S. (2002) Visualization of irreparable ischemic damage in brain by selective labeling of double-strand blunt-ended DNA breaks. *Mol. Med.* **8**, 818–823.
39. Didenko, V.V., Ngo, H., James, W., and Baskin, D.S. (2003) Early necrotic DNA degradation: presence of blunt ended DNA breaks, 3′ and 5′ overhangs in apoptosis but only 5′ overhangs in necrosis. *Am. J. Pathol.* **162**, 1571–1578.
40. Sikorska, M., and Walker, P.R. (1998) Endonuclease activities and apoptosis, in *When Cells Die* (Lockshin, R.A., Zakeri, Z., and Tilly, J.L., eds.), Wiley-Liss, New York, pp.211–242.
41. Barry, M.A., and Eastman, A. (1993) Identification of deoxyribonuclease II as an endonuclease involved in apoptosis. *Arch. Biochem. Biophys.* **300**, 440–450
42. Didenko, V.V. (2008) New in apoptosis imaging: dual detection of self-execution and waste-management. *Cellutions* **1**, 13–15.
43. Grosse, F., and Manns, A. (1993) Terminal deoxyribonucleotidyl transferase (EC 2.7.7.31), in *Enzymes of Molecular Biology* (Burrell, M.M. ed.), Humana, Totowa, NJ, pp.95–105.
44. Maunders, M.J. (1993) DNA and RNA ligases (EC 6.5.1.1, EC 6.5.1.2, and EC 6.5.1.3), in *Enzymes of Molecular Biology* (Burrell, M.M. ed.), Humana, Totowa, NJ, pp.213–230.
45. Didenko, V.V., Ngo, H., and Baskin, D.S. (2002) In situ detection of double-strand DNA breaks with terminal 5′OH groups, in *In Situ Detection of DNA Damage: Methods and Protocols* (Didenko, V.V. ed.), Humana, Totowa, NJ, pp.153–159.
46. Didenko, V.V., Minchew, C.L., Shuman, S., and Baskin, D.S. (2004) Semi-artificial fluorescent molecular machine for DNA damage detection. *Nano Lett.* **4**, 2461–2466.
47. Didenko, V.V. (2006) Oscillating probe for dual detection of 5′PO$_4$ and 5′OH DNA breaks in tissue sections, in *Fluorescent Energy Transfer Nucleic Acid Probes: Methods and Protocols* (Didenko, V.V. ed.), Humana, Totowa, NJ, pp.59–69.
48. Al-Lamki, R.S., Skepper, J.N., Loke, Y.W., King, A., and Burton, G.J. (1998) Apoptosis in the early human placental bed and its discrimination from necrosis using the in-situ DNA ligation technique. *Hum. Reprod.* **13**, 3511–3519.
49. Donath, S., Li, P., Willenbockel, C., Al-Saadi, N., Gross, V., Willnow, T., Bader, M., Martin, U., Bauersachs, J., Wollert, K.C., Dietz, R., and von Harsdorf, R. (2006) Apoptosis repressor with caspase recruitment domain is required for cardioprotection in response to biomechanical and ischemic stress. *Circulation* **7**, 1203–1212.
50. Durand, E., Mallat, Z., Addad, F., Vilde, F., Desnos, M., Guérot, C., Tedgui, A., and Lafont, A. (2002) Time courses of apoptosis and cell proliferation and their relationship to arterial remodeling and restenosis after angioplasty in an atherosclerotic rabbit model. *J. Am. Coll. Cardiol.* **39**, 1680–1685.
51. Kockx, M.M., Muhring, J., Knaapen, M.W.M., and De Meyer, G.R.Y. (1998) RNA synthesis and splicing interferes with DNA in situ end labeling techniques used to detect apoptosis. *Am. J. Pathol.* **152**, 885–888.
52. Zhu, C., Wang, X., Hagberg, H., and Blomgren, K. (2000) Correlation between caspase-3 activation and three different markers of DNA damage in neonatal cerebral hypoxia-ischemia. *J. Neurochem.* **75**, 819–829.
53. Watanabe, M., Hitomi, M., van der Wee, K., et al. (2002) The pros and cons of apoptosis assays for use in the study of cells, tissues, and organs. *Microsc. Microanal.* **8**, 375–391.
54. Murata, I., Takemura, G., Asano, K., Sano, H., Fujisawa, K., Kagawa, T., Baba, K., Maruyama, R., Minatoguchi, S., Fujiwara, T., and Fujiwara, H. (2002) Apoptotic cell loss following cell proliferation in renal glomeruli of Otsuka Long-Evans Tokushima Fatty rats, a model of human type 2 diabetes. *Am. J. Nephrol.* **22**, 587–595.
55. Zavitsanou, K., Nguyen, V., Greguric, I., Chapman, J., Ballantyne, P., and Katsifis, A. (2007) Detection of apoptotic cell death in the thymus of dexamethasone treated rats using [^{123}I]annexin V and in situ oligonucleotide ligation. *J. Mol. Histol.* **38**, 313–319.

56. Mahaney, B.L., Meek, K., and Lees-Miller, S.P. (2009) Repair of ionizing radiation-induced DNA double-strand breaks by non-homologous end-joining. *Biochem. J.* **417**, 639–650.

57. Staley, K., Blaschke, A.J., and Chun, J. (1997) Apoptotic DNA fragmentation is detected by a semi-quantitative ligation-mediated PCR of blunt DNA ends. *Cell Death Differ.* **4**, 66–75.

Chapter 6

In Situ Ligation Simplified: Using PCR Fragments for Detection of Double-Strand DNA Breaks in Tissue Sections

Vladimir V. Didenko

Abstract

The simplified *in situ* ligation procedure is described. All reagents for the assay can be easily obtained in any molecular or cell biology laboratory. The technique uses ligation of double-stranded, PCR-derived DNA fragments labeled with digoxigenin or fluorophores for highly selective detection of apoptotic cells in paraffin-embedded tissue sections. Two types of DNA fragments prepared by PCR are employed. The fragment synthesized by Taq polymerase contains single-base 3′ overhangs, whereas the Pfu polymerase-made fragment is blunt ended. Both fragments can be used as specific, sensitive and cost-effective DNA damage probes. After ligation to apoptotic nuclei in tissue sections, they indicate the presence of double-strand DNA breaks with single-base 3′ overhangs as well as blunt ends.

Key words: Blunt-ended DNA breaks, Double-strand DNA breaks, Single-base overhangs, PCR fragments, Apoptosis detection, DNA damage detection, *In situ* labeling, *In situ* ligation, Tissue sections

1. Introduction

The *in situ* ligation technique is often used for specific labeling of apoptotic cells in tissue sections. The technique employs the enzyme T4DNA ligase for the attachment of oligoprobes to the ends of double-strand DNA breaks in fixed cells (1–3). These probes are hairpin-shaped oligonucleotides bearing a single fluorophore or biotin. During labeling of DNA breaks, a single fluorophore is placed at the end of each break. The sensitivity of such labeling, when using a nonconfocal fluorescent microscope for observation, is approximately 45,000 FITC molecules in the area occupied by a nucleus 0.01 mm in diameter (4). This is sufficient for labeling apoptotic cells because the amount of breaks in apoptosis rises from about 50,000 per genome, at the initial high molecular weight DNA degradation at the earliest stages of apoptosis, to 3×10^6 during the later internucleosomal DNA fragmentation stage (5). Consequently, the molecular probes, which carry a single fluorophore per probe, permit visualization of the full spectrum of breaks present in apoptosis from the initial early stage, to the massive DNA cleavage stage.

However, probes with a signal which is orders of magnitude higher than that of hairpin-shaped oligos would be advantageous in several cases. This refers to the described instances where apoptosis occurs with very few DNA breaks, and to cases where small numbers of apoptotic cells are present in a tissue. Probes with stronger signal can definitively visualize even solitary apoptotic cells, including those at the earlier stages of apoptosis.

One way to increase the intensity of signal is by placing multiple tags into each *in situ* ligation probe. This can be accomplished by using PCR-derived probes instead of hairpin probes. The additional attractive feature of these probes is that they are inexpensive and can be easily produced when needed in any molecular biology laboratory.

PCR-derived probes were introduced in our first paper describing *in situ* ligation (6). We used 226-bp double-stranded DNA fragments containing digoxigenin or Texas Red to label DNA breaks. In our subsequent work, we also used longer 460-bp probes and a variety of fluorescent and conventional labels.

Soon thereafter, we introduced a different type of probe based on hairpin-shaped oligonucleotides, but the PCR-based probes still remained popular (7–10). There are several reasons for this popularity. The probes are very sensitive, easy to produce and very inexpensive. Here, we present the detailed protocol for their generation and application for apoptosis detection in the tissue section format.

Generally, two forms of double-stranded PCR-derived DNA fragments are used as probes for *in situ* ligation. A Taq polymerase-synthesized fragment has single-base 3' overhangs, whereas a Pfu polymerase-synthesized fragment has blunt-ends. Both types of fragments can be ligated to DNA breaks in apoptotic nuclei, indicating in such nuclei the presence of double-strand DNA breaks with either single-base 3' overhangs or blunt ends.

Compared to hairpin probes, the PCR-labeled fragments generate a much stronger signal. This signal is proportional to the probe length because longer probes contain higher numbers of labeled nucleotides.

A wide range of PCR probes was successfully used by different laboratories. PCR fragments as short as 146 bp (9, 11) and as long as 441 bp (12) were all successfully employed. Our initial PCR probes were 226 bp-long (6) and, in most cases, the PCR probes used in other laboratories stayed close to 200 bp. Some examples include PCR probes of 174 bp (13, 14); 200 bp (10) and 245 bp (7, 8, 15), etc. This is because the much longer probes are impractical as they can break during purification and storage, producing smaller fragments with nonspecific configurations of ends.

In general, digoxigenin-11-dUTP or biotin-dUTP are incorporated with high efficiency if the ratio of substituted-dUTP to dTTP in the labeling mixture is 1:2 (16). Therefore, usually ~35% of a dTTP in the PCR reaction mix is substituted by its tagged analog. This is expected to translate into the substitution of ~35% of thymidines in our 226-bp probe by tagged nucleotides and would result in the insertion of 26 tags.

In this chapter, we present the *in situ* ligation procedure which uses PCR-derived probes for detection of two types of DNA breaks in fixed tissue sections. We also present the PCR protocols for probes preparation. The described probes are applicable for fluorescence and conventional detection of both blunt-ended DNA breaks and breaks with 3′ overhangs for selective labeling of apoptotic versus necrotic cells (6).

2. Materials

2.1. PCR Probe Preparation and Labeling (see Note 1)

1. PCR primers. Dissolve in distilled water to 100 pmoles/µL.
   Primer 1 – (27-mer): 5′-GCT GGT CTG CCG CCG TTT TCG ACC CTG-3′
   Primer 2 – (21-mer): 5′-TGG CCT GCC CAA GCT CTA CCT-3′
2. TE buffer: 10 mM Tris–HCl, pH 7.5, 1 mM EDTA.
3. Uncut pBluescript-bSDI 1 plasmid (see Note 1).
   Dissolve in TE to 100 pg/µL (see Note 2).
4. Pfu DNA polymerase 2.5 U/µL (Stratagene, La Jolla, CA).
5. Taq DNA polymerase 5 U/µL (Invitrogen, Carlsbad, CA).
6. 10× buffer for Pfu polymerase (Stratagene, La Jolla, CA): 200 mM Tris–HCl, pH 8.8, 100 mM KCl, 100 mM ammonium sulfate, 20 mM $MgSO_4$, 1% Triton X-100, 1 mg/mL BSA.
7. 10× buffer for Taq polymerase: 100 mM Tris–HCl, pH 8.35, 500 mM KCl.
8. 8.25 mM $MgCl_2$ stock solution.
9. Digoxigenin-11-dUTP (1 mM) (Roche Molecular Biochemicals, Indianapolis, IN) (see Note 3).
10. dATP, dGTP, dCTP, dTTP in separate vials 100 mM stock solutions (Roche Molecular Biochemicals, Indianapolis, IN).

Make 20× dNTPs-dig-dUTP labeling mix containing 1 mM dATP; dGTP, dCTP; 0.65 mM dTTP, 0.33 mM digoxigenin-11-dUTP (see Note 4). Store at –20°C.

11. High Pure™ PCR Product Purification Kit (Roche Molecular Biochemicals, Indianapolis, IN).
12. Agarose.
13. Ethidium bromide at 0.5 µg/mL.
14. Spectrophotometer.
15. PCR cycler.

2.2. In Situ Ligation in Tissue Sections

1. The 5–6 µm-thick sections cut from paraformaldehyde-fixed, paraffin-embedded tissue blocks onto ProbeOn™ Plus charged and precleaned slides (Fisher Scientific, Pittsburgh, PA). Other slide brands can also be used if they retain tissue well.
2. Xylene.
3. 80 and 96% Ethanol.
4. T4 DNA ligase 5 U/µL (Roche Molecular Biochemicals, Indianapolis, IN) (see Note 5).
5. 10× reaction buffer for T4 DNA ligase: 660 mM-Tris–HCl, 50 mM $MgCl_2$, 10 mM dithioerythritol, 10 mM ATP, pH 7.5 (20°C) (Roche Molecular Biochemicals, Indianapolis, IN) (see Note 6).
6. 30% (w/v) solution of PEG-8000 (Sigma, St. Louis, MO) in bidistilled water (see Note 7).
7. Proteinase K (Roche Molecular Biochemicals, Indianapolis, IN) 20 mg/mL stock solution in distilled water. Store at –20°C. In the reaction use 50 µg/mL solution in PBS, prepared from the stock. Do not reuse (see Note 8).
8. Phosphate-buffered saline (1× PBS): dissolve 9 g NaCl, 2.76 g $NaH_2PO_4 \cdot H_2O$, 5.56 g $Na_2HPO_4 \cdot 7H_2O$ in 800 mL of distilled water. Adjust to pH 7.4 with NaOH, and fill to 1 L with distilled water.
9. Washing buffer: 100 mM Tris–HCl, pH 7.5, 100 mM NaCl.
10. Blocking solution: 10% sheep serum in 100 mM Tris–HCl, pH 7.5, 100 mM NaCl.
11. 1-Step™ NBT/BCIP (Pierce) – ready-to-use substrate for alkaline phosphatase that yields a very intense purple signal. If background caused by endogenous phosphatases is a problem, then 1-Step NBT/BCIP plus Suppressor can be used. This substrate formulation additionally contains 1 mM levamisole to inhibit endogenous phosphatase activity.
12. Anti-digoxigenin antibody – alkaline-phosphatase conjugate (Roche Molecular Biochemicals, Indianapolis, IN). Fab fragments from an anti-digoxigenin antibody from sheep,

conjugated with alkaline phosphatase for the detection of digoxigenin-labeled compounds.

13. 22×22 mm or 22×40 mm glass or plastic coverslips. Plastic coverslips are preferable as they are easier to remove from the section.

14. Aqueous slide mounting media such as Aqua-Mount (Thermo Fisher Scientific, Waltham, MA) or ImmunoHistoMount (Santa Cruz Biotechnology, Santa Cruz, CA).

15. Microscope with appropriate filters and objectives.

3. Method

3.1. Preparation of PCR-Labeled In Situ Ligation Probes Containing 3′ A-Overhangs

1. Set up the PCR reaction mix in thin-wall PCR tubes on ice (100 μL volume):

   Combine in this order:

   75 μL – distilled water

   10 μL – 10× buffer for Taq polymerase

   6 μL – $MgCl_2$ (25 mM stock) (1.5 mM final concentration)

   5 μL – 20× dNTPs-dig-dUTP labeling mix (final concentrations dA, dC, dG – 50 μM, dT – 32.5 μM, dig-dUTP – 16.5 μM) (see Note 4)

   1 μL – pBluescript-bSDI1plasmid (10–100 pg/μL) (see Note 2)

   1 μL – Primer 1 (100 pmoles/μL)

   1 μL – Primer 2 (100 pmoles/μL)

   1 μL – Taq polymerase (2.5 U/μL)

2. Perform PCR labeling for 40 cycles using the following cycle profile

   95°C – 20 s

   61°C – 20 s

   74°C – 2 min

   Postamplification extension at 74°C – 5 min

   Store at 4°C.

3. Take 2 μL from the PCR reaction mix and run on a 1% agarose gel to verify successful labeling

4. Purify using High Pure™ PCR Product Purification Kit. Elute in 50 μL TE (see Note 9). Store at −20°C. This solution can be directly used for *in situ* labeling

5. Measure probe concentration using spectrophotometer. Dilute in TE to 0.5 μg/μL

3.2. Preparation of PCR-Labeled Blunt-Ended Probes for In Situ Ligation

1. Set up the PCR reaction mix in thin-wall PCR tubes on ice (100 μL volume):

   Combine in this order:

   81 μL – distilled water

   10 μL – 10× buffer for Pfu polymerase

   5 μL – 20× dNTPs-dig-dUTP labeling mix (final concentrations dA, dC, dG – 50 μM, dT – 32.5 μM, dig-dUTP – 16.5 μM) (see Note 10)

   1 μL – pBluescript-bSDI1plasmid (10–100 pg/μL) (see Note 2)

   1 μL – Primer 1 (100 pmoles/μL)

   1 μL – Primer 2 (100 pmoles/μL)

   1 μL – Pfu turbo polymerase (2.5 U/μL)

2. Perform PCR labeling for 40 cycles using the following cycle profile

   Postamplification extension at 74°C – 5 min

   Store at 4°C

   95°C – 20 s

   61°C – 20 s

   74°C – 2 min

3. Take 2 μL from the PCR reaction mix and run on a 1% agarose gel to verify successful labeling

4. Purify using High Pure™ PCR Product Purification Kit. Elute in 50 μL TE (see Note 9). Store at –20°C. This solution can be directly used for *in situ* labeling

5. Measure probe concentration using spectrophotometer. Dilute in TE to 0.5 μg/μL

3.3. In Situ Ligation in Tissue Sections

1. Place the sections in a slide rack and dewax in xylene for 15 min, transfer to a fresh xylene bath for an additional 5 min.

2. Rehydrate by passing through graded ethanol concentrations: 96% Ethanol – 2×5 min; 80% Ethanol – 5 min; water – 2×5 min.

3. Digest section with Proteinase K. Use 100 μL of a 50 μg/mL solution per section. Incubate for 15 min at room temperature in a humidified chamber (see Note 11).

4. Rinse in distilled water for 4×2 min.

5. Apply 100 μL per section of the preincubation solution. Incubate for 15 min at room temperature (23°C). The preincubation solution consists of a 1× T4 DNA ligase reaction buffer supplemented with PEG-8000 to the final concentration

of 15%. It contains 66 mM Tris–HCl, pH 7.5, 5 mM $MgCl_2$, 1 mM dithioerythritol, 1 mM ATP, and 15% polyethylene glycol (see Note 12).

6. Aspirate the preincubation solution and apply the full ligase reaction mix with the probe (750 ng/µL stock) and T4 DNA ligase (see Notes 13 and 14).

   In situ ligation labeling solution (20 µL per section):

   Prepare on ice in this order:

   6 µL – bidistilled water.

   10 µL – PEG8000 (30% stock solution).

   2 µL – 10× buffer for T4 DNA ligase.

   1 µL – PCR labeled probe (0.5 µg/µL) (for smaller sections can be 0.5 µL).

   1 µL – T4 DNA ligase (5 U/µL) (see Note 5).

   The total volume of the labeling solution can be scaled up to accommodate bigger sections.

   Incubate overnight at room temperature (23°C) (see Note 15) in a humidified chamber with a plastic coverslip. The incubation time can be reduced to 4 h, if a faster detection is needed, this, however, decreases the signal and has to be compensated by longer color development.

7. Remove coverslips by gently immersing the slides vertically in a Coplin jar containing water at room temperature. Then wash sections 3 × 10 min in distilled water.

8. Apply 100 µL per section of the blocking solution. Incubate for 15 min at room temperature (23°C) in humidified chamber. The blocking solution contains 10% sheep serum in 100 mM Tris–HCl, pH 7.5, 100 mM NaCl.

9. Aspirate the preincubation solution and apply 50 µL per section of solution of anti-digoxigenin antibody – alkaline-phosphatase conjugate diluted 1:100 in the blocking solution. Incubate 10 min at room temperature (23°C) in humidified chamber.

10. Wash 2 × 10 min in 100 mM Tris–HCl, pH 7.5, 100 mM NaCl.

11. Apply 1-Step™ NBT/BCIP substrate solution for alkaline phosphatase. Keep in dark. Monitor color development. The color usually develops in 5–20 min.

12. Stop reaction by rinsing sections in water. Add water-soluble mounting media such as Aqua-Mount or ImmunoHistoMount and coverslip.

4. Notes

1. This protocol describes Taq and Pfu polymerase-based labeling of 226-bp fragments, used as *in situ* ligation probes. The fragment is amplified from pBluescript-bSDI 1 plasmid, which can be obtained from Dr. Peter Hornsby, San Antonio, TX but almost any other plasmid or a DNA fragment, for which appropriate primers are available to amplify a 200–500 bases-long fragment, can be used instead.

2. Although the pBluescript-bSDI1plasmid is, possibly, the most used and tested template for preparation of the PCR-derived probes for *in situ* ligation (6, 17–21), other commercially-available plasmids can be used instead with similar results. The examples of the plasmids and primers used by other laboratories include.

    (a) Bluescript II KS$^{+/-}$ (Stratagene, La Jolla, CA) with primers complementary to the T3 and T7 promoters in the multicloning site, resulting in 146 bp probes (9, 11).

    (b) pOCME1B plasmid with primers 5′-ATGCTCTTCA GTTCGTGTGT-3′ and 5′-CTGACTTGGCAGGCT TGAGG-3′ amplifying 174 bp fragment (13, 14).

    (c) pBK-CMV plasmid (Stratagene, La Jolla, CA), as a source of a 200-bp DNA fragment corresponding to its multiple cloning site (10).

    (d) pBluescript SK(–) plasmid with primers 5′-CTCATAG CTCACGCTGTAGG-3′ and 5′-AGTGTAGCCGTA GTTAGGCC-3′ amplifying a 245 bp fragment (7, 8, 15).

    (e) pBluescript SK(–) plasmid with primers 5′-TCGAG GTCGACGGATTCGATG-3′ and 5′-CCGCTCTAGA ACTAGTGGATC-3′ amplifying a 441 bp fragment (12).

3. Texas Red-12-dUTP, FITC-12-dUTP, biotin-dUTP and other modified nucleotide analogs can be used instead. However, sometimes probes labeled with these tags had higher background staining. In our experiments, the probes labeled with digoxigenin-11-dUTP were less prone to non-specific background staining.

4. We found that significantly more labeled nucleotides can be inserted into the synthesized fragment during the Taq-driven PCR labeling when different formulations of dNTPs mixes are used. The successful labeling occurs when the concentration of digoxigenin-11-dUTP in the mixes is kept constant and is 16.5 µM, whereas, concentration of dTTP is reduced from the commonly used 32.5–24.5 µM or to 16.5–8 µM or

even to 0 µM. In the last case, every dTTP in the PCR fragment is substituted by digoxigenin-11-dUTP. With our PCR fragments the efficiency of PCR labeling reaction was not visibly affected in all of these cases. Using FITC-12-dUTP instead of digoxigenin-11-dUTP in this reaction had similar results.

Although the fragments generated in these PCR reactions contain more tags, it is untested whether they produce stronger signal as probes. This is due to the possibility of steric hindrance or, in case of fluorophores, of self-quenching, preventing detection of the closely-located labels.

5. T4 DNA ligase in lower stock concentrations (1 U/µL) can also be used. The labeling reaction is only marginally less efficient at lower concentrations of ligase in labeling solution (125–250 U/mL). However, the highly concentrated (5 U/µL) (Roche Molecular Biochemicals, Indianapolis, IN) ligase preparation gives the best signal.

6. ATP in the ligase reaction buffer is easily destroyed in repetitive cycles of thawing-freezing. Aliquot the buffer in small 15–20 µL portions and store at –20°C. Use once.

7. 15% PEG-8000 in the ligation mix strongly stimulates the ligation reaction increasing the effective concentrations of the probe and ligase by volume exclusion (22). Use water of the highest quality for all solutions, which come in contact with T4 DNA ligase. If nonspecific background staining is a problem, reducing PEG concentration or complete elimination of PEG from labeling solution significantly reduces nonspecific signal. However, it also weakens the labeling reaction.

8. Proteinase K is a very stable enzyme, when stored at concentrations higher than 1 mg/mL. However, autolysis of the enzyme occurs in aqueous solutions at low concentrations (~10 µg/mL) (23).

9. If not purified, the probe is prone to producing background staining.

10. We found that significantly more labeled nucleotides can be inserted into the synthesized fragment during the Pfu-driven PCR labeling when different formulations of dNTPs mixes are used. The successful labeling occurs when the concentration of digoxigenin-11-dUTP in the mixes is kept constant and is 16.5 µM, whereas concentration of dTTP is reduced from the commonly used 32.5–24.5 µM or to 16.5–8 µM or even to 0 µM. In the last case, every dTTP in the PCR fragment is substituted by digoxigenin-11-dUTP. With our PCR fragments the efficiency of PCR labeling reaction was not visibly affected in all of these cases. Using FITC-12-dUTP instead of digoxigenin-11-dUTP in this reaction had similar results.

Although the fragments generated in these PCR reactions contain more tags, it is untested whether they produce stronger signal as probes. This is due to the possibility of steric hindrance or, in case of fluorophores, of self-quenching, preventing detection of the closely-located labels.

11. Proteinase K digestion time may need adjustment depending on the tissue type. Times of 15–25 min are usually used, though hard tissues might require longer digestion. Insufficient digestion may result in a weaker signal. Overdigestion, on the other hand, results in signal disappearance and section disruption.

12. Preincubation with ligation buffer ensures even saturation of the section prior to addition of the enzyme and the probe, and was shown to increase the ligation efficiency. Prepare preincubation solution by adding 10 μL of the 10× ligase buffer (Roche Molecular Biochemicals, Indianapolis, IN) to 40 μL of bidistilled water and 50 μL of 30% PEG-8000.

    Sometimes, in the case of very high background staining, the 1 h 30 min preblock with 20% BSA instead of ligation buffer solves the problem.

13. A mock reaction is recommended as a regular control in order to rule out nonspecific background staining. In a mock reaction solution, an equal volume of 50% glycerol in water is substituted for T4 DNA ligase.

14. The probe has exclusively 5′OH groups and is not self-ligatable because T4DNA ligase can only ligate the 5′PO_4 end of one probe to the 3′OH of another (22).

15. Lowering of the temperature to 16°C reduces the signal; a temperature increase to 37°C completely eliminates the signal.

References

1. Didenko, V.V., Tunstead, J.R., Hornsby, P.J. (1998) Biotin-labeled hairpin oligonucleotides. Probes to detect double-strand breaks in DNA in apoptotic cells. *Am J Pathol* **152**, 897–902.
2. Didenko, V.V., Boudreaux, D.J., Baskin, D.S. (1999) Substantial background reduction in ligase-based apoptosis detection using newly designed hairpin oligonucleotide probes. *Biotechniques* **27**, 1130–1132.
3. Didenko, V.V. (2002) Detection of specific double-strand DNA breaks and apoptosis *in situ* using T4 DNA ligase, in *In Situ Detection of DNA Damage: Methods and Protocols* (Didenko, V.V. ed.), Humana, Totowa, NJ, pp.143–151.
4. Didenko, V.V., Minchew, C.L., Shuman, S., Baskin, D.S. (2004) Semi-artificial fluorescent molecular machine for DNA damage detection. *Nano Lett* **27**, 1130–1132.
5. Walker, P.R., Leblanc, J., Carson, C., Ribecco, M., Sikorska, M. (1999) Neither caspase-3 nor DNA fragmentation factor is required for high molecular weight DNA degradation in apoptosis. *Ann N Y Acad Sci* **887**, 48–59.
6. Didenko, V.V., and Hornsby, P.J. (1996) Presence of double-strand breaks with single-base 3′ overhangs in cells undergoing apoptosis but not necrosis. *J Cell Biol* **135**, 1369–1376.
7. Koda, M., Takemura, G., Kanoh, M., Hayakawa, K., Kawase, Y., Maruyama, R., Li, Y., Minatoguchi, S., Fujiwara, T., Fujiwara, H. (2003) Myocytes positive for *in situ* markers for DNA breaks in human hearts which are

hypertrophic, but neither failed nor dilated: a manifestation of cardiac hypertrophy rather than failure. *J Pathol* **199**, 229–236.

8. Okada, H., Takemura, G., Koda, M., Kanoh, M., Kawase, Y., Minatoguchi, S., Fujiwara, H. (2005) Myocardial apoptotic index based on in situ DNA nick end-labeling of endomyocardial biopsies does not predict prognosis of dilated cardiomyopathy. *Chest* **128** (2), 1060–1062.

9. Schoppet, M., Al-Fakhri, N., Franke, F.E., Katz, N., Barth, P.J., Maisch, B., Preissner, K.T., Hofbauer, L.C. (2004) Localization of osteoprotegerin, tumor necrosis factor-related apoptosis-inducing ligand, and receptor activator of nuclear factor-kappaB ligand in Mönckeberg's sclerosis and atherosclerosis. *J Clin Endocrinol Metab* **89** (8), 4104–4112.

10. Audo, I., Darjatmoko, S.R., Schlamp, C.L., Lokken, J.M., Lindstrom, M.J., Albert, D.M., Nickells, R.W. (2003) Vitamin D analogues increase p53, p21, and apoptosis in a xenograft model of human retinoblastoma. *Invest Ophthalmol Vis Sci* **44** (10), 4192–4199.

11. Al-Fakhri, N., Chavakis, T., Schmidt-Woll, T., Huang, B., Cherian, S.M., Bobryshev, Y.V., Lord, R.S.A., Katz, N., Preissner, K.T. (2003) Induction of apoptosis in vascular cells by plasminogen activator inhibitor-1 and high molecular weight kininogen correlates with their anti-adhesive properties. *J Biol Chem* **384**, 423–435.

12. Matsuoka, R., Ogawa, K., Yaoita, H., Naganuma, W., Maehara, K., Maruyama, Y. (2002) Characteristics of death of neonatal rat cardiomyocytes following hypoxia or hypoxia-reoxygenation: the association of apoptosis and cell membrane disintegrity. *Heart Vessels* **16** (6), 241–248.

13. Guerra, S., Leri, A., Wang, X., Finato, N., Di Loreto, C., Beltrami, C.A., Kajstura, J., Anversa, P. (1999) Myocyte death in the failing human heart is gender dependent. *Circ Res* **85** (9), 856–866.

14. Leri, A., Claudio, P.P., Li, Q., Wang, X., Reiss, K., Wang, S., Malhotra, A., Kajstura, J., Anversa, P. (1998) Strech-mediated release of angiotensin II induces myocyte apoptosis by activating p53 that enhances the local renin-angiotensin system and decreases the Bcl-2 to Bax protein ratio in the cell. *J Clin Invest* **101**, 1326–1342.

15. Murata, I., Takemura, G., Asano, K., Sano, H., Fujisawa, K., Kagawa, T., Baba, K., Maruyama, R., Minatoguchi, S., Fujiwara, T., Fujiwara, H. (2002) Apoptotic cell loss following cell proliferation in renal glomeruli of Otsuka Long-Evans Tokushima Fatty rats, a model of human type 2 diabetes. *Am J Nephrol* **22** (5–6), 587–595.

16. Emanuel, J.R. (1991) Simple and efficient system for synthesis of non-radioactive nucleic acid hybridization probes using PCR. *Nucleic Acids Res* **19**, 2790.

17. Bozkurt, B., Kribbs, S.B., Clubb, F.J. Jr., Michael, L.H., Didenko, V.V., Hornsby, et al. (1998) Pathophysiologically relevant concentrations of tumor necrosis factor-alpha promote progressive left ventricular dysfunction and remodeling in rats. *Circulation* **97**, 1382–1391.

18. Frustaci, A., Chimenti, C., Setoguchi, M., Guerra, S., Corsello, S., Crea, F., et al. (1999) Cell death in acromegalic cardiomyopathy. *Circulation* **99**, 1426–1434.

19. Li, B., Setoguchi, M., Wang, X., Andreoli, A.M., Leri, A., Malhotra, A., Kajstura, J., Anversa, P. (1999) Insulin-like growth factor-1 attenuates the detrimental impact of nonocclusive coronary artery constriction on the heart. *Circ Res* **84**, 1007–1019.

20. Leri, A., Liu, Y., Li, B., Fiordaliso, F., Malhotra, A., Latini, R., Kajstura, J., Anversa, P. (2000) Up-regulation of AT_1 and at_2 receptors in postinfarcted hypertrophied myocytes and stretch-mediated apoptotic cell death. *Am J Pathol* **156**, 1663–1672.

21. Fiordaliso, F., Li, B., Latini, R., Sonnenblick, E.H., Anversa, P., Leri, A., Kajstura, J. (2000) Myocyte death in streptozotocin-induced diabetes in rats is angiotensin II-dependent. *Lab Invest* **80**, 513–527.

22. Maunders, M.J. (1993) DNA and RNA ligases (EC 6.5.1.1, EC 6.5.1.2, EC 6.5.1.3), in *Enzymes of Molecular Biology* (Burrell, M.M. ed.), Humana, Totowa, NJ, pp. 213–230.

23. Sweeney, P.J., and Walker, J.M. (1993) Proteinase K (EC 3.4.21.14), in *Enzymes of Molecular Biology* (Burrell, M.M. ed.), Humana, Totowa, NJ, pp. 305–311.

Chapter 7

5′OH DNA Breaks in Apoptosis and Their Labeling by Topoisomerase-Based Approach

Vladimir V. Didenko

Abstract

Recently, the concept of apoptotic cell elimination was expanded and programed cell death is no longer viewed as an individual cellular event. The complete description of the apoptotic process now includes two phases: the self-driven cell disassembly and the externally-controlled elimination of apoptotic cell corpses by phagocytizing cells. The second, phagocytic phase is essential, highly conserved, and is even more important than the internal phase of cell disassembly. This is because it ensures the complete degradation of the dying cell's DNA, preventing the release of pathological, viral and tumor DNA, and self-immunization. In different cells and species from mammals to flies, a single conserved enzyme – DNase II is responsible for the elimination of cellular DNA in the second "mopping up" phase of apoptosis. Here, we present an assay for the selective detection of the phagocytic phase of apoptosis. The technology capitalizes on the fact that phagocytic DNase II produces identifiable signature DNA breaks, which can be labeled by vaccinia topoisomerase. The assay permits labeling of the previously underestimated phase of apoptotic execution and is a useful tool in the apoptosis detection arsenal.

Key words: Apoptosis labeling, DNA breaks, *In situ* detection, Vaccinia topoisomerase I, DNase II-type breaks, Blunt-ended DNA breaks, 5′OH DNA breaks; phagocytizing cells, Apoptotic cell corpse elimination

1. Introduction

1.1. Cell-Autonomous and Lysosomal (Phagocytic) Nucleases in Apoptosis

Based on their role in cell corpse elimination, all apoptotic nucleases are divided into two categories: cell-autonomous and waste-management nucleases (1). These two groups of nucleases have very different functions. Cell-autonomous nucleases cleave the DNA within a cell as it undergoes apoptosis. The lysosomal (phagocytic) nucleases take part in the engulfment-mediated DNA degradation – the "cleaning up" of corpses of cells that have died by apoptosis. While cell-autonomous DNA degradation is important for carrying out an individual cell death program, the waste-control nucleases are essential for the life of other cells in the organism. The proper disposal of postapoptotic corpses is critically important because their uncleaned debris pose danger to other cells (1).

Even though the cell-autonomous nucleases are important, they are dispensable in many cell types. This is because in the organism, after a cell has died, its corpse is destroyed and engulfed by other cells. Therefore, even if a cell has not degraded its own DNA, its neighbors will do it instead. However, there is no such "plan B" in the case of waste-management nucleases. If they fail to clean the corpses, the organism would be poisoned with nondegraded DNA. Thus, lysosomal nucleases are essential for life (1, 2).

Tissue section assays which can label the activity of these two categories of enzymes permit comprehensive imaging of apoptotic degradation.

Here, we present an assay which labels the phagocytic phase of apoptosis, i.e. phagocytosis of DNA from apoptotic cell corpses by the "mop up" cells. The assay addresses the challenges created by a very recent shift in the apoptotic paradigm which now views apoptosis not as a "private matter" of a dying cell, but as a broad reaction analogous to the immune response, continuing beyond the individual cell program and requiring the participation of other specialized cells (1, 3). The assay takes into consideration this novel broad perspective of programed cell death which is necessary for better understanding of apoptosis-related pathologies.

The technology capitalizes on the fact that phagocytic nuclease activities in apoptosis are highly conserved. They produce characteristic and highly specific DNA breaks. The assay uses the enzymatic action of vaccinia topoisomerase I to label signature DNA breaks produced by waste-management nucleases.

1.2. DNase I- and DNase II-Types of DNA Breaks

The ability to accurately label apoptotic cells and visualize the progression of apoptosis in the tissue section format is critical for many branches of biomedical science. However, the task of choosing an appropriate apoptotic marker is complicated by the fact that apoptotic pathways are remarkably complex. They not only vary in different cell types and tissues, but can change within the same cell. Nucleolytic activities are critical for apoptotic disassembly as they degrade cellular DNA, preventing the release of pathological, viral and tumor DNA, and self-immunization (1, 3).

Every apoptotic cell goes through sequential steps ensuring its complete disassembly and disappearance. Massive and systematic DNA fragmentation is one of the characteristic features of this process. Not surprisingly, it is often used as specific marker in apoptosis detection. When apoptotic DNA is cut, the result is an abundant amount of DNA fragments. These DNA fragments are not random. Mainly they possess blunt ends and some have ends with short, single-nucleotide staggers with one of the strands protruding slightly (4–6).

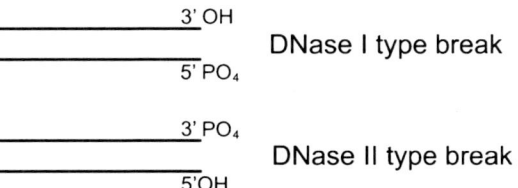

Fig. 1. DNase I type and DNase II type DNA breaks.

Different functional groups can be present at those ends. A double-stranded break forms when a DNA duplex is cut through. It exposes the 3′ and 5′ ends of the two DNA strands. These ends can carry either a phosphate (PO_4) group or a hydroxyl (OH) group (Fig. 1). The distribution of these groups at the ends provides important information about the enzyme which did the cutting.

Based on the distribution of functional groups, the two types of cuts are identified. The cuts with 3′OH/5′PO_4 configuration are DNase I type cuts. The cuts with the inverse distribution of functional groups – 3′PO_4/5′OH, are of DNase II type (7, 8) (Fig. 1).

The cuts received their names because they match the cleavage patterns of the two major nucleases – DNase I and DNase II. The terms are used for convenience and do not imply any specific relationship between these enzymes and the apoptotic cutters. The actual apoptotic nuclease might or might not be related to DNase I and II.

So far, in the majority of apoptotic cells in mammals, DNase I type fragmentation was detected (3′OH/5′PO_4 at the ends) (1, 9, 10).

DNase II type was detected in many important cases too (2, 3, 11, 12). Although in general, this type of cleavage is less prevalent than the self-inflicted apoptotic DNA fragmentation. The underdevelopment of techniques for the selective labeling of DNase II type cleavage might have contributed to this situation. In fact, the enzymatic assays for labeling apoptosis in tissue sections focus exclusively on DNase I type cleavage.

Different features of DNase I cleavage are used by various apoptosis assays. For example, the TUNEL assay specifically detects one marker of DNase I cleavage and labels the 3′OH groups with the help of the enzyme Terminal Deoxyribonucleotidyl Transferase (TdT) (13). The other assay, *in situ* nick translation, labels nicks or single-stranded DNA breaks with 3′OH using DNA polymerase I (14). Yet another technique, *in situ* ligation, selectively detects double-stranded DNA breaks (10, 15). It relies on the attachment of double-stranded DNA probes with blunt ends, or short 3′ overhangs, to the ends of such breaks. The ligation reaction is carried out by the enzyme T4 DNA ligase, which needs terminal 5'PO$_4$ in the breaks to attach the probe. Consequently, the assay detects exclusively 5' phosphorylated double-stranded breaks and does not label DNase II type breaks with 5'OH. Therefore, in all of these assays, the cells with DNase II type cleavage go undetected.

In the meantime, important changes occurred in the very concept of apoptotic cell elimination which significantly increased the value of detecting DNase II type breaks. It was demonstrated that DNase II plays a critical role in apoptosis at a level different from executioner nucleases.

1.3. Role of DNase II in Elimination of Apoptotic Cell Corpses

The apoptotic paradigm has recently changed to incorporate the notion that the apoptotic process is not an internal cellular event (1, 3). Instead, it continues beyond the individual cell reaction and requires the participation of surrounding cells. The complete apoptotic process now includes two phases: the self-driven cell disassembly and the externally-controlled elimination of apoptotic cell corpses by phagocytizing waste-management cells (1). This externalized waste-control phase is essential, highly conserved, and is considered to be even more important than the internal phase of cell disassembly (1, 2, 12). This is because it ensures complete degradation of the dying cell's DNA, preventing the release of pathological, viral and tumor DNA, and self-immunization (see Note 1).

DNase II plays a fundamental role in engulfment-mediated DNA degradation during the waste-management phase of apoptosis (1, 3). DNase II is present in lysosomes, the sac-like organelles inside cells that contain digestive enzymes that break down cellular components. The enzyme destroys DNA of apoptotic cells after their corpses are engulfed by waste-management cells, such as tissue macrophages and many other tissue cells capable of phagocytosis.

Consequently, novel approaches for apoptosis detection are needed which will take this paradigm change into account. Several attempts were made to accomplish this task.

Initially to resolve this problem, the T4 DNA kinase-based technique was introduced for the detection of 5'OH bearing DNA breaks (8). However, in this case, the 5'PO$_4$ breaks must be labeled first so that in the subsequent reaction with T4 DNA kinase all remaining 5'OH breaks are converted to the detectable 5' phosphate format. As a result, the approach requires two successive overnight labeling reactions with multiple controls, making it complicated and time consuming.

We have subsequently developed a new approach for labeling of 5'OH double-stranded DNA breaks. The approach was an offshoot of our work in bionanotechnology developing an oscillating nano-size device, which contained a vaccinia virus encoded protein linked with a dual-labeled DNA part (16) (also featured in (17)).

The construct exemplified a practical approach to the design of molecular devices and machines, and illustrated our notion that nano-size constructs that use mechanisms developed in the evolution of biological molecules are simpler and uniquely suitable for nanoscale environments. To make it into a useful assay, we adapted this development for the practical usage as an oscillating double-hairpin oligonucleotide probe which uses vaccinia DNA topoisomerase I for the simultaneous detection of two specific types of DNA damage *in situ* (16, 18).

However, the later work permitted us to significantly simplify the assay and make it faster and more cost-effective. The improvements resulted in the development of a new much shorter probe of a different configuration, which substitutes for the previously used oscillating double-hairpin. This significantly increased the speed of detection which now takes only minutes for completion and makes the assay significantly more robust and more sensitive.

Here, we describe this new and improved technique. The assay selectively detects blunt-ended DNA breaks with terminal 5′OH groups. It labels the waste-management phase of apoptotic DNA degradation in tissue sections. This technique takes into account recent changes in the apoptotic paradigm that views apoptosis as a broad reaction continuing beyond the individual cell program and requiring participation of other cells.

1.4. Principle of the Assay (Fig. 2)

The assay uses an oligonucleotide probe and vaccinia DNA topoisomerase I. The assay utilizes the unique enzymatic properties of vaccinia DNA topoisomerase I, which can join two DNA molecules employing a mechanism different from those of ligases. This enzyme binds to duplex DNA having the CCCTT3′ recognition sequence, and creates a nick at its 3′ end (2). In normal conditions, the enzyme then seals the nick, religating the strand back to the acceptor DNA end with 5′OH.

In the assay, the topoisomerase CCCTT3′ recognition sequence is located in a hairpin oligonucleotide probe which has a 12-base-long single-stranded region on its 3′ end. The enzyme recognition site is positioned at the end of the duplex-forming part of the probe and on the edge of an unhybridized 12-base overhang (Fig. 2). When topoisomerase attaches to the probe, it cleaves the strand just after the recognition sequence. This cuts the 12-base long part of the oligonucleotide which then permanently separates, leaving vaccinia topoisomerase attached to the 3′ end of the blunt-ended hairpin. Now, the oligonucleotide has a topoisomerase molecule strongly attached to its 3′ end. It can label DNase II type breaks because vaccinia topoisomerase I, which remains bound to the CCCTT motif, will religate the oligonucleotide to a double-stranded DNA break possessing a complementary 5′OH blunt-end. Therefore, the breaks of DNase II type are detected specifically and directly.

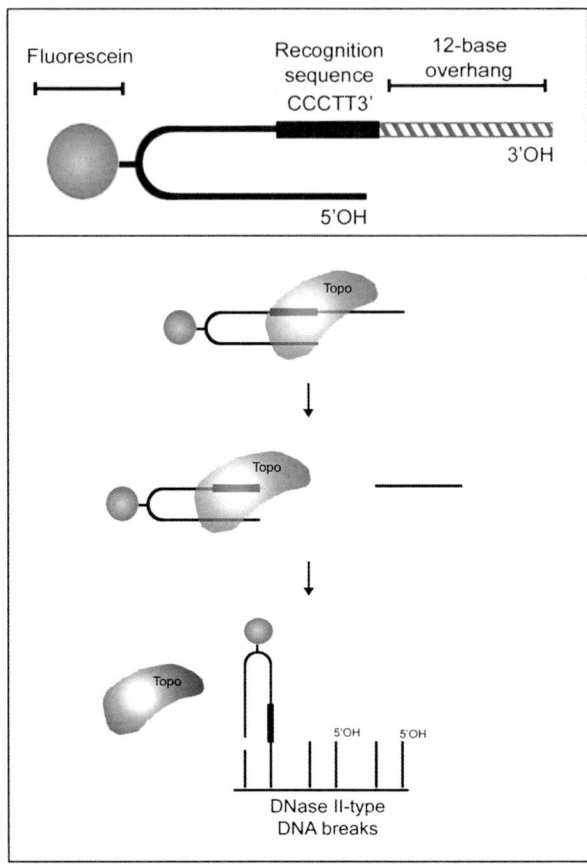

Fig. 2. Principle of the assay for DNA damage detection *in situ* using vaccinia topoisomerase I. Vaccinia topoisomerase I (TOPO) binds to the oligoprobe and cleaves it at the 3′ end of the recognition sequence. The TOPO-activated FITC-labeled portion of the probe then religates to the blunt-ended DNA breaks with 5′OH in tissue section.

The 12-bases overhang on the probe is required because the enzyme will not cut a shorter strand (19) and will therefore be unable to attach to the probe and activate its 3′end.

Although the topoisomerase-based assay can be used on its own, it can also be combined with *in situ* ligation. In this case, the T4 DNA ligase-based assay will label breaks with DNase I architecture, bearing 5′PO$_4$ groups (Fig. 1). The combined detection, using both topoisomerase and ligase, is especially informative in the tissue section format because it can detect both phases of apoptotic cell disassembly: the self-driven cell disassembly and externally-controlled elimination of apoptotic cell corpses by phagocytizing cells. When tested in tissue sections of dexamethazone-treated rat thymus, such a combined assay successfully detected both the primary DNase I-like cleavage in apoptotic thymocyte nuclei and the DNase II-like breaks in the cytoplasm of cortical macrophages ingesting apoptotic cells (16, 18).

The sensitivity of the assay depends on the method of signal registration and with the right microscope can potentially visualize very small numbers of DNase II-type breaks in individual cells (see Note 2).

In this chapter, we present the complete protocol for topoisomerase-based detection applicable for fixed tissue sections.

2. Materials

1. 5–6 μm-thick sections cut from paraformaldehyde-fixed, paraffin-embedded tissue blocks. Use slide brands which retain sections well, such as ProbeOn™ Plus charged and precleaned slides (Fisher Scientific, Pittsburgh, PA) or similar product. Apoptotic tissue sections, such as dexamethazone treated rat thymus are recommended as controls (see Note 3).

2. Xylene.

3. 80% and 96% Ethanol.

4. Oligoprobe 1 was synthesized by IDT (Integrated DNA Technologies, Inc., Coralville, IA). However, it can be synthesized by many commercial oligonucleotide producers. PAGE or HPLC purification is recommended. Dilute with bidistilled water to 450 ng/μL stock concentration. Store at –20°C protected from light. The probe is a labeled with a single fluorescein and detects blunt-ended DNA breaks with 5′OH:

   5′-AAG GGA CCT GCF GCA GGT CCC TTG ATA CGA TTC TA -3′ **F** – FITC-dT

5. Vaccinia DNA topoisomerase 1 – 3,000 U/μL (Millipore) (see Note 4).

6. 50 mM Tris–HCl, PH 7.4.

7. Proteinase K (Roche Molecular Biochemicals, Indianapolis, IN) 20 mg/mL stock solution in distilled water. Store at –20°C. In the reaction, use 50 μg/mL solution in PBS, prepared from the stock. Do not reuse (see Note 5).

8. Vectashield with DAPI (Vector Laboratories, Burlingame, CA).

9. 100 ng/μL DNase I (Roche, Indianapolis, IN) in 50 mM Tris–HCl, pH 7.4, and 10 mM $MgCl_2$. DNase I and DNase II are needed for controls if verification of labeling specificity is required (see Note 6).

10. 500 ng/μL DNase II in the buffer supplied with the enzyme (Sigma Chemicals, St. Louis, MO). Make the DNase II stock

by diluting DNase II powder in water to the concentration of 1 mg/mL, aliquot in small volumes and keep the stock at −20°C. To run a reaction, dilute the stock solution 1:1 with the DNase II reaction buffer supplied with the enzyme.

11. Phosphate-buffered saline (1× PBS): dissolve 9 g NaCl, 2.76 g $NaH_2PO_4 \cdot H_2O$, and 5.56 g $Na_2HPO_4 \cdot 7H_2O$ in 800 mL of distilled water. Adjust to pH 7.4 with NaOH, and fill to 1 L with distilled water.

12. Sodium bicarbonate buffer: 50 mM $NaHCO_3$, 15 mM NaCl, pH 8.2.

13. 22 × 22 mm or 22 × 40 mm glass or plastic coverslips. Plastic coverslips are preferable as they are easier to remove from the section.

14. Fluorescent microscope with appropriate filters and objectives.

15. Video or photo camera for documentation.

3. Method

3.1. Labeling 5′OH blunt-ended DNA breaks in Tissue Sections

1. Place the sections in a slide rack and dewax in xylene for 15 min, transfer to a fresh xylene bath for an additional 5 min.

2. Rehydrate by passing through graded ethanol concentrations: 96% Ethanol – 2 × 5 min; 80% Ethanol – 5 min; water – 2 × 5 min.

3. Digest section with Proteinase K. Use 100 µL of a 50 µg/mL solution per section. Incubate 10 min at room temperature (23°C) in a humidified chamber (see Note 7).

4. Rinse in distilled water for 2 × 10 min.

5. While sections are rinsed in water, combine in a vial: 100 pmol of Oligoprobe 1 and 100 pmol (3.3 µg) of vaccinia topoisomerase I in solution of 50 mM Tris–HCL, pH 7.4 (see Note 8). Incubate at room temperature for 15 min to allow for probe activation. Use 25 µL of this reaction solution per section.

6. Aspirate water from sections and apply the reaction solution containing Oligoprobe 1 and vaccinia topoisomerase I.

7. Incubate for 15 min at room temperature (23°C) in a humidified chamber with a plastic coverslip, protected from light.

8. Remove coverslips by gently immersing the slides vertically in a coplin jar containing water at room temperature. Then, wash section 3 × 10 min in distilled water.

9. Rinse with sodium bicarbonate buffer (see Note 9).

10. Cover section with an antifading solution (Vectashield with DAPI), coverslip and analyze the signal using a fluorescent microscope. Double-stranded DNA breaks with 5′OH will fluoresce green (see Note 10).

4. Notes

1. So far, the only disease with suggested links to a disruption in apoptotic corpse disassembly is systemic lupus erythematosus (SLE). In SLE, the apoptotic cells are not properly eliminated due to failures of the clearance system in the phase II of apoptotic DNA degradation. Cell corpses become the source of autoantigens responsible for the initiation and progression of the disease (20, 21).

2. The visualization of smaller numbers of breaks generated without apoptosis might require a biotin-labeled probe with enzymatic amplification of signal or a confocal microscope.

3. To make apoptotic thymus, inject Sprague-Dawley rats (150 g) subcutaneously with 6 mg/kg dexamethasone (Sigma) dissolved in 30% dimethyl sulfoxide in water. Animals have to be sacrificed 24 h post injection. Fix the thymus tissue by incubating 18 h in 4% paraformaldehyde. Then, pass it through graded alcohols to 100% ethanol, place overnight in chloroform and embed in paraffin.

4. Highly concentrated vaccinia topoisomerase I, which works well with the described assay, can be obtained from Millipore, sold as a part of the ApopTag® ISOL Dual Fluorescence Apoptosis Detection Kit. Our initial enzyme preparation was obtained directly from Dr. Stewart Shuman (Sloan-Kettering Institute, New York) who originally described its purification (22). We have also received highly concentrated vaccinia topoisomerase I (3,000 U/μL i.e. 0.2 μg /μL ~6 pmol/μL) from Epicentre Biotechnologies (Madison, WI) as a gift. The company's regular preparation of 10 U/μL is too weak for the assay (1 pmol of the concentrated enzyme equals approximately 500 U).

5. Proteinase K is a very stable enzyme, when stored at concentrations higher than 1 mg/mL. However, autolysis of the enzyme occurs in aqueous solutions at low concentrations (~10 μg/mL) (23).

6. DNase I and II treated sections can be used as positive and negative controls. Use sections of tissues with no preexisting DNA damage, such as normal bovine adrenal or rat heart.

Not all normal tissues provide good controls. Normal rat brain sections, after the treatment with DNase II, display high nonspecific background levels. Sections of such organs as normal thymus or small intestine contain cells with DNA cleavage.

The sections should be washed in water (2 × 10 min), and treated with 100 ng/μL of DNase I (Roche) in 50 mM Tris–HCL, pH 7.4, and 10 mM $MgCl_2$ overnight at 37°C, or with 500 ng/μL DNase II in the buffer supplied with the enzyme (Sigma) for 30 min at 37°C. After washing in water (3 × 10 min), and preblocking with 2% BSA (15 min, 23°C) the sections are ready for labeling.

Mock reactions without enzymes are also recommended as regular controls in order to assess nonspecific background staining.

Sometimes, additional controls might be needed to rule out possible contamination of vaccinia topoisomerase preparations with nucleases. For this, the pretreatment of control sections with vaccinia topoisomerase I for 2 h at 37°C is recommended followed by DNA breaks detection.

7. Proteinase K digestion time may need adjustment depending on the tissue type. Hard tissues might require longer digestion. Times of 10 min are usually used. Omitting the digestion step results in a weaker signal. On the other hand, overdigestion, results in signal disappearance and section disruption.

8. In the initial experiments, we used 215 pmol (7.1 μg) of the enzyme per every 25 μL of the reaction mix. However, the topoisomerase concentration can be significantly reduced without any loss of sensitivity. We later used four times less of the enzyme per section (1.76 μg in 25 μL of the reaction mix per section) with similar results. Reducing the amount of enzyme to 880 ng (in 25 μL of the reaction mix) resulted in a weaker signal and 8.8 ng of enzyme produced no signal.

9. Alkaline solution rinse enhances FITC fluorescence, which is pH sensitive and is significantly reduced below pH 7. This step is not needed if other non pH-sensitive fluorophores are used in the probe.

10. Topoisomerase-based labeling and combined topoisomerase-ligase labeling can be performed using the ApopTag® ISOL Dual Fluorescence Kit (Millipore).

Acknowledgment

I am grateful to Candace Minchew for her outstanding technical assistance.

References

1. Samejima, K., and Earnshaw, W.C. (2005) Trashing the genome: role of nucleases during apoptosis. *Nat. Rev. Mol. Cell Biol.* **6**, 677–688.
2. Kawane, K. *et al.* (2003) Impaired thymic development in mouse embryos deficient in apoptotic DNA degradation. *Nat. Immunol.* **4**, 138–144.
3. Parrish, J.Z., and Xue, D. (2006) Cuts can kill: the roles of apoptotic nucleases in cell death and animal development. *Chromosoma* **115**, 89–97.
4. Didenko, V.V., and Hornsby, P.J. (1996) Presence of double-strand breaks with single-base 3′ overhangs in cells undergoing apoptosis but not necrosis. *J. Cell Biol.* **135**, 1369–1376.
5. Widlak, P., Li, P., Wang, X., and Garrard, W.T. (2000) Cleavage preferences of the apoptotic endonuclease DFF40 (caspase-activated DNase or nuclease) on naked DNA and chromatin substrates. *J. Biol. Chem.* **275**, 8226–8232.
6. Staley, K., Blaschke, A.J., and Chun, J. (1997) Apoptotic DNA fragmentation is detected by a semi-quantitative ligation-mediated PCR of blunt DNA ends. *Cell Death Differ.* **4**, 66–75.
7. Weir, A.F. (1993) Deoxyribonuclease I and II, in *Enzymes of Molecular Biology* (Burrell, M.M. ed.), Humana, Totowa, NJ, pp. 7–16.
8. Didenko, V.V., Ngo, H., and Baskin, D.S. (2002) In situ detection of double-strand DNA breaks with terminal 5′OH groups, in *In Situ Detection of DNA Damage: Methods and Protocols* (Didenko, V.V. ed.) Humana, Totowa, NJ, pp. 143–151.
9. Sikorska, M., and Walker, P.R. (1998) Endonuclease activities and apoptosis, in *When Cells Die* (Lockshin, R.A., Zakeri, Z., and Tilly, J.L., eds.), Wiley-Liss, New York, pp. 211–242.
10. Didenko, V.V. (2002) Detection of specific double-strand DNA breaks and apoptosis in situ using T4 DNA ligase, in *In Situ Detection of DNA Damage: Methods and Protocols* (Didenko, V.V. ed.), Humana, Totowa, NJ, pp. 143–151.
11. Chahory, S., Padron, L., Courtois, Y., and Torriglia, A. (2004) The LEI/L-DNase II pathway is activated in light-induced retinal degeneration in rats. *Neurosci. Lett.* **367**, 205–209.
12. Krieser, R.J., MacLea, K.S., Longnecker, D.S., Fields, J.L., Fiering, S., and Eastman, A. (2002) Deoxyribonuclease IIalpha is required during the phagocytic phase of apoptosis and its loss causes perinatal lethality. *Cell Death Differ* **9**, 956–962.
13. Loo, D.T. (2011) In situ detection of apoptosis by the TUNEL assay: an overview of techniques, in *DNA Damage Detection In Situ, Ex Vivo, and In Vivo: Methods and Protocols* (Didenko, V.V. ed.), Humana, Totowa, NJ, pp. This volume.
14. Thiry, M. (2002) In situ nick translation at the electron microscopic level, in *In Situ Detection of DNA Damage: Methods and Protocols* (Didenko, V.V. ed.), Humana, Totowa, NJ, pp. 121–130.
15. Hornsby, P.J., and Didenko, V.V. (2011) In situ oligo ligation (ISOL): A decade and a half of experience, in *DNA Damage Detection In Situ, Ex Vivo, and In Vivo: Methods and Protocols* (Didenko, V.V. ed.), Humana, Totowa, NJ, pp. This volume.
16. Didenko, V.V., Minchew, C.L., Shuman, S., and Baskin, D.S. (2004) Semi-artificial fluorescent molecular machine for DNA damage detection. *Nano Lett.* **4**, 2461–2466.
17. Holmes, B. (2004) Colour-coded tags show DNA damage. *New Sci.* **2477**, 23.
18. Didenko, V.V. (2006) Oscillating probe for dual detection of 5′PO_4 and 5′OH DNA breaks in tissue sections, in *Fluorescent Energy Transfer Nucleic Acid Probes: Methods and Protocols*. (Didenko, V.V. ed.), Humana, Totowa, NJ, pp. 59–69.
19. Shuman, S. (1992) Two classes of DNA end-joining reactions catalyzed by vaccinia topoisomerase I. *J. Biol. Chem.* **267**, 16755–16758.
20. Gabler, C., Blank, N., Winkler, S., Kalden, J.R., and Lorenz, H.M. (2003) Accumulation of histones in cell lysates precedes expression of apoptosis-related phagocytosis signals in human lymphoblasts. *Ann. N. Y. Acad. Sci.* **1010**, 221–224.
21. Cocca, B.A., Cline, A.M., and Radic, M.Z. (2002) Blebs and apoptotic bodies are B cell autoantigens. *J. Immunol.* **169**, 159–166.
22. Shuman, S., Golder, M., and Moss, B. (1988) Characterization of vaccinia virus DNA topoisomerase I expressed in Escherichia coli. *J. Biol. Chem.* **263**, 16401–16407.
23. Sweeney, P.J., and Walker, J.M. (1993) Proteinase K (EC 3.4.21.14), in *Enzymes of Molecular Biology* (Burrell, M.M. ed.), Humana, Totowa, NJ, pp. 305–311.

Part II

Detection in Cell Cultures

Chapter 8

Detection of DNA Strand Breaks in Apoptotic Cells by Flow- and Image-Cytometry

Zbigniew Darzynkiewicz and Hong Zhao

Abstract

Extensive DNA fragmentation that generates a multitude of DNA double-strand breaks (DSBs) is a hallmark of apoptosis. A widely used approach to identify apoptotic cells relies on labeling DSBs *in situ* with fluorochromes. Flow or image cytometry is then used to detect and quantify apoptotic cells labeled this way. We developed several variants of the methodology that is based on the use of exogenous terminal deoxynucleotidyl transferase (TdT) to label 3′-OH ends of the DSBs with fluorochromes, defined as the TUNEL assay. This chapter describes the variant based on DSBs labeling using 5-Bromo-2′-deoxyuridine-5′-triphosphate (BrdUTP) as a TdT substrate and the incorporated BrdU is subsequently detected immunocytochemically with anti-BrdU antibody. We also describe modifications of the protocol that allow using other than BrdUTP deoxyribonucleotides to label DSBs. Concurrent differential staining of cellular DNA and multiparameter analysis of cells by flow- or image cytometry enables one to correlate the induction of apoptosis with the cell cycle phase. Examples of the detection of apoptotic cells in cultures of human leukemic cell lines treated with TNF-α and DNA topoisomerase I inhibitor topotecan are presented. The protocol can be applied to the cells growing *in vitro*, treated *ex vivo* with cytotoxic drugs as well as to clinical samples.

Key words: Apoptosis, DNA damage, Flow cytometry, Laser scanning cytometry, Cell cycle, Immuno-fluorescence, BrdU

1. Introduction

During apoptosis, DNA undergoes extensive fragmentation at the internucleosomal sections which generates a multitude of DNA double-strand breaks (DSBs) (1). Their presence is considered to be one of the most characteristic markers of apoptotic cells. A widely used approach to identify apoptotic cells, thus, relies on labeling DSBs *in situ* either with fluorochromes (2–4) or absorption dyes (5). We have developed several variants of the methodology that is based on the use of exogenous terminal deoxynucleotidyl transferase (TdT) to label 3′-OH termini of the DSBs either indirectly or directly with fluorochrome-tagged deoxyribonucleotides, commonly defined as the TUNEL assay (2–4, 6–8). In this chapter, we describe the variant based on DSBs labeling with 5-Bromo-2′-deoxyuridine-5′-triphosphate (BrdUTP) that subsequently is detected immunocytochemically with BrdU antibody (Ab). The BrdUTP labeling assay offers much greater sensitivity than other TUNEL variants (8). However, modifications of the protocol that allow one to use deoxyribonucleotides other than BrdUTP also are described. Concurrent staining of cellular DNA with propidium iodide (PI) or 4′,6-diamidino-2-phenylindole (DAPI) and multiparameter analysis of cells by flow or image cytometry enables one to correlate the induction of apoptosis with the cell cycle phase (9). The protocol can be applied to cells growing *in vitro*, treated *ex vivo* with cytotoxic drugs as well as to clinical samples (see Note 1).

The method presented in this chapter can be applied to cells measured by flow cytometry, and its modification, also included, to cells attached on microscope slides. The latter can be analyzed by image cytometry, e.g., using an instrument, such as the laser scanning cytometer (LSC). LSC is the microscope-based cytofluorometer that allows one to measure rapidly, with high sensitivity and accuracy, fluorescence of individual cells (10). Cells staining on slides eliminates their loss that otherwise occurs during repeated centrifugations in sample preparation for flow cytometry. Therefore, the procedure offers an advantage when applied to samples with paucity of cells, such as fine needle aspirate or spinal fluid tap (see Note 2). Another advantage of LSC is that it offers a possibility of electronic selection (gating) of cells of interest during the initial measurement for their subsequent analysis by imaging or staining with other fluorochromes. Imaging and visual examination are of particular importance because the characteristic changes in cell morphology are considered the gold standard for positive identification of apoptotic cells (3, 4). Furthermore, the cell attributes measured by LSC on live cells can be correlated with the attributes that to be measured require cell fixation (11). For example, the activation of caspases (11–13), DNA replication (14), translocation of Bax to mitochondria (15), or activation of NF-κB transcription factor (16), and the key events associated with apoptosis can be correlated, in the very same cells, with the presence of apoptosis-associated DSBs.

Fixation and permeabilization of the cells are the initial essential steps to successfully label DSBs. Cells are briefly fixed with the cross-linking fixative formaldehyde and then permeabilized by suspending in ethanol or using detergents in the subsequent rinses. By cross-linking small DNA fragments to other cell constituents, formaldehyde prevents their extraction, which otherwise occurs during repeated centrifugations and rinses (17). The 3′OH-termini of the DSBs serve as primers and become labeled in this procedure with BrdU when incubated with BrdUTP in a reaction catalyzed by exogenous TdT (2). The incorporated BrdU is immunocytochemically detected by BrdU Ab conjugated to fluorochromes of a desired emission wavelength (8). The BrdU Ab is a widely available reagent, also used in studies of cell proliferation to detect BrdU incorporated during DNA replication (18). The sensitivity of DSBs detection is higher and overall cost of reagents is significantly lower when BrdUTP is used, as compared to the alternative labeling with biotin- (or digoxigenin-) (2) or directly fluorochrome-tagged deoxyribonucleotides (7). The alternate procedures, utilizing digoxigenin, biotin, or directly labeled deoxynucleotides, however, are also described in the chapter.

2. Materials

2.1. Reagents and Glassware

1. Phosphate-buffered saline (PBS), pH 7.4.
2. 1% Formaldehyde (methanol-free, "ultrapure") (Polysciences, Warrington, PA), in PBS, pH 7.4.
3. 70% Ethanol.
4. TdT (Roche Diagnostics, Indianapolis, IN). Supplied in storage buffer: (60 mM potassium phosphate at pH 7.2, 150 mM KCl, 1 mM 2-mercaptoethanol and 0.5% Triton X-100, 50% glycerol). The 5× TdT reaction buffer (Roche Diagnostics) contains: 1 M potassium (or sodium cacodylate), 125 mM HCl, pH 6.6, and 1.25 mg/mL bovine serum albumin (BSA).
5. BrdUTP stock solution (50 µL): 2 mM BrdUTP (Sigma) in 50 mmM Tris–HCl, pH 7.5.
6. 10 mM $CoCl_2$ (Sigma).
7. Rinsing buffer: 0.1% Triton X-100 (Sigma) and 5 mg/mL BSA dissolved in PBS.
8. Alexa Fluor 488-conjugated anti-BrdU monoclonal antibody (mAb): Dissolve 1.0 µg of Alexa Fluor 488-conjugated anti-BrdU Ab in 100 µL of PBS containing 0.3% Triton X-100 and 1% (w/v) BSA. Alternatively, use Fluorescein- (FITC)- or Alexa Fluor 647-conjugated anti-BrdU Ab. These Abs are available from Phoenix Flow System (San Diego, CA) or from Invitrogen/Molecular Probes (Eugene, OR).
9. PI staining buffer: 5 µg/mL PI (Invitrogen/Molecular Probes) and 10 µg/mL of RNase A (DNase-free) (Sigma) in PBS. Alternatively, use 1 µg/mL solution of DAPI (Invitrogen/Molecular Probes) in PBS.

10. Microscope slides or single- or multichambered Falcon CultureSlides (BD Biosciences) (to be used in conjunction with analysis by LSC/iCys$^{(R)}$ Research Imaging Cytometer).

11. Coplin jars (to be used in conjunction with analysis by LSC/iCys$^{(R)}$ Research Imaging Cytometer).

12. Parafilm "M" (American National Can, Greenwich, CT) (to be used in conjunction with LSC/iCys).

13. Glycerol (to be used in conjunction with analysis by LSC/iCys$^{(R)}$ Research Imaging Cytometer).

2.2. Commercial Kits

Several kits for labeling DSBs are commercially available. The APO-BRDU kit (Phoenix Flow Systems, San Diego, CA) uses a BrdUTP methodology similar to that described in this chapter. As mentioned, this methodology offers the most sensitive means of DNA strand break detection (8). The APO-DIRECT kit (also from Phoenix) offers a single-step labeling of DNA strand breaks with the fluorochrome-tagged deoxynucleotide. Its virtue is simplicity, but it is less sensitive than the APO-BRDU. Of importance, the positive and negative control cells are supplied with each of these Phoenix kits. It should be noted that the kits developed by Phoenix Flow Systems are provided also by other vendors. The kits utilizing biotin- or digoxigenin-tagged dUTP are also commercially available.

2.3. Instrumentation

Flow cytometers of different types, offered by several manufacturers, can be used to measure cell fluorescence following staining according to the procedures described below. The manufacturers of the most common flow cytometers are Coulter/Beckman Corporation (Miami, FL), BD Biosciences (formerly Becton Dickinson Immunocytometry Systems; San Jose, CA), iCyt (Urbana-Champain, IL) and PARTEC GmbH (Zurich, Switzerland). The multiparameter LSC (iCys$^{(R)}$ Research Imaging Cytometer) is available from CompuCyte, Inc., (Westwood, MA). Cytospin centrifuge, which is used in conjunction with LSC/iCys, is provided by Shandon (Pittsburgh, PA).

The software to deconvolute the DNA content frequency histograms, to analyze the cell cycle distributions, is available from Phoenix Flow Systems or Verity Software House (Topham, MA).

3. Methods

3.1. DNA Strand Break Labeling with BrdUTP for Analysis by Flow Cytometry

1. Suspend $1–2 \times 10^6$ cells in 0.5 mL PBS. With a Pasteur pipette transfer this suspension into a 5 mL polypropylene tube (see Note 2) containing 4.5 mL of ice-cold 1% formaldehyde (see Note 3). Keep the tube for 15 min on ice.

2. Centrifuge at 300×g for 5 min and resuspend cell pellet in 5 mL of PBS. Centrifuge again and resuspend cell pellet in 0.5 mL of PBS. With a Pasteur pipette transfer the suspension to a tube containing 4.5 mL of ice-cold 70% ethanol. The cells can be stored in ethanol, at −20°C for several weeks.

3. Centrifuge at 200×g for 3 min, remove ethanol, resuspend cells in 5 mL of PBS, and centrifuge at 300×g for 5 min.

4. Resuspend the pellet in 50 µL of a solution containing:
   - 10 µL TdT 5× reaction buffer.
   - 2.0 µL of BrdUTP stock solution.
   - 0.5 µL (12.5 units) TdT.
   - 5 µL of 10 mM $CoCl_2$ solution.
   - 33.5 µL distilled H_2O.

5. Incubate the cells in this solution for 40 min at 37 °C (see Notes 4 and 5).

6. Add 1.5 mL of the rinsing buffer, and centrifuge at 300×g for 5 min.

7. Resuspend cell pellet in 100 µL of Alexa Fluor 488-conjugated anti-BrdU Ab solution. (Alternatively you may use the Ab conjugated either with fluorescein (FITC) or Alexa Fluor 647).

8. Incubate at room temperature for 1 h.

9. Add 1 mL of PI staining solution (alternatively you may add 1 mL of the DAPI staining solution).

10. Incubate for 30 min at room temperature, or 20 min at 37 °C, in the dark.

11. Analyze cells by flow cytometry.
    - Illuminate with blue light (488 nm laser line or BG12 excitation filter).
    - Measure green fluorescence of FITC (or Alexa Fluor 488)-conjugated anti-BrdU Ab at 530±20 nm.
    - Measure intensity of red fluorescence of PI at >600 nm. Alternatively, if DNA was stained with DAPI instead of PI use UV light as an excitation source and measure the intensity of blue fluorescence (480±20 nm).

The bivariate (DSBs versus cellular DNA content) distributions (scatterplots) illustrating the cell populations containing a fraction of apoptotic cells labeled according to the method described in the protocol and analyzed by flow cytometry are shown in Fig. 1 and analyzed by LSC (iCys[R] Research Imaging Cytometer) are shown in Fig. 2. A correlation between the induction of apoptosis and cell position in the cell cycle is clearly evident: in the case of topotecan treated HL-60 cells, nearly all

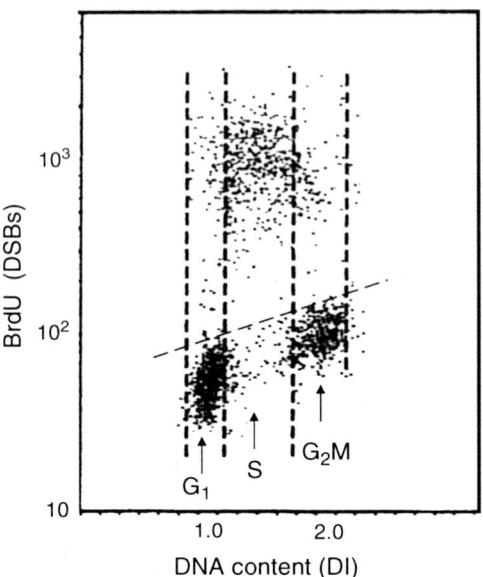

Fig. 1. Detection of apoptotic cells after DSBs labeling with BrdUTP and fluorescence analysis by flow cytometry. To induce apoptosis leukemic HL-60 cells were treated in culture with DNA topoisomerase I inhibitor topotecan (0.15 μM) for 4 h. The cells were then subjected to DSBs labeling with BrdUTP as described in the protocol using fluorescein-tagged BrdU Ab and staining DNA with PI. Cellular fluorescence was measured by flow cytometry. The data are presented as the bivariate distributions (scatterplots) illustrating cellular DNA content (DNA index, DI) versus DSBs labeled with BrdU Ab. Note that essentially only S-phase cells underwent apoptosis as shown by high intensity of their BrdU-associated fluorescence, above the control level marked by the skewed dashed line. The leukemic cells treated with topoisomerase I inhibitors topotecan or camptothecin for 3–5 h present a convenient experimental model to assess whether the protocol of DSBs labeling is effective because in the same cell population there are DSBs positive (S-phase) and negative (G_1 and G_2M) cells.

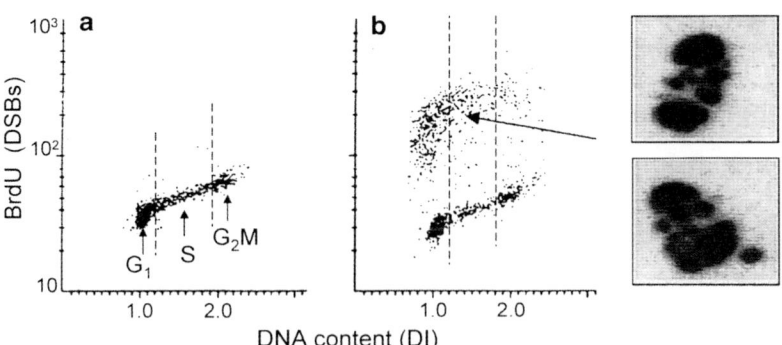

Fig. 2. Detection of apoptotic cells after DSBs labeling with BrdUTP and analysis by LSC. U-937 cells were untreated (**a**) or treated with tumor necrosis factor-α (TNF-α) in the presence of cycloheximide (**b**, (22, 23)). The cells were then subjected to DNA strand break labeling and DNA staining as described in the protocol using fluorescein-tagged BrdU Ab and staining DNA with PI. Cell fluorescence was measured by iCys(R) Research Imaging Cytometer. The bivariate distributions (scatterplots) allow one to identify apoptotic

apoptotic cells are S-phase cells (Fig. 1) while the apoptotic U-932 cells treated with TNF-α are predominantly G_1-cells.

3.2. DSBs Labeling with Other Markers for Analysis by Flow Cytometry

As mentioned in the Subheading 1 DNA strand breaks can be labeled with deoxynucleotides tagged with variety of other fluorochromes. For example, the Molecular Probes, Inc., catalog lists several types of dUTP conjugates, including BODIPY dyes (e.g., BODIPY-FL-X-dUTP), fluorescein, Cascade Blue and Texas Red. Several cyanine dyes conjugates (e.g., CY-3-dCTP) are available from Biological Detection Systems (Pittsburgh, PA). Indirect labeling, via biotinylated- or digoxigenin-conjugated deoxyribonucleotides offers a multiplicity of commercially available fluorochromes (fluorochrome-conjugated avidin or streptavidin, as well as digoxigenin antibodies) with different excitation and emission characteristics. DNA strand breaks, thus, can be labeled with a dye of any desired fluorescence color and excitation wavelength.

The procedure described in Subheading 3.1 can be adopted to utilize any of these fluorochromes. In the case of the direct labeling (7), the fluorochrome-conjugated deoxynucleotide is included in the reaction solution (0.25–0.5 nmol per 50 μL) instead of BrdUTP, as described in step 4 of Subheading 3.1. Following the incubation step (step 5), omit steps 6–8, and stain cells directly with PI (step 9). In the case of the indirect labeling, instead of BrdUTP, digoxygenin- or biotin-conjugated deoxynucleotides are included into the reaction buffer (0.25–0.5 nmol per 50 μL) at step 4. The cells are then incubated either with the fluorochrome-conjugated antidigoxigenin MAb (0.2–0.5 μg per 100 μL of PBS containing 0.1% Triton X-100 and 1% BSA), or with fluorochrome conjugated avidin or streptavidin (0.2–0.5 μg per 100 μL, as above) at step 7 and then processed through steps 8–10 as described in the protocol. Analysis by flow cytometry is carried out with excitation and emission wavelengths appropriate to the used fluorochrome.

3.3. DNA Strand Break Labeling for Analysis by LSC (iCys(R) Research Imaging Cytometer)

1. Transfer 300 μL of cell suspension (in tissue culture medium, with serum) containing approximately 20,000 cells into a cytospin chamber. Cytocentrifuge at 1,000 rpm (~150×*g*) for 6 min to deposit the cells on a microscope slide. (Alternatively, to analyze cells that grow attached to surface maintain them in cultures in single- or multichambered

---

Fig. 2. (continued) cells as the cells with labeled DSBs (strong green fluorescence intensity), and also reveal the cell cycle position of cells in either apoptotic or nonapoptotic population. Note predominance of G_1 cells among apoptotic cells. The cells with strong DSBs labeling were relocated, imaged, and their representative images are presented as shown. These cells show nuclear fragmentation and chromatin condensation, the typical features of apoptosis (3, 4).

Falcon CultureSlides (BD Biosciences). When harvested, remove the walls of the chambers, rinse cells with PBS and fix in formaldehyde as described in step 2 of Subheading 3.3.).

2. Without allowing the cytospin to completely dry, prefix the cells by transferring the slides for 15 min to a Coplin jar containing 1% formaldehyde in PBS, cooled to ice temperature.

3. Rinse the slides in PBS and transfer to 70% ethanol; fix in ethanol for at least 1 h; the cells can be stored in ethanol for weeks at −20°C.

4. Follow steps 4–8 of Subheading 3.1 as described for flow cytometry. Carefully layer small volumes (approximately 100 μL) of the respective buffers, rinses or staining solutions on the cytospin area (or over the sites of individual chambers if the cells were grown on Chamber Slides) of the horizontally placed slides. At appropriate times, remove these solutions with Pasteur pipette (or vacuum suction pipette). To prevent drying, place a 2 × 4 cm strip of Parafilm over the site where the cells are present atop of the solutions used for cell incubations (see Note 8).

5. Replace the PI staining solution with a drop of a mixture of glycerol and PI staining solution (9:1) and mount under the coverslip. To preserve the specimen for longer period of time or transport, seal the coverslip with nail polish or melted paraffin.

6. Measure cell fluorescence on LSC.

   - Excite fluorescence with 488 nm laser line.
   - Measure green fluorescence of Alexa Fluor 488 or fluorescein anti BrdU Ab at 530 ± 20 nm.
   - Measure red fluorescence of PI at >600 nm.

   (Alternatively, if DSBs are labeled with Alexa Fluor 647 excite fluorescence with red diode laser and measure fluorescence intensity at far-red wavelength. If DAPI is used to stain DNA, excite DAPI fluorescence with UV laser and measure fluorescence intensity at 480 ± 20 wavelength).

   The typical results are shown in Figs. 1 and 2 (see Notes 6 and 7).

3.4. Controls

The procedure of DNA strand break labeling is rather complex and involves many reagents. Negative results, therefore, may not necessarily mean the absence of DNA strand breaks (see Note 7) but may be due to methodological problems, such as the loss of TdT activity, degradation of BrdUTP, etc. It is necessary, therefore, to include the positive and negative control. An excellent control is to use HL-60 cells treated (during their exponential growth) for 3–4 h with 0.2 μM of the DNA topoisomerase I inhibitor camptothecin (CPT). Because CPT under these conditions induces apoptosis selectively during S phase, cells in G_1 and G_2/M may serve as negative control populations, while the S phase cells in the same sample, represent the positive control.

Another negative control consist cells processed identically as described in Subheading 3.1 except that TdT is excluded from step 4.

4. Notes

1. This method is also useful for clinical material, such as that obtained from in leukemias, lymphomas, and solid tumors (19, 20), and that can be combined with surface immunophenotyping. In such instance, the cells are first immunophenotyped, then fixed with 1% formaldehyde (which stabilizes the antibody bound on the cell surface) and subsequently subjected to the DSBs detection assay using other color fluorochrome (see Subheading 3.1) than the one used for immunophenotyping. The percent of apoptotic (DSBs-positive) cells is then estimated within the gated-immunophenotype cell population.

2. If the sample initially contains small number of cells, cell loss during repeated centrifugations is a problem. To minimize cell loss, polypropylene, or siliconized glass tubes are recommended. Since transferring cells from one tube to another one results in electrostatic attachment of a large fraction of cells to the surface of each new tube all steps of the procedure (including fixation) should be done in the same tube. Addition of 1% (w/v) BSA into rinsing solutions also decreases cell loss. When the sample contains very few cells, the carrier cells, which later can be recognized based on differences in DNA content (e.g., chick erythrocytes) may be included. Because there is no cell loss during processing for analysis by LSC the samples with paucity of cells can easily be measured.

3. Cell pre-fixation with a cross-linking agent, such as formaldehyde is required to prevent the extraction of the fragmented DNA from apoptotic cells (17). This ensures that despite the repeated cell washings during the procedure, the DNA content of apoptotic cells (and with it the number of DSBs) is not markedly diminished.

4. Alternatively, incubate at room temperature overnight.

5. Control cells may be incubated in the same solution, but without TdT.

6. It is generally easy to identify apoptotic cells, due to their intense labeling with Alexa Fluor 488, fluorescein, or Alexa Fluor 647-conjugated anti-BrdU Ab. The high fluorescence intensity often requires the use of the exponential scale (logarithmic amplifiers of the flow cytometer or LSC) for data acquisition and display (Figs. 1 and 2). As it is evident in these figures, because cellular DNA content of each, apoptotic and nonapoptotic cell population is measured, the cell cycle distribution and/or DNA ploidy of these both populations can be estimated.

7. While the presence of extensive DNA breakage, detected following strand break labeling, by the strong fluorescence, is a very characteristic feature of apoptosis, a weak fluorescence may not necessarily mean the lack of apoptosis. In some cell systems, DNA fragmentation stops at 300–50 kb size DNA fragments and does not progress into internucleosomal sections (21).

8. It is essential that the incubations are carried out in a humidified chamber. Even minor drying produces severe artifacts.

Acknowledgment

Supported by NCI grant RO1 28 704.

References

1. Nagata, S. (2000) Apoptotic DNA fragmentation. *Exp. Cell Res.*, **256**, 12–18.
2. Gorczyca, W., Bruno, S., Darzynkiewicz, R.J, Gong, J., and Darzynkiewicz, Z. (1992) DNA strand breaks occurring during apoptosis: Their early in situ detection by the terminal deoxynucleotidyl transferase and nick translation assays and prevention by serine protease inhibitors. *Int. J. Oncol.*, **1**, 639–648.
3. Darzynkiewicz, Z., Bruno, S., Del Bino, G., Gorczyca, W., Hotz, M.A., Lassota, P., and Traganos, F. (1992) Features of apoptotic cells measured by flow cytometry. *Cytometry*, **13**, 795–808.
4. Darzynkiewicz, Z., Juan, G., Li, X., Gorczyca, W., Murakami, T., and Traganos, F. (1997) Cytometry in cell necrobiology: analysis of apoptosis and accidental cell death (necrosis). *Cytometry*, **27**, 1–20.
5. Gavrieli, Y. Sherman, Y., and Ben-Sasson, S.A. (1992) Identification of programmed cell death in situ via specific labeling of nuclear DNA fragmentation. *J. Cell Biol.*, **119**, 493–501.
6. Gorczyca, W., Tuziak, T., Kram, A., Melamed, M.R., and Darzynkiewicz, Z. (1994) Detection of apoptosis in fine-needle aspiration biopsies by in situ end-labeling of fragmented DNA. *Cytometry*, **15**, 169–175.
7. Li, X., Traganos, F., Melamed, M.R., and Darzynkiewicz, Z. (1995) Single step procedure for DNA strand break labeling. Detection of apoptosis and DNA replication. *Cytometry*, **20**, 172–180.
8. Li, X., and Darzynkiewicz, Z. (1995) Labelling DNA strand breaks with BrdUTP. Detection of apoptosis and cell proliferation. *Cell Prolif.*, **28**, 571–579.
9. Gorczyca, W., Gong, J., Ardelt, B., Traganos, F., and Darzynkiewicz, Z. (1993) The cell cycle related differences in susceptibility of HL-60 cells to apoptosis induced by various antitumor drugs. *Cancer Res.*, **53**, 3186–3192.

10. Pozarowski, P., Holden, E., and Darzynkiewicz, Z. (2006) Laser scanning cytometry: Principles and applications. *Methods Mol. Biol.*, **319**, 165–192.
11. Li, X., and Darzynkiewicz, Z. (1999). The Schrödinger's cat quandary in cell biology: integration of live cell functional assays with measurements of fixed cells in analysis of apoptosis. *Exp. Cell Res.*, **249**, 4–412.
12. Li, X., Du, L., and Darzynkiewicz, Z. (2000) Caspases are activated during apoptosis independently of dissipation of mitochondrial electrochemical potential. *Exp. Cell Res.*, **257**, 290–297.
13. Huang, X., Okafuji, M., Traganos, F., Luther, E., Holden, E., and Darzynkiewicz, Z. (2004) Assessment of histone H2AX phosphorylation induced by DNA topoisomerase I and II inhibitors topotecan and mitoxantrone and by DNA crosslinking agent cisplatin. *Cytometry A*, **58A**, 99–110.
14. Li, X., Melamed, M.R., and Darzynkiewicz, Z. (1996) Detection of apoptosis and DNA replication by differential labeling of DNA strand breaks with fluorochromes of different color. *Exp. Cell Res.*, **222**, 28–37.
15. Bedner, E., Li, X., Kunicki, J., and Darzynkiewicz, Z. (2000) Translocation of Bax to mitochondria during apoptosis measured by laser scanning cytometry. *Cytometry*, **41**, 83–88.
16. Deptala, A., Bedner, E., Gorczyca, W., and Darzynkiewicz, Z. (1998) Simple assay of activation of nuclear factor kappa B (NF-κB) by laser scanning cytometry (LSC). *Cytometry*, **33**, 376–382.
17. Gong, J., Traganos, F., and Darzynkiewicz, Z. (1994) A selective procedure for DNA extraction from apoptotic cells applicable for gel electrophoresis and flow cytometry. *Anal. Biochem.*, **218**, 314–319.
18. Dolbeare, F., and Selden, J.R. (1994) Immunochemical quantitation of bromodeoxyuridine: Application to cell cycle kinetics. *Methods Cell Biol.*, **41**, 297–316.
19. Gorczyca, W., Bigman, K., Mittelman, A., Ahmed, T., Gong, J., Melamed, M.R., and Darzynkiewicz, Z. (1993) Induction of DNA strand breaks associated with apoptosis during treatment of leukemias. *Leukemia*, 7, 659–670.
20. Li, X., Gong, J., Feldman, E., Seiter, K., Traganos, F., and Darzynkiewicz, Z. (1994) Apoptotic cell death during treatment of leukemias. *Leuk. Lymph.*, **13**, 65–72.
21. Oberhammer, F., Wilson, J.W., Dive, C., Morris, I.D., Hickman, J.A., Wakeling, A.E., Walker, P.R., and Sikorska, M. (1993) Apoptotic death in epithelial cells: Cleavage of DNA to 300 and 50 kb fragments prior to or in the absence of internucleosomal degradation of DNA. *EMBO J.*, **12**, 3679–3684.
22. Li, X., and Darzynkiewicz, Z. (2000) Cleavage of poly(ADP-ribose) polymerase measured in situ in individual cells: relationship to DNA fragmentation and cell cycle position during apoptosis. *Exp. Cell Res.*, **255**, 125–132.
23. Bedner, E., Smolewski, P., Amstad, P., and Darzynkiewicz, Z. (2000) Activation of caspases measured in situ by binding of fluorochrome-labeled inhibitors of caspases (FLICA): correlation with DNA fragmentation. *Exp. Cell Res.*, **260**, 308–313.

Chapter 9

Fluorochrome-Labeled Inhibitors of Caspases: Convenient *In Vitro* and *In Vivo* Markers of Apoptotic Cells for Cytometric Analysis

Zbigniew Darzynkiewicz, Piotr Pozarowski, Brian W. Lee, and Gary L. Johnson

Abstract

Activation of caspases is a hallmark of apoptosis. Several methods, therefore, were developed to identify and count the frequency of apoptotic cells based on the detection of caspases activation. The method described in this chapter is based on the use of *f*luorochrome-*l*abeled *i*nhibitors of *ca*spases (FLICA) applicable to fluorescence microscopy, and flow- and image-cytometry. Cell-permeant FLICA reagents tagged with carboxyfluorescein or sulforhodamine when applied to live cells *in vitro* or *in vivo*, exclusively label cells that are undergoing apoptosis. The FLICA labeling methodology is simple, rapid, robust, and can be combined with other markers of cell death for multiplexed analysis. Examples are presented on FLICA use in combination with a vital stain (propidium iodide), detection of the loss of mitochondrial electrochemical potential, and exposure of phosphatidylserine on the outer surface of plasma cell membrane using Annexin V fluorochrome conjugates. Following cell fixation and stoichiometric staining of cellular DNA, FLICA binding can be correlated with DNA ploidy, cell cycle phase, DNA fragmentation, and other apoptotic events whose detection requires cell permeabilization. The "time window" for the detection of apoptosis with FLICA is wider compared to that with the Annexin V binding, making FLICA a preferable marker for the detection of early phase apoptosis and more accurate for quantification of apoptotic cells.

Key words: FLICA, Caspases, Apoptosis, Flow cytometry, Laser scanning cytometry, Cell death, Mitochondrial potential, Annexin V binding

1. Introduction

Caspases are *c*ysteine-*a*spartic acid-*s*pecific prote*ases* that are activated in response to different cell death-inducing stimuli (1). Their activation initiates specific cleavage of the respective target proteins and, therefore, is considered to be a marker of the irreversible steps leading to cell demise. Although there are exceptions (2), caspase activation is considered a characteristic event of apoptosis, and identification of apoptotic cells often relies on detection of their activity (3, 4). Caspases specifically recognize a four-amino acid sequence on their substrate proteins; the carboxyl end of aspartic acid within this sequence is the target for cleavage. Several approaches have been developed to detect the process of caspases activation. Because the activation involves the transcatalytic cleavage of the zymogen pro-caspases (1), the cleavage products having lower molecular weight than the zymogen can be revealed electrophoretically and identified in Western blots using caspase-specific antibodies. Another approach utilizes the fluorogenic (or chromogenic) substrates of caspases. Peptide substrates were developed, which are colorless or non-fluorescent, but upon cleavage, generate colored or fluorescing products (5, 6). These two approaches, however, are primarily used to detect caspases activation in cell extracts, thereby providing no information on their activation in individual cells that otherwise is needed to reveal heterogeneity of cell populations or to correlate caspases activation with other cell attributes such as cell cycle position.

Among the approaches that can be applied to study activation of caspases *in situ* are the methods based on immunocytochemical detection of the epitope that is characteristic of the caspases' active form. Antibodies that react only with the activated caspases are now available and have been used in cytometric assays (7, 8). Activation of caspases also can be detected indirectly, by immunocytochemical identification of the specific cleavage products, e.g., the p89 fragment of poly(ADP-ribose) polymerase, and this method has been adapted to cytometry as well (9).

The method described here relies on the use of the *f*luorochrome-*l*abeled *i*nhibitors of *ca*spases (FLICA; (10, 11)). The principle of this methodology was introduced long ago in the studies of the esterases and proteases utilizing radio-labeled specific inhibitors that bound to the active centers of these enzymes and were detected by autoradiography (12). In the case of caspases, the ligands that bind to their active centers are carboxyfluorescein (FAM)- or fluorescein (FITC)-tagged peptide-fluoromethyl ketone (FMK) inhibitors. These ketone reagents penetrate the plasma membrane of live cells and are relatively nontoxic to the cells, at least in short-term incubations (13). Actually, in some cell systems by virtue of caspase inhibition, FLICA slow down the process of apoptosis and prevent total cell disintegration, making it possible to quantify the frequency of apoptotic cells more accurately (apoptotic index; AI) (14). The recognition peptide moieties of these reagents were expected to provide some level of specificity of their binding to particular caspases. Thus the FLICA containing VDVAD, DEVD, VEID, YVAD, LETD, LEHD, and AEVD peptide moieties were expected to preferentially bind to activated caspases-2, -3, -6, -1, -8, -9, and -10, respectively. On the contrary, the FAM-VAD-FMK containing the valylalanyl aspartic acid sequence was designed to be pan-caspase reactive, binding to activated caspases-1, -3 -4, -5, -7, -8, and -9 (15). It was later observed, however, that because of the FMK group reactivity, likely with thiol groups of intracellular proteins that become available upon cleavage by caspases, binding of the sequence is not as caspase specific as initially thought (7, 16). Despite the apparent lack of specificity in labeling caspases, the FLICA probes have consistently shown themselves to be highly reliable reporters of caspases activation and convenient markers of apoptotic cells.

Exposure of live cells to FLICA results in the rapid uptake of these reagents followed by their covalent binding to apoptotic cells containing activated caspase enzymes. Unbound FLICA are removed from the nonapoptotic cells by rinsing with wash-buffer. The protocols given below describe labeling cells that contain activated caspases using FAM-VAD-FMK. The same protocol can be applied to other FLICAs. Concurrent exposure of cells to propidium iodide (PI) strongly labels all cells with a compromised plasma membrane integrity that cannot exclude this cationic dye; the loss of membrane integrity is a key feature of mid-late apoptotic cells or cells undergoing necrosis. On the contrary, simultaneous staining with the mitochondrial electrochemical potential probe chloromethyl-X-rosamine (CMXRos; MitoTracker Red) (17) and FLICA allows one to discriminate between two sequential events of apoptosis: dissipation of the inner mitochondrial membrane potential and activation of the caspase enzyme cascade, respectively. Likewise, simultaneous staining with FLICA and Annexin V conjugated to red fluorescing dyes such as Cy5 reveals the relationship between caspases activation and exposure of phosphatidylserine (PS) residues on the external surface of plasma membrane during apoptosis (7, 18). Cells labeled with FLICA and PI, CMXRos, or Annexin V–Cy5 can be examined by fluorescence microscopy or subjected to quantitative analysis by flow or image cytometry such as laser scanning cytometry (LSC). By virtue of the ability to measure large cell populations rapidly to analyze cell images, LSC is particularly useful in studies of apoptosis (19).

2. Materials

1. *Cells to be analyzed*: can be grown on slides (see Subheading 3.1) or in suspension.
2. *Microscope slides, coverslips*: see Subheading 3.1

3. *Instrumentation*: Fluorescence microscope, or multiparameter Laser Scanning Cytometer (LSC; iCys$^{(R)}$ Research Imaging Cytometer), available from CompuCyte, Inc. (Westwood, MA). Flow cytometers of different types, offered by several manufacturers, can be used to measure cell fluorescence following staining according to the procedures described below. The manufacturers of the most common flow cytometers are Beckman/Coulter Corporation (Miami, FL), BD Biosciences (formerly Becton Dickinson Immunocytometry Systems; San Jose, CA), iCyt (Urbana-Champain, IL), and PARTEC GmbH (Zurich, Switzerland).

   The software to deconvolute the DNA content frequency histograms, to analyze the cell cycle distributions, is available from Phoenix Flow Systems (San Diego, CA) or Verity Software House (Topham, MA).

4. Phosphate-buffered saline (PBS).

5. Dimethylsulfoxide (DMSO; Sigma Chemical Co., (St. Louis, MO)).

6. *Stock solution of PI*: dissolve 1 mg of PI (Invitrogen/Molecular Probes, Eugene, OR) in 1 mL of distilled water. This solution can be stored at 4°C in the dark for several months.

7. *Stock FLICA solution*: dissolve lyophilized FLICA (e.g., FAM-VAD-FMK; available as a component of the FLICA™ Apoptosis Detection ("Green FLICA") kit from Immunochemistry Technologies LLC, Bloomington, MN) in dimethylsulfoxide (DMSO) as specified in the kit to obtain 150× concentrated (stock) solution of this inhibitor. Also available from this vendor are caspase-2 (VDVAD), caspase-3 (DEVD), caspase-6 (VEID), caspase-1 (YVAD), caspase-8 (LETD), caspase-9 (LEHD), and caspase-10 (AEVD) FLICA. Aliquots of FLICA solution may be stored at −20°C in the dark for 6 months.

8. *Intermediate (30× concentrated) FLICA solution*: prepare a 30× concentrated solution of FAM-VAD-FMK by diluting the stock solution 1:5 in PBS. Mix the vial until the contents become transparent and homogenous. This solution should be made freshly. Protect all FLICA solutions from light.

9. *FLICA staining solution*: just prior to the use, add 3 μL of 30× concentrated FAM-VAD-FMK solution into 100 μL of culture medium.

10. *Rinsing solution*: 1% (w/v) BSA in PBS.

11. *Staining solution of PI*: add 10 μL of stock solution of PI to 1 mL of the rinsing solution.

12. MitoTracker Red (Invitrogen/Molecular Probes).

13. Annexin V–Cy5 conjugate and the binding buffer (Abcam, Cambridge, MA, or Alexis Biochemicals, San Diego, CA).

3. Methods

3.1. Attachment of Cells to Slides (Cells to be Analyzed by Microscopy and/or LSC)

The procedure requires incubation of live (unfixed, not permeabilized) cells with solutions of FLICA. A variety of adherent cells are available for growth in cell culture flasks. Such cells can be attached to microscope slides by culturing them on slides or coverslips. Culture vessels having microscope slides at the bottom of the chamber are commercially available (e.g., "Chamberslide," Nunc, Inc., Naperville, IL, or single- or multi-chambered Falcon CultureSlides, BD Biosciences). The cells growing in these chambers spread and attach to the slide surface after incubation at 37°C for several hours. Alternatively, the cells can be grown on coverslips, e.g., placed on the bottom of Petri dishes. The coverslips are then inverted over shallow (<1 mm) wells on the microscope slides. The wells can be prepared by constructing the well walls (~2 × 1 cm size) either with a pen that deposits a hydrophobic barrier ("Isolator," Shandon Scientific), nail polish, or melted paraffin. The wells also may be made by preparing a strip of Parafilm "M" (American National Can, Greenwich, CT) of the size of the slide, cutting a hole ~2 × 1 cm in the middle of this strip, placing the strip on the microscope slide, and heating the slide on a warm plate until the Parafilm starts to melt. It should be stressed, however, that because the cells detach during late stages of apoptosis, these cells may be selectively lost if the analysis is limited to attached cells.

Cells that normally grow in suspension can be attached to glass slides by electrostatic forces. This is due to the fact that sialic acid residues that cover the cell surface have a net negative charge in contrast to the glass surface which is positively charged. Incubation of cells on microscope slides in the absence of any serum or serum proteins (which otherwise neutralize the charge), thus, leads to their attachment. The cells taken from culture should be rinsed in PBS in order to remove serum proteins contained in the cell culture media and then resuspended in PBS at a concentration of 2×10^5 cells/mL. An aliquot (50–100 µL) of this suspension should be deposited within a shallow well (prepared as described above) on the horizontally placed microscope slide. To prevent drying, a small piece (~2 × 2 cm) of a thin polyethylene foil or Parafilm may be placed atop the cell suspension drop. A short (15–20 min) incubation of such cell suspension at room temperature in a closed box containing wet tissue or filter paper that provides 100% humidity is adequate to ensure that most cells firmly attach to the slide surface. Cells attached in this manner remain viable for several hours and can be subjected to surface immunophenotyping, viability tests, or intracellular enzyme kinetics assays. Such preparations can then be fixed (e.g., in formaldehyde) without significant loss of cells from the slide. However, as in the case of cell growth on glass, late apoptotic cells have a tendency to detach even after the initial attachment.

It should be stressed that the microscope slide to which the cells are going to be attached electrostatically should be extra clean. Fingerprints leave oils on the slide that interfere with cell attachment. To remove possible contamination of the glass surface that may interfere with cell attachment, soak the microscope slides in a household detergent, then rinse in water and 100% ethanol, respectively. The slides should be allowed to air dry and used the same day they were cleaned.

3.2. Cell Staining and Analysis by Microscopy or LSC (iCys(R) Research Imaging Cytometer)

1. Attach the cells to the microscope slide as described in Subheading 3.1. Keep the cells immersed in the culture medium by adding 100 µL of the medium (with 10% serum) into the well on the microscope slide to cover the area with the cells. In the case of cells growing on microscope slide chambers, move directly to step 2.
2. Remove the medium and replace it with 100 µL of 1× FLICA (e.g., FAM-VAD-FMK) staining solution.
3. Place ~2 × 4 cm strip of Parafilm atop the staining solution to prevent drying. Incubate the slides horizontally for 1 h at 37°C in a closed box with wet tissue or filter paper to ensure 100% humidity, in the dark.
4. Remove the staining solution with Pasteur pipette. Rinse twice with the rinsing solution, each time adding a new aliquot, gently mixing, and after 2 min replacing with the next rinse.
5. Apply 100 µL of the PI staining solution atop the cells deposited on the slide. Cover with a coverslip and seal the edges to prevent drying (see Notes 1–4).
6. During the subsequent 30 min after the addition of PI solution, observe the cells under fluorescence microscope (blue light illumination) or measure cell fluorescence with LSC/iCys. Use the argon ion laser (488 nm) of LSC/iCys to excite fluorescence, contour on light scatter, and measure green fluorescence of FLICA at 530 ± 20 nm and red fluorescence of PI at >600 nm (Fig. 1) (See Note 5).

3.3. Cell Staining and Analysis by Flow Cytometry

1. Suspend 5×10^5–10^6 cells in 0.3 mL of complete culture medium (with 10% serum) in a centrifuge tube.
2. Add 10 µL of the 30× concentrated ("intermediate") FLICA solution to this cell suspension. Mix the cell suspension by flicking the tube.
3. Incubate for 60 min at 37°C in atmosphere of air with 5% CO_2, at 100% humidity, in the dark.

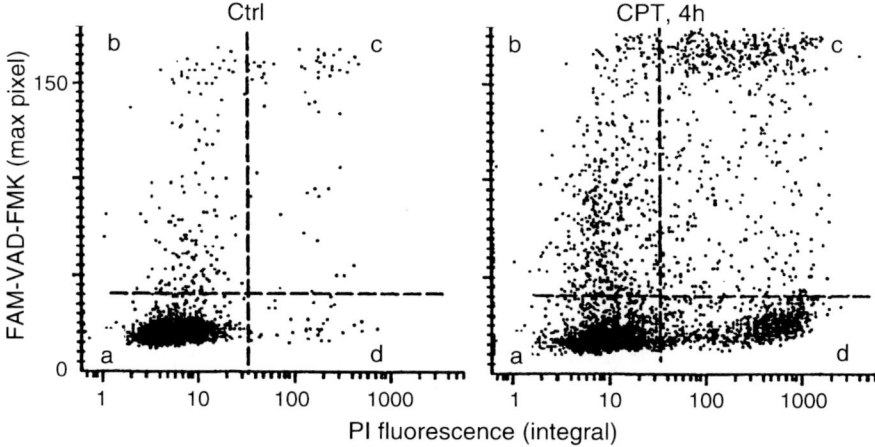

Fig. 1. The bivariate distributions (scatterplots) of FAM-VAD-FMK (FLICA; green maximal pixel) vs. PI (red integral) fluorescence of the control (Ctrl) and CPT-treated HL-60 cells. To induce apoptosis, the cells were treated for 4 h with DNA topoisomerase I inhibitor camptothecin (0.15 µM; CPT). The cells were then stained according to the protocol (Subheading 3.2) and their fluorescence was measured by iCys(R) Research Imaging Cytometer. The live non-apoptotic cells, which are predominant in Ctrl, are unlabeled (quadrant a). Early apoptotic cells have increased FLICA fluorescence but minimal fluorescence of PI (quadrant b). The cells that are more advanced in apoptosis show variable degrees of both, FLICA and PI, fluorescence (quadrant c). At very late stages of apoptosis ("necrotic stage"), caspases either leak out of cells or are not reactive with FLICA, and the cells become FLICA negative/PI positive (quadrant d) (see Notes 6–11).

4. To the cell suspension with FLICA, add 5 mL of the rinsing solution (PBS with BSA) or 1× "wash buffer" provided with the FLICA kit and gently mix the cell suspension.

5. Centrifuge at $300 \times g$ for 5 min at room temperature and remove supernatant by aspiration.

6. Resuspend cell pellet in 2 mL of the rinsing solution or in 1× "wash buffer."

7. Centrifuge at $300 \times g$ for 5 min and aspirate the supernatant (See Note 5).

8. Resuspend cells in 1 mL of the PI staining solution. Place the tube on ice (see Notes 2–4).

9. Measure cell fluorescence by flow cytometry:
   - Excite cell fluorescence with 488-nm laser line.
   - Measure green fluorescence of FLICA at 530 ± 20 nm.
   - Measure red fluorescence of PI at >600 nm.

4. Notes

1. Protect cells from light throughout the procedure.
2. Staining with PI is optional. It allows us to identify the cells that have compromised cell membrane integrity features.

Cells bearing compromised cell membrane integrity structure (necrotic and mid-late apoptotic cells, cells with mechanically damaged membranes, isolated cell nuclei will stain red as a result of their inability to exclude PI (Fig. 1).

3. If concurrently with FLICA cells are stained with CMXRos (MitoTracker Red), instead of adding PI solution as described above, add 100 µL of PBS containing 0.2 µM CMXRos. Alternatively, the CMXRos potentiometric dye can be replaced by a tetramethylrhodamine derivative dye such as tetramethylrhodamine methyl ester perchlorate (TMRM, available from Sigma, Invitrogen/Molecular Probes, or in kit form from Immunochemistry Technologies). Cells can be stained with 0.4 µM of TMRM for 15 min at 37°C. After addition of either the CMXRos or TMRM potentiometric dyes, rinse the cells with rinsing solution, suspend in rinsing solution, and measure their green (FLICA) and red (CMXRos or TMRM) fluorescence the same way as in the case of cells stained with FLICA and PI (Fig. 2).

4. If concurrently with FLICA cells are stained with Annexin V–Cy5 conjugate, instead of adding PI solution, add 100 µL of the 1× binding buffer containing Annexin V–Cy5 at the concentration suggested by the vendor. Incubate for 15 min, rinse once with the binding buffer, and suspend in rinsing

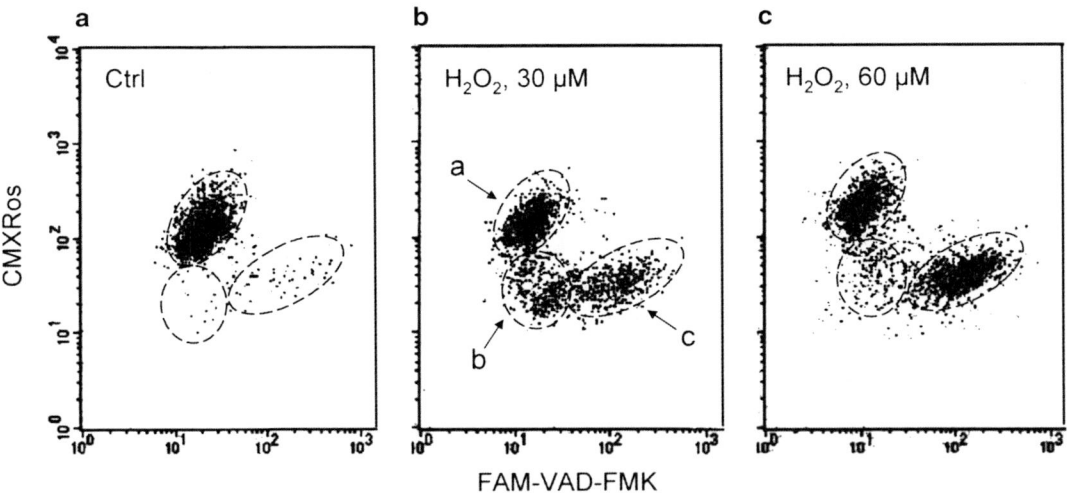

Fig. 2. Concurrent detection of caspases activation by FLICA (FAM-VAD-FMK) and the loss of mitochondrial electrochemical potential ($\Delta\psi_m$) during apoptosis. To induce apoptosis, human leukemic Jurkat cells were treated with 30 or 60 µM H_2O_2 for 7 h (7). The cells were then subjected to staining with FLICA and CMXRos (MitoTracker Red), and their fluorescence was measured by flow cytometry as described in the protocol (Subheading 3.3 and Note 3). Based on differential binding of FLICA and CMXRos, one can distinguish three cell subpopulations: (**a**) live, non-apoptotic cells; (**b**) cells having decreased mitochondrial potential (CMXRos binding), prior to caspases activation, and (**c**) cells showing both decreased mitochondrial potential and increased caspases activation (**c**). The data illustrate that the reduction of mitochondrial potential precedes FLICA binding. Treatment with 60 µM H_2O_2 accelerates the process of apoptosis as evidenced by an increase in frequency of FLICA-positive cells. (For more details, see ref. 7) (see Note 9).

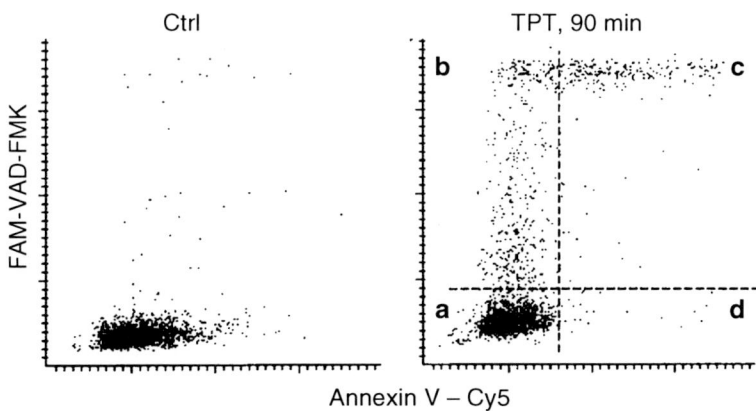

Fig. 3. Relationship between caspases activation as detected by FLICA binding and externalization of phosphatidylserine revealed by Annexin V binding during apoptosis. Apoptosis of HL-60 cells was induced by treatment with the DNA topoisomerase I inhibitor topotecan (TPT, 0.15 μM, 90 min). The cells were then labeled with FLICA (FAM-VAD-FMK), subsequently with Annexin V–Cy5 conjugate, and their fluorescence was measured by iCys(R) Research Imaging Cytometer as described in the protocol (see Subheading 3.2 and Note 4). Four quadrants shown in the right panel identify cells that are (**a**) non-apoptotic, (**b**) early apoptotic cells that bind FLICA but do not bind Annexin V–Cy5, (**c**) apoptotic cells that bind both FLICA and Annexin V–Cy5, and (**d**) very late apoptotic or necrotic cells that are FLICA negative. The data show that the "time window" to detect apoptosis is wider for the FLICA assay as the early apoptotic cells are FLICA positive (**b**) and not detectable by the Annexin V–Cy5 assay (see Notes 6–11).

buffer. The calcium containing binding buffer must be used to maintain the affinity interaction between the Annexin and PS residues on the apoptotic cell membrane surface. Excite cell fluorescence with both 488 nm and red diode (647 nm) lasers and measure the emission of Cy5 at far red (>650 nm) wavelength (Fig. 3). This procedure, however, can be combined with staining with PI. High intensity of PI fluorescence at red wavelength (590–620 nm, excited at 488 nm) allows one to identify and gate out the cells with a damaged plasma membrane (PI positive) and analyze the PI negative cells with respect to their Annexin V–Cy5 and FLICA fluorescence.

5. After step 4 (Subheading 3.2) or step 7 (Subheading 3.3), the cells may be fixed in 1% formaldehyde (10 min on ice) followed by 70% ethanol and then subjected to staining with PI in the presence of RNase A or stained with 7-aminoactinomycin D. In this format, PI is used to stain the DNA of all cells to facilitate the assessment of DNA ploidy parameters. Analysis of the FLICA vs. PI fluorescence by LSC or flow cytometry allows for the correlation of caspases activation with cellular DNA content, i.e., the cell cycle position or DNA ploidy.

Details of this procedure are provided in reference (7). Alternatively, when two-laser excitation is available and one of the lasers produces UV light the cellular DNA may be counterstained with Hoechst 33342 or with 4′,6-diamidino-2-phenylindole (DAPI), both available from Invitrogen/ Molecular Probes.

6. One has to keep in mind that FLICA are not passive reagents that mark the activated caspases, but react directly with the caspase by covalent interaction with the active site of the enzyme. This inhibits caspase activity, suppressing the process of apoptosis. Thus, the rate of apoptosis progression and all the events related to caspases activity are suppressed by FLICA (14).

7. Another problem that should be taken into an account when using FLICA to mark the activated caspases in live cells pertains to fragility of apoptotic cells. The flow cytometric assay requires incubation of live cells with these reagents followed by repeated rinsing to remove unbound FLICA from the non-apoptotic cells. Apoptotic cells, particularly at late stages of apoptosis, are fragile and may be preferentially lost during the centrifugations. A certain degree of stability is derived from the presence of serum (up to 20% v/v) or BSA (up to 2% w/v) in the rinsing buffers. Also, the cells should be sedimented with minimal g force and short centrifugation time. Because of a possibility of a selective loss of apoptotic cells, one has to be careful when drawing conclusions about the absolute frequency of apoptosis based on the percentage of FLICA positive cells in the samples assayed by flow cytometry. In the case of cell analysis by LSC/iCys$^{(R)}$ Research Imaging Cytometer , the propensity of apoptotic cells to detach in cultures and thus escape from analysis should also be taken into account, as it may favor a downward bias in the estimation of the apoptotic index as well. The above technical difficulties in the analysis of frequency of apoptosis pertain to any cytometric methodology, not only by FLICA, and are discussed in extent elsewhere (20).

8. It is difficult to assess the specificity of *in situ* bound individual FLICA sequences designed to be markers for their respective caspases in the light of the evidence of a nonspecific component in FLICA binding (7, 16). While activation of caspases is definitely associated with FLICA binding (7), it is likely that products of cleavage of other proteins by caspases have exposed thiol groups reactive with FLICA as well (16).

9. Also available are FLICA red fluorescing reagents containing sulforhodamine (SR) instead of FAM. Their availability extends multiparameter FLICA applications in combination with other markers.

10. A combination of FLICA with SYTO dyes offers an attractive marker of apoptosis concurrently revealing activation of caspases and condensation of nucleic acids (21).

11. It should be stressed that FLICA reagents at concentrations used to label apoptotic cells do not show toxicity to the non-apoptotic cells, at least in short-term (up to 48 h) *in vitro* experiments (13). Therefore, they are ideally suited to serve as *in vivo* markers of apoptotic cells. Defined as FLIVO™ *in vivo* apoptosis detection and imaging reagents (Immunochemistry Technologies, LLC), these reagents were successfully used to image and mark apoptotic cells in rats and mice following induction of apoptosis by different reagents (22–24). Increased levels of tumor cell apoptosis resulting from suppression of gp78 protein in sarcoma cells were assessed using *in vivo* injections of the SR-VAD-FMK (Red FLIVO) probe (24). These FLIVO probes were also used successfully *in vivo*, to measure increased levels of apoptosis in rat liver ischemia studies (25, 26).

Acknowledgment

This study was supported by NCI grant RO1 28704.

References

1. Salvesen G.S., and Riedl, S.J. (2008) Caspase mechanisms. *Adv. Exp. Med. Biol.*, **615**, 13–23.
2. Kroemer, G., Galluzzi, L., Vandenabeele, P., Abrams, J., Alnemri, E.S., Baehrecke, E.H., Blagosklonny, M.V., El-Deiry, W.S., Golstein, P., Green, D.R., Hengartner, M., Knight, R.A., Kumar, S., Lipton, S.A., Malorni, W., Nunez, G., Peter, M.E., Tschopp, J., Yuan, J., Piacentini, M., Zhivotovski, B., and Melino, G. (2008) Classification of cell death: recommendations of the nomenclature committee on cell death 2009. *Cell Death Differ.*, **16**, 3–11.
3. Earnshaw, W.C., Martins, L.M., and Kaufmann, S.H. (1999) Mammalian caspases: structure, activation, substrates, and functions during apoptosis. *Annu. Rev. Biochem.*, **68**, 383–424.
4. Nicholson, D.W. (1999) Caspase structure, proteolytic substrates and function during apoptotic cell death. *Cell Death Differ.*, **6**, 1028–1042.
5. Komoriya, A., Packard, B.Z., Brown, M.J., Wu, M.L., and Henkart, P.A. (2000) Assessment of caspase activities in intact apoptotic thymocytes using cell-permeable fluorogenic caspase substrates. *J. Exp. Med.*, **191**, 1819–1828.
6. Lee, B.W., Johnson, G.L., Hed, S.A., Darzynkiewicz, Z., Talhouk, J.W., and Mehrota, S. (2003) DEVDase detection in intact apoptotic cells using the cell permeant fluorogenic substrate, (z-DEVD)$_2$-cresyl violet. *BioTechniques*, **35**, 1080–1085.
7. Pozarowski, P., Huang, X., Halicka D.H., Lee, B., Johnson, G., and Darzynkiewicz, Z. (2003) Interactions of fluorochrome-labeled caspase inhibitors with apoptotic cells. A caution in data interpretation. *Cytometry A*, **55A**, 50–60.
8. Huang, X., Okafuji, M., Traganos, F., Luther, E., Holden, E., and Darzynkiewicz, Z. (2004) Assessment of histone H2AX phosphorylation induced by DNA topoisomerase I and II inhibitors topotecan and mitoxantrone and by

9. Li, X., and Darzynkiewicz, Z. (2000) Cleavage of poly(ADP-ribose) polymerase measured in situ in individual cells: relationship to DNA fragmentation and cell cycle position during apoptosis. *Exp. Cell Res.*, **255**, 125–132.

10. Bedner, E., Smolewski, P., Amstad, P., and Darzynkiewicz, Z. (2000) Activation of caspases measured in situ by binding of fluorochrome-labeled inhibitors of caspases (FLICA): correlation with DNA fragmentation. *Exp. Cell Res.*, **260**, 308–313.

11. Smolewski, P., Bedner, E., Du, L., Hsieh, T.-C., Wu, J.M., Phelps, D.J., and Darzynkiewicz, Z. (2001) Detection of caspases activation by fluorochrome-labeled inhibitors: multiparameter analysis by laser scanning cytometry. *Cytometry*, **44**, 73–82.

12. Darzynkiewicz, Z., and Barnard, E.A. (1966) Specific proteases of mast cells. *Nature*, **212**, 1198–1203.

13. Smolewski, P., Grabarek, J., Lee, B.W., Johnson, G.L., and Darzynkiewicz, Z. (2002) Kinetics of HL-60 cell entry to apoptosis during treatment with TNF-α or camptothecin assayed by the stathmo-apoptosis method. *Cytometry*, **47**, 143–149.

14. Smolewski, P., Grabarek, J., Phelps, D.J., and Darzynkiewicz, Z. (2001) Stathmo-apoptosis: arresting apoptosis by fluorochrome-labeled inhibitor of caspases. *Int. J. Oncol.*, **19**, 657–663.

15. Thornberry, N.A., Rano, T.A., Peterson, E.P., Rasper, D.M., Timkey, T., Garcia-Calvo, M., Houtzager, V.M., Nordstrom, P.A., Roy, S., Valliancourt, J.P., Chapman, K.T., and Nicholson, D.W. (1997) A combinatorial approach defines specificities of members of the caspase family and granzyme B. *J. Biol. Chem.*, **272**, 17907–17911.

16. Darzynkiewicz, Z., and Pozarowski, P. (2007) All that glitters is not gold: all that binds FLICA is not caspase. A caution in data interpretation – and new opportunities. *Cytometry A*, **71A**, 536–537.

17. Pendergrass, W., Wolf, and Poot, M. (2004) Efficacy of MitoTracker Green and CMXRosamine to measure changes in mitochondrial membrane potentials in living cells and tissues. *Cytometry*, **61**, 162–169.

18. Koopman, G., Reutelingsperger, C.P.M., Kuijten, G.A.M., Kechnen, R.M.J., Pals, S.T., and van Oers, M.H.H. (1994) Annexin V for flow cytometric detection of phosphatidylderine expression of B cells undergoing apoptosis. *Blood*, **84**, 1415–1420.

19. Pozarowski, P., Holden, E., and Darzynkiewicz, Z. (2005) Laser scanning cytometry. Principles and applications. *Methods Mol. Biol.*, **319**, 165–192.

20. Darzynkiewicz, Z., Bedner, E., and Traganos, F. (2001) Difficulties and pitfalls in analysis of apoptosis. *Methods Cell Biol.*, **63**, 527–559.

21. Wlodkowic, D., Skommer, J., Hillier, C., and Darzynkiewicz, Z. (2008) Multiparameter detection of apoptosis using red-excitable SYTO probes. *Cytometry A*, **73A**, 563–569.

22. Griffin R.J., Williams, B.W., Bischof, J.C., Olin, M., Johnson, G.L., and Lee, B.W. (2007) Use of fluorescently labeled poly-caspase inhibitor for in vivo detection of apoptosis related to vascular-targeting agent arsenic trioxide for cancer therapy. *Technol. Cancer Res. Treat.*, **6**, 651–654.

23. Lee, B.W., Olin, M.R., Johnson, G.L., and Griffin, R.J. (2008) In vitro and in vivo apoptosis detection using membrane permeant fluorescent-labeled inhibitors of caspases. *Methods Mol. Biol.*, **414**, 109–135.

24. Tsai, Y.C., Mendoza, A., Mariano, J.M., Zhou, M., Kostova, Z., Chen, B., Veenstra, T., Hewitt, S.M., Helman, L.J., Khanna, C., and Weissman, A.M. (2007) The ubiquitin ligase gp78 promotes sarcoma metastasis by targeting KAI1 for degradation. *Nat. Med.*, **13**, 1504–1509.

25. Cursio, R., Colosetti, P., Auberger, P., and Gugenheim, J. (2008) Liver apoptosis following normothermic ischemia-reperfusion: in vivo evaluation of caspase activity by FLIVO assay in rats. *Transplant. Proc.*, **40**, 2038–2041.

26. Cursio, R., Miele, C., Filippa, N., Colosetti, P., Auberger, P., Van Obberghen, E., and Gugenheim, J. (2009) Tyrosine phosphorylation of insulin receptor substrates during ischemia/reperfusion-induced apoptosis in rat liver. *Langenbecks Arch. Surg.*, **394**, 123–131.

Chapter 10

Combining Fluorescent *In Situ* Hybridization with the Comet Assay for Targeted Examination of DNA Damage and Repair

Sergey Shaposhnikov, Preben D. Thomsen, and Andrew R. Collins

Abstract

The comet assay is a simple and sensitive method for measuring DNA damage. Cells are embedded in agarose on a microscope slide, lysed, and electrophoresed; the presence of strand breaks allows the DNA to migrate, giving the appearance of a comet tail, the percentage of DNA in the tail reflecting the break frequency. Lesion-specific endonucleases extend the usefulness of the method to investigate different kinds of damage. DNA repair can be studied by treating cells with damaging agent and monitoring the damage remaining at intervals during incubation. An important feature of the assay is that damage is detected at the level of individual cells. By combining the comet assay with fluorescent *in situ* hybridization (FISH), using labeled probes to particular DNA sequences, we can examine DNA damage and repair at the level of single genes or DNA sequences. Here we provide protocols for the comet assay and the FISH modification, answer some technical questions, and give examples of applications of the technique.

Key words: Comet assay, Fluorescent *in situ* hybridization, DNA damage, DNA repair

1. Introduction

1.1. The Comet Assay

The alkaline comet assay is a common method for measuring DNA damage, whether as a biomarker in lymphocyte samples from human population or intervention studies, or in *in vitro* or *in vivo* genotoxicity testing, or in investigating the mechanisms of DNA damage and repair (1). Briefly, cells are embedded in low melting point agarose on a microscope slide, and lysed in a solution containing Triton X-100, which breaks membranes, and high salt, which strips off histones. This leaves the DNA in a nucleosome-free form, which however remains attached to the nuclear matrix as loops retaining the supercoiling that was present in the nucleosomes. On electrophoresis, DNA is attracted towards the anode, but migration occurs only if breaks are present.

The migrating DNA from each cell forms a comet-like structure when viewed (usually) by fluorescence microscopy with a suitable stain, and the fraction of DNA in the tail is proportional to the frequency of breaks. The advantages of the assay are its simplicity, speed, sensitivity, applicability to different cell types, avoidance of radioactive labeling, and – perhaps most significantly – the fact that damage is assessed at the level of individual cells.

The assay can be combined with lesion-specific endonucleases, so that damage other than simple strand breaks can be measured. Formamidopyrimidine DNA glycosylase (FPG) is used to detect oxidized purines (mainly 8-oxoGua), endonuclease III for oxidized pyrimidines, and T4 endonuclease V for UV-induced cyclobutane pyrimidine dimers.

Repair of different kinds of DNA damage can be monitored with the comet assay, simply by incubating cells after treating them with a specific damaging agent, and measuring the damage remaining (e.g. strand breaks, or enzyme-sensitive sites) at intervals during the incubation.

As stated above, the special feature of the comet assay is the ability to look at damage in individual cells. By combining comets with fluorescent *in situ* hybridization (FISH) using labeled probes to particular DNA sequences, we can examine damage and repair at an even finer level of resolution, i.e. at the level of individual genes and DNA sequences. This potential was recognized early on, but relatively few FISH-comet papers have been published to date.

Figure 1 illustrates general principles of the comet assay combined with FISH.

Studying DNA damage and repair with FISH combined with the comet assay is difficult but potentially very informative. In comparison with standard FISH on chromosome spreads, additional care is necessary as comet DNA preparations are embedded in fragile agarose gels. On the other hand, this feature can be useful as it allows investigation of DNA preparations in 3-D space. In this chapter, we provide protocols for the comet assay and the FISH modification, answer some technical questions, and give examples of applications of the technique. FISH with comets should be useful to researchers interested in the structural organization of DNA and chromatin, the localization of DNA damage, and the kinetics of repair of damage.

1.2. Probes for FISH with Comets

Depending on the types of target sequences to be detected, a variety of probes have been developed and applied in standard FISH techniques and, subsequently, with comets. The most widely used repetitive sequences used as FISH probes are centromere, telomere, and ribosomal DNA repeats, as well as interspersed repetitive elements (SINEs, LINEs). These types of probes are technically convenient for use as they give strong,

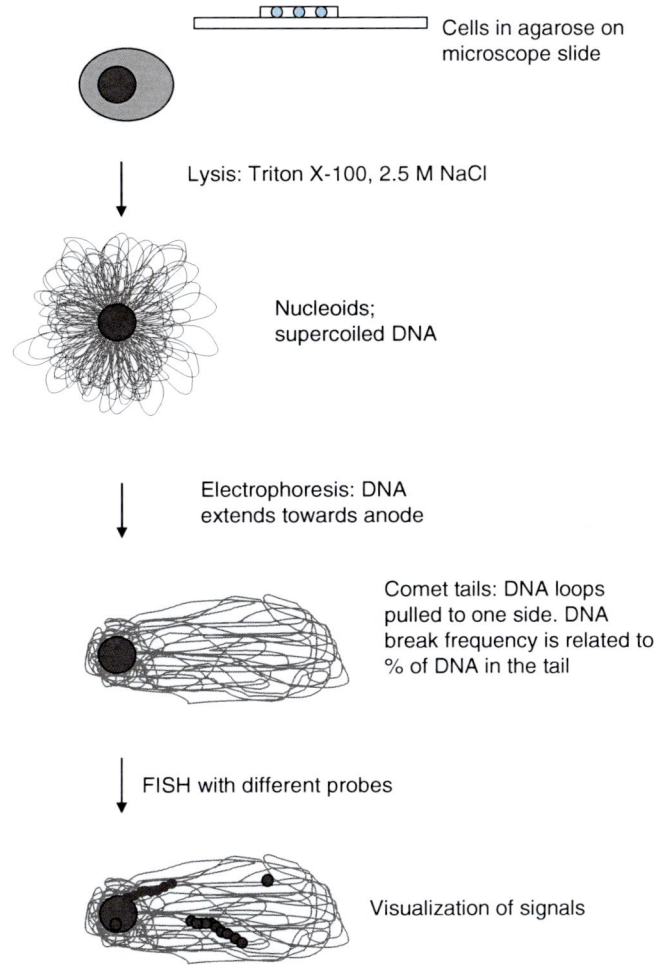

Fig. 1. General principles of the comet assay combined with FISH.

easily detectable signals. Other widely used nongene-specific probes producing strong signals in comets are chromosome arm- or band-specific painting probes (DNA from micro-dissected chromosomes), as well as chromosome painting probes (DNA from flow-sorted chromosomes). The types of probes mentioned are usually produced commercially. Specific DNA sequences are studied using unique probes: PCR products, cDNAs, and genomic DNA cloned in cosmids, P1-artificial chromosomes (PACs), bacterial artificial chromosomes (BACs), or yeast artificial chromosomes (YACs). Oligonucleotide probes can also be applied. Many of the large unique probes are not commercially available but can be prepared for FISH using standard molecular biology techniques. Thus, genomic DNA clones (cosmids, PACs, BACs, YACs) can be obtained from various institutions dealing with genome mapping and sequencing. Of these, PAC and BAC clones become especially valuable as they have been used to construct well-characterized and sequenced genomic libraries, thus allowing accurate and precise selection of the target sequences. These clones are easy to work with and their insert size can be greater than 700 kb, with usual insert size of 100–350 kb. These large genomic DNA fragments, however, can often contain a fraction of repetitive sequences, which may result in a high background. In some cases, it is possible to avoid this problem by using more specific chemically synthesized oligonucleotide probes. As they are small, probe molecules that are not bound to the specific target DNA are readily removed on washing, so ensuring a low background. However, the signals from these probes are weak.

Peptide nucleic acid (PNA) probes are synthetic analogs of DNA in which the bases are attached via methylene carbonyl bonds to repeating units of N-(2-aminoethyl) glycine in place of the phosphodiester backbone (2). The lack of charged phosphate groups means that PNA-DNA duplexes have enhanced stability when PNA is used as a FISH probe.

Padlock probes are linear oligonucleotides that are designed so that the two end segments, connected by a linker region, are complementary to adjacent target sequences (3). Upon hybridization, the two probe ends become juxtaposed and can be joined by a DNA ligase, thereby circularizing the padlock probe and rendering it physically catenated to the target sequence. The reaction is stable and highly specific, requiring a perfect match between the probe ends and the target sequence. By amplifying circularized padlock probes through rolling circle DNA synthesis *in situ*, specifically reacted probes can be distinguished from nonspecifically bound probe molecules and other detection reagents. When used for FISH on comets, the essential feature of these probes is that padlock reaction steps are performed at 37°C – the "normal" comet assay temperature, which ensures gel conservation.

1.3. Visualizing FISH Comets

When analyzing FISH comets results, visualization and scoring depend entirely on direct observation. In most cases, there is no possibility to score the signals automatically because of the complexity of the preparations. Figure 2 illustrates different appearances of the signals: a linear array (Fig. 2a) or separate spots (Fig. 2b). Investigators have to look at each signal to judge whether to count it. However, rapid developments of automated systems for detection of various objects at the microscopic level give hope for automated scoring of at least certain kinds of FISH signals in the near future.

An important difference between visualization of the standard chromosomal FISH experiments and FISH comets is the dimensions of the preparations and, subsequently, of the observed signals. In standard FISH studies, DNA preparations that are fixed or adsorbed onto microscope slides are normally used.

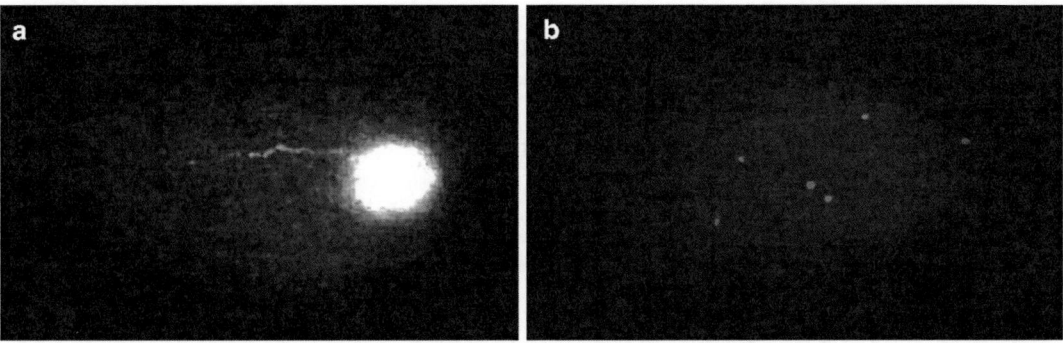

Fig. 2. FISH with biotin-labeled probes from the LCAT gene cluster on human chromosome 16q22.1 to neutral and alkaline comets prepared from U-2 OS cells treated with 0.1 mM H_2O_2. (**a**) PAC clone RPCI1-213H19 (8), neutral comet (reproduced from *Electrophoresis* 2008 (8), with permission. Copyright Wiley-VCH Verlag GmbH & Co. KGaA). The probe appears as a linear array, consistent with extension from a fixed point on the nuclear core or matrix. (**b**) Cosmid probe Odin (8), alkaline comet. The probe appears as a series of spots in the head and the tail of the comet.

Thus, 2-D preparations are under investigation. In contrast, comet preparations are organized in 3-D space. Electrophoresis may alter the original chromatin organization, but the signals from FISH are still in a 3-D matrix. On the one hand, this 3-D organization of the DNA in comet preparations can be a great advantage, as it reflects the real organization of chromatin in the living cell, but on the other hand, it produces difficulties in the process of visualizing and scoring the signals and makes it impossible to obtain realistic images in photographs that record only one (2-D) plane.

1.4. Applications of the FISH Comet Technique

1.4.1. Information About Chromatin Structure and Organization

Although the potential for combining FISH with the comet assay was apparent from early on, it was a decade before Santos et al. published the first successful results (4). Employing a simple neutral comet assay, their aim was to investigate aspects of the structural organization of the nucleus in human interphase lymphocytes – specifically, how centromeric and telomeric DNAs behave under electrophoresis. They used probes to all centromeres, all telomeres, and chromosome-specific centromere and telomere DNA, plus three segments of the gene *MGMT* (coding for the repair enzyme O^6-methylguanine methyltransferase). Telomere probes appeared as discrete spots, localized mostly over the head, which is consistent with their known attachment to the nuclear membrane. The size of telomeric sequences is in the region of 15–50 kb. In contrast, centromere DNA (1,000 kb in size) gave signals that were arranged as long strings of dots, extending well into the comet tail. The *MGMT* signals were found in the head as well as the tail of comets, the three segments generally appearing close together in a linear array.

Arutyunyan et al. (5, 6) were also interested in telomeric DNA, and its sensitivity to damage induced by drugs used in cancer therapy.

They used telomere-specific PNA probes with comets developed from human leukocytes by the alkaline method. After bleomycin treatment, the appearance of telomere probes in the tail paralleled total DNA migration, i.e. there was no evidence for preferential damage to telomere DNA.

1.4.2. Information Relating to the Comet Assay Itself

FISH can be used to answer questions relating to how DNA behaves in the comet assay. A common question is whether damage to mitochondrial DNA (mtDNA) can be detected. We identified mtDNA using specific padlock probes (7). In preparations of cells embedded in agarose, before lysis, signals were thickly clustered around the counterstained nucleus. After lysis even for a few minutes, signals had moved away from the nucleus, and after electrophoresis they were randomly distributed over the gel. In contrast, FISH with probes to *Alu* sequences, which are widespread throughout the genomic DNA, showed no tendency to disperse, but remained colocalized with tail DNA.

Recently, we compared the signals obtained by applying probes to DNA from neutral and alkaline electrophoresis of cells with experimentally induced DNA single strand breaks (8). The sequence probed – with a length of about 100 kb – was equivalent to a substantial fraction of the tail length. After neutral electrophoresis, signals appeared as linear chains of dots, consistent with the idea that the tail consists of DNA molecules extending out from the nucleoid core towards the anode. In contrast, after alkaline electrophoresis, signals were mostly isolated dots; few chains were seen. A possible explanation was that the DNA had undergone further fragmentation. When differently labeled probes to adjacent target regions (with a combined length of about 200 kb) were applied to DNA from alkaline comets, the two colors used to visualize the probes generally appeared as pairs, in close association – arguing against further fragmentation, which would have separated them.

The number of signals (mostly dots) seen after alkaline electrophoresis was approximately double the number of signals (mostly chains) after neutral electrophoresis, as expected since the denatured DNA provides two strands for hybridization. Furthermore, the average numbers of signals seen per cell corresponded closely with the numbers expected according to the gene copy number in a random interphase cell population. The frequency of breaks was so low that there was only a small probability of breaks occurring within the targeted region (see Note 1).

1.4.3. Gene-Specific DNA Repair

The overall process of cellular DNA damage and repair is followed with the comet assay by measuring tail intensity at intervals during incubation following DNA-damaging treatment. Damage to a specific gene (more accurately, of the DNA region containing the gene) and its repair can be monitored by following the "retreat" of the gene-specific signals from tail to head during the incubation period. The kinetics of overall genomic and gene-specific repair can thus be compared.

McKenna et al. (9) examined strand break repair in human tumor cells following 5 Gy of γ-irradiation, using a probe for the *TP53* gene. They found an increased number of signals (mostly in the comet tails) immediately after irradiation and a decrease in number over the first 15 min at which time most were in comet heads. By 60 min, the number of signals was back to the normal level, while the percentage of tail DNA (representing total DNA and its repair) was still elevated. Thus *TP53* repair was faster than repair of the overall DNA. Kumaravel et al. (10) and Horvathova et al. (11) also reported preferential repair of *TP53*, after ionizing radiation or H_2O_2 treatment.

Conventional methods for examining gene-specific DNA repair (12, 13) require high doses of UV and operate at the level of resolution of a restriction fragment of the gene. With the comet assay, in contrast, very low damage doses are employed, and the level of resolution is the DNA loop containing the gene.

2. Materials

2.1. The Comet Assay

2.1.1. Equipment and Supplies

1. Staining jars (vertical or horizontal).
2. Glass slides (frosted end)
3. Coverslips (20 × 20 mm and 22 × 22 mm).
4. Parafilm.
5. Metal plate.
6. Incubator (37°C/55°C).
7. Moist chamber (e.g. glass or plastic box with platform for slides above a layer of water).
8. Eletrophoresis tank (horizontal).
9. Electrophoresis power supply.
10. Fluorescence microscope.

2.1.2. Reagents, Solutions

1. Normal melting point agarose (NMP agarose): 1% in H_2O. A few hundred milliliters is enough for several hundred slides.
2. Low melting point agarose (LMP agarose): 1% in PBS. Store at 4°C in 10 mL aliquots.
3. Phosphate Buffered Saline (PBS).
4. Reagents for cell culture.
5. Histopaque 1077 (Sigma) or Lymphoprep (Nycomed).

6. Lysis solution: 2.5 M NaCl, 0.1 M EDTA, 10 mM Tris. Prepare 1 L. Set pH to 10 with 10 M NaOH solution. (Add 35 mL of NaOH straight away to dissolve EDTA, and then add dropwise to pH 10). Add 1 mL Triton X-100 per 100 mL immediately before use.

7. Alkaline electrophoresis solution: 0.1 M NaOH, 1 mM EDTA.

8. Neutral electrophoresis solution (1× TBE): 0.09 M Tris-borate, 0.002 M EDTA. Set to pH 7.5 by adding concentrated HCl.

9. Enzyme reaction buffer: 40 mM HEPES, 0.1 M KCl, 0.5 mM EDTA, 0.2 mg/mL BSA, pH 8.0 with KOH (can be made as 10× stock, adjusted to pH 8.0 and frozen at −20°C).

2.1.3. Enzymes

Endonuclease III (endo III), formamidopyrimidine glycosylase (FPG) and T4 endonuclease V are commercially available in purified form, or may be obtained from a laboratory producing them. They are isolated from bacteria containing over-producing plasmids. Because such a high proportion of protein is the enzyme, a crude extract is satisfactory, as nonspecific nuclease activity is not significant at the concentrations employed. The final dilution of the working solution will vary from batch to batch. Follow supplier's instructions, or use the following as a guide (assuming a final dilution of 3,000×) (see Note 2).

FPG:

1. On receipt, dispense the stock solution into 5 µL aliquots and refreeze at −80°C. This is to minimize repeated freezing and thawing.

2. Take one of these aliquots and dilute to 0.5 mL using the enzyme reaction buffer – with the addition of 10% glycerol. Dispense into 10 µL aliquots (label as "100× diluted") and freeze at −80°C.

3. For use, dilute one of these 10 µL aliquots to 300 µL with buffer (no glycerol) and keep on ice until you add it to the gels; do not refreeze this working solution.

Endo III and T4 endonuclease V:

1. Dispense the stock solution into 5 µL aliquots and refreeze at −80°C.

2. Take one of these aliquots and dilute to 0.5 mL using the enzyme reaction buffer (no need to add glycerol as these enzymes are more stable). Dispense this into 10 µL aliquots (label as "100× diluted") and freeze at −80°C.

3. For use, dilute one of these 10 µL aliquots to 300 µL with buffer (no glycerol) and keep on ice until you add it to the gels.

2.2. FISH on Comets

2.2.1. Labeling of DNA Probes by Nick Translation

1. DNase I (Roche); store at −20°C.
2. 10× nick translation buffer: 0.5 M Tris-HCl pH 7.5, 0.1 M $MgSO_4$, 1 mM DTT, 0.5 mg/mL bovine serum albumin. Aliquot and store at −20°C.
3. Stock solutions of deoxyribonucleoside triphosphates (dNTPs), 100 mM each (Roche); store at −20°C.
4. Mixture of dNTPs (0.5 mM each) containing no dCTP (dNTP-C): mix 1 µL of dATP, 1 µL of dGTP and 1 µL of dTTP with 197 µL of H_2O. Store at −20°C.
5. Mixture of dNTPs (0.5 mM each) containing no dATP (dNTP-A): mix 1 µL of dCTP, 1 µL of dGTP and 1 µL of dTTP with 197 µL of H_2O. Store at −20°C.
6. Mixture of dNTPs (0.5 mM each) containing no dTTP (dNTP-T): mix 1 µL of dATP, 1 µL of dCTP and 1 µL of dGTP with 197 µL of H_2O. Store at −20°C.
7. Biotin-14-dCTP (0.4 mM) (Invitrogen); store at −20°C.
8. Digoxigenin-11-dUTP (1 mM), (Roche), store at −20°C.
9. Cot-I DNA (Invitrogen, from human placental DNA) (see Note 3) Concentrate by ethanol precipitation to 10 µg/µL:
   (a) add 0.05 vol. of 5 M NaCl and 2.5 vol. of 96% ethanol, keep at −20°C overnight or at −80°C for 2 h
   (b) spin down for 10 min at 14,000×g at 4°C, and carefully remove the supernatant
   (c) add 0.5 mL 70% ethanol, spin down for 10 min at 14,000×g at 4°C, carefully remove the supernatant, repeat this step one more time
   (d) dry the pellet (let the tube stand open for several minutes, do not allow to overdry)
   (e) dissolve in H_2O, store at −20°C
10. DNA polymerase I, *Escherichia coli* (Invitrogen), store at −20°C.
11. Salmon sperm DNA (Sigma), store at −20°C.

2.2.2. Hybridization, Signal Detection and Visualization

1. MM 2.1 hybridization buffer:
   (a) mix 0.5 mL 20× SSC, 5.5 mL formamide and 1 mL H_2O
   (b) add 1 g dextran sulfate, dissolve for 2–3 h at 70°C, set pH to 7.0 with either HCl or NaOH (dextran sulfate has a pH 6–8, and formamide becomes acidic with time)
   (c) store at −20°C
2. Formamide (Sigma)
3. 9 mm round glass coverslips

4. 20 × 20 mm glass coverslips
5. "Fixo gum" rubber cement (Marabuwerke GmbH & Co)
6. 20× SSC
7. Tween-20 (Sigma)
8. Nonfat milk powder (Sigma)
9. Antibody dilution solution: 4× SSC, 0.05% Tween-20, 5% nonfat milk powder
10. Cy3-conjugated streptavidin (Jackson Immuno Research Laboratories)
11. Biotinylated anti-avidin D (Vector Laboratories)
12. Fluorescent antibody enhancer set for digoxigenin detection (Roche Applied Science):
    (a) solution 1: anti-DIG monoclonal antibody against digoxigenin, mouse IgG1
    (b) solution 2: anti-mouse-Ig-DIG, F(ab')2, fragment from sheep
    (c) solution 3: anti-DIG-Fluorescein, Fab fragments from sheep

3. Methods

3.1. The Comet Assay

3.1.1. Slide Preparation for the Comet Assay (Precoating)

1. The slides for precoating should be grease-free; clean if necessary by soaking in alcohol and then wiping dry with a clean tissue.
2. Dip slides in a vertical staining jar of melted 1% standard agarose in H_2O.
3. Drain off excess agarose, wipe the back clean and dry by leaving on a clean bench overnight.

3.1.2. Embedding Cells in Agarose (Work Quickly as the Agarose Sets Quickly at Room Temperature!)

1. Melt the LMP agarose by heating carefully (with cap loose!) in microwave oven. Then place in water bath at 37°C and allow time for it to equilibrate to this temperature.
2. Suspend cells (after gentle trypsinization in the case of attached cells in culture, or after disaggregation of tissue) at 10^6/mL in PBS.
3. Place an aliquot of 40 μL in a microcentrifuge tube.
4. Quickly add 140 μL of 1% LMP agarose at 37°C and mix by aspirating agarose up and down (once) with pipette.
5. Place two 70 μL aliquots (use same pipettor tip) on one slide.
6. Cover each drop with a 20 × 20 mm coverslip. Leave slides in fridge for 5 min.

3.1.3. Lysis

1. Add 1 mL Triton X-100 to 100 mL of lysis solution (4°C).
2. Remove coverslips from slides and place in this solution in a (horizontal) staining jar.
3. Leave at 4°C for 1 h.

3.1.4. Enzyme Treatment

1. Prepare 300 mL of enzyme reaction buffer. Put aside 1 mL for enzyme dilutions.
2. Wash slides in three changes of this buffer (4°C) in staining jar, for 5 min each.
3. Meanwhile, prepare dilutions of enzyme. The final dilution of the working solution will vary from batch to batch.
4. Remove slides from last wash, and dab off excess liquid with tissue.
5. Place 40–50 µL of enzyme solution (or buffer alone, as control) onto gel, and cover with a coverslip.
6. Put slides into moist box (to prevent desiccation) and incubate at 37°C for 30 min.

3.1.5. Alkaline Electrophoresis (see Note 4)

1. Gently place slides on platform in tank, and immerse in cold (4°C) alkaline electrophoresis solution, forming complete rows (gaps filled with blank slides).
2. Make sure that tank is level and gels are just covered. Leave 40 min. Alkaline treatment and electrophoresis are carried out at 4°C.
3. For most tanks (i.e. of standard size), run at 25 V (constant voltage setting) for 30 min.
4. If there is too much electrolyte covering the slides, the current may be so high that it exceeds the maximum – so set this at a higher level than you expect to need. If necessary, i.e. if 25 V is not reached, remove some solution. Normally the current is around 300 mA but this is not crucial (see Note 5).
5. Neutralize by washing for 10 min with PBS in staining jar at 4°C, followed by 10 min wash in water.
6. Proceed to staining while gels are wet: OR prepare for FISH as described below.
7. Dry (room temperature) for storage.

3.1.6. Neutral Electrophoresis (see Note 4)

1. Rinse the slides two times each in jars with 1× TBE buffer at 4°C.
2. Gently place on platform in tank, immersed in cold (4°C) 1× TBE buffer, forming complete rows (gaps filled with blank slides).
3. Make sure that tank is level and gels are just covered.
4. Electrophorese at 25 V for 30 min at 4°C.

5. Proceed to staining while gels are wet: OR prepare for FISH as described below.

6. Dry (room temperature) for storage.

3.2. FISH on Comets

When using standard FISH on comets, special care is required to avoid possible problems (see Note 6). After the electrophoresis step of the comet assay, prepare slides for FISH by incubating in 96% ethanol for 30 min at 4°C, then in 0.5 M NaOH for 25 min and finally dehydrate through 70, 84, and 96% ice-cold ethanol (see Note 7).

3.2.1. Probes

A number of commercial FISH probes are available from different companies. If these probes are used, general protocols recommended by the supplier are followed with some modifications – described in the technical notes. However, for some applications, there are no commercial probes available, and so you must prepare them yourself.

Before carrying out experiments with comet DNA, it is advisable to check (by hybridization with human metaphase spreads) that each probe produces low unspecific hybridization (to ensure a high signal-to-noise ratio).

3.2.2. Labeling of DNA Probes by Nick Translation

Human Cot-I DNA and purified large insert bacterial clones (cosmids, PAC or BAC clones) containing human genomic DNA inserts can be labeled with botin-14-dCTP or digoxigenin-11-dUTP by conventional nick translation. To label approximately 1–4 µg DNA in a final volume of 100 µL (adjusted with sterile distilled water), the following steps are performed (scaling up or down is possible):

1. Prepare the reaction mix by adding the reagents in order. Work on ice.

    (a) Biotin-14-dCTP reaction mix:
    - distilled water (calculate the amount for the final volume of 100 µL)
    - 10× nick translation buffer: 10 µL
    - DNA probe: 1–4 µg
    - dNTP-C: 7.5 µL (0.5 mM each)
    - biotin-14-dCTP: 6.25 µL
    - DNA polymerase I: 40 units

    (b) Digoxigenin-11-dUTP reaction mix:
    - distilled water (calculate the amount for the final volume of 100 µL)
    - 10× nick translation buffer: 10 µL
    - DNA probe: 1–4 µg

- (dNTP-T): 6 μL (0.5 mM each)
- (dNTP-A): 1.5 μL (0.5 mM each)
- digoxigenin-11-dUTP: 2 μL
- DNA polymerase I: 40 units

2. Mix, and centrifuge using bench microcentrifuge.
3. Add 4 μL of DNase I (using the optimal dilution established by titration, see Subheading 3.2.3). Mix, and spin down using bench microcentrifuge.
4. Place the tube with reaction mix in water bath at 13–14°C for 30 min.
5. Immediately after that add 10 μL of 0.5 M EDTA and put the reaction mix on ice.
6. Add 5 μL of 5 M NaCl and 300 μL 96% ethanol, keep at –20°C overnight or at –80°C for 2 h.
7. Centrifuge for 10 min at 4°C at $14,000 \times g$.
8. Remove supernatant and add 500 μL of cold 70% ethanol.
9. Centrifuge for 5 min at 4°C at $14,000 \times g$, remove supernatant.
10. Repeat the ethanol wash step.
11. Remove the supernatant, briefly dry the pellet, dissolve in 15 μL H_2O, store at –20°C until use.

3.2.3. DNase Titration

1. Make the following dilutions of DNase I in 50% glycerol/1.5 M NaCl: 1:10, 1:50, 1:100, 1:250, 1:500, and 1:1,000.
2. For each dilution, make a separate reaction mix (work on ice):
   (a) 5 μL nick translation buffer
   (b) 1 μg genomic DNA (total genomic DNA from different organisms as well as large-insert genomic clones can be used)
   (c) 2 μL DNase I
   (d) distilled water up to 50 μL.
3. Place the reaction mixes in a water bath at 13–14°C for 30 min.
4. Check the length of the resulting DNA fragments by agarose gel electrophoresis; select the dilution producing fragments of 100–500 bp for further use in nick translation experiments. The dilutions can be stored at –20°C and retitrated if the fragment size in nick translation reactions is increased or if a much stronger background appears when FISH signals are visualized.

3.2.4. Preparation of Hybridization Mix

Hybridization reaction mix contains labeled DNA probe (or probes), salmon sperm DNA, Cot-I DNA (not labeled, when it is used as a blocking agent rather than a probe), 1× SSC, 0.1 mg/mL dextran sulfate, 55% formamide.

(a) To prepare 15 µL of hybridization mix with genomic DNA clones used as probes, mix:
- 100 ng labeled DNA probe
- 10.5 µL MM 2.1 buffer
- 20 µg Cot-I DNA
- 10 µg salmon sperm DNA
- adjust the volume to 15 mL with distilled water. Scaling up or down is possible.

(b) To prepare 15 µL of hybridization mix with labeled Cot-I DNA used as probes, mix:
- 100 ng labeled Cot-I DNA
- 10.5 µL MM 2.1 buffer
- 1 µg salmon sperm DNA (optional)
- adjust the volume to 15 mL with distilled water. Scaling up or down is possible.

3.2.5. Probe Denaturation and Start of Hybridization

1. Denature probe in a water bath at 70°C for 10 min.
2. Transfer to 37°C for preannealing for 1 h (this step is omitted when labeled Cot-I DNA is used as a probe).
3. Pipette 3–4 µL of the probe mix onto slide, cover with 9 mm round coverslip, seal with rubber cement. More of the probe mix can be used under bigger coverslips, e.g. up to 30 µL of the mix is used under a 20×20 mm coverslip. Plastic coverslips cut from a standard overhead transparency film can be also used as they are flexible and easier to remove after hybridization is finished.
4. Incubate in a humid box overnight at 37°C. A shorter time of incubation (down to 1 h) is possible when Cot-I DNA is used as a probe.

3.2.6. Posthybridization Washes (see Note 8)

1. Prepare three jars containing 50% formamide/2× SSC and two jars with 2× SSC.
2. Warm the jars to 42°C in a water bath. Remember that the temperature in the jars will be a couple of degrees lower than in the water bath.
3. Carefully remove the rubber cement from the slides and place them in a jar with 2× SSC at room temperature for 10 min or until the coverslips are washed off.
4. Transfer the slides into one of the jars with 50% formamide/2× SSC at 42°C, incubate for 5 min.

5. Incubate for 5 min each in the second and third jar with 50% formamide/2× SSC.

6. Wash the slides for 5 min each in the two jars of 2× SSC at 42°C.

7. Transfer the slides to a jar with 4× SSC, 0.05% Tween-20 for 10 min at room temperature.

3.2.7. Preparation of Antibody Detection Solutions

1. Prepare 500 mL (or more, depending of the number of slides used) wash buffer (WB): 4× SSC, 0.05% Tween-20.

2. Prepare antibody dilution solution: 4× SSC, 0.05% Tween-20, 5% nonfat milk powder.

3. Depending on the probe's label, prepare detection solutions as described below (Subheading 3.2.7, steps 1–3).

4. The amount of detection solution per one slide per incubation is given and should be scaled up in accordance with the number of slides used.

5. After all the components of the solutions are mixed, spin the tubes down using a bench microcentrifuge, let stand for 10 min in a dark place at room temperature, centrifuge for 10 min at 14,000×g at room temperature and, finally, transfer the supernatant to fresh tubes. Keep in a dark place before use.

6. The solutions should be used the same day.

3.2.7.1. Antibody Solutions for Detection of Probes Labeled with Biotin

1. Detection solutions 1 and 3: mix 2 μL of Cy-3 conjugated streptavidin with 98 μL of the antibody dilution solution.

2. Detection solution 2: mix 1 μL antiavidin-D with 99 μL of the antibody dilution solution.

3.2.7.2. Antibody Solutions for Detection of Probes Labeled with Digoxigenin

1. Detection solution 1: mix 4 μL of solution 1 from the fluorescent antibody enhancer set for digoxigenin detection with 96 μL of the antibody dilution solution.

2. Detection solution 2: mix 4 μL of solution 2 from the fluorescent antibody enhancer set for digoxigenin detection with 96 μL of the antibody dilution solution.

3. Detection solution 3: mix 4 μL of solution 3 from the fluorescent antibody enhancer set for digoxigenin detection with 96 μL of the antibody dilution solution.

3.2.7.3. Antibody Solutions for Simultaneous Detection of Probes Labeled with Biotin and Probes Labeled with Digoxigenin

For simultaneous detection of probes labeled with biotin and probes labeled with digoxigenin in the same hybridization reaction, combine the solutions of antibodies as follows:

1. Detection solution 1: mix 2 μL of Cy-3 conjugated streptavidin and 4 μL of solution 1 from the fluorescent antibody enhancer set for digoxigenin detection with 94 μL of the antibody dilution solution.

2. Detection solution 2: mix 1 μL antiavidin-D and 4 μL of solution 2 from the fluorescent antibody enhancer set for digoxigenin detection with 95 μL of the antibody dilution solution.

3. Detection solution 3: mix 2 μL of Cy-3 conjugated streptavidin and 4 μL of solution 3 from the fluorescent antibody enhancer set for digoxigenin detection with 94 μL of the antibody dilution solution.

3.2.8. Signal Detection

Perform all the incubations in a humid box at 37°C under parafilm coverslips. Do not allow slides to dry at any stage of the detection process.

1. Drain the slides and apply 100 μL of the antibody dilution solution under parafilm coverslips

2. Incubate for 30 min at 37°C in a humid box, remove the parafilm and let the solution drain onto a tissue.

3. Apply 100 μL of antibody detection solution 1, cover with parafilm and incubate for 40–60 min at 37°C in a humid box.

4. Remove the parafilm and wash three times in WB at room temperature.

5. Apply 100 μL of antibody detection solution 2, cover with parafilm and incubate for 40-60 min at 37°C.

6. Repeat steps 4, 3 and 4 (in that order).

7. Dehydrate the slides in 70, 80, and 95% ethanol for 5 min each and air dry.

3.2.9. Signal Visualization

1. Stain the gels with 20 μL of DAPI prepared in Vectashield (Vector Laboratories).

2. Alternative stains: Propidium iodide (2.5 μg/mL), Hoechst 33258 (0.5 μg/mL), SYBR Gold (0.1 μL/mL) or ethidium bromide (20 μg/mL) can be used in place of DAPI for the visualization of comet DNA.

3. Visualize and record the signals using appropriate filters; overlay the recorded images.

4. Notes

1. Theoretical note: though the comet assay is used by many laboratories, it is still not completely understood what happens when damaged DNA is electrophoresed. In conventional comet experiments, it is common to describe DNA in the tail as damaged and DNA in the head as undamaged. However, FISH experiments have shown very clearly that there are two factors determining where a particular sequence

of DNA will be located, i.e. whether it is able to "escape" into the tail: this will depend not only on the presence or absence of damage, but also on the organization of the chromatin.

2. The buffer in which enzyme is stored may contain β-mercaptoethanol to preserve the enzyme. However, inclusion of sulfhydryl reagents in the reaction buffer would significantly increase background DNA breakage.

3. The Cot value of a particular DNA sequence is defined as the product of the nucleotide concentration (Co, in moles of nucleotides per liter), its reassociation time (t, in seconds), and an appropriate buffer factor based upon cation concentration. Cot-I DNA is the fraction of total DNA with low Cot value consisting largely of highly repetitive sequences.

4. Electrophoresis solution should be cooled before use, e.g. by pouring into the electrophoresis tank in the cold room an hour or so before it is needed.

5. The voltage depends on tank dimensions. 0.8 V/cm is recommended, calculated on the basis of the distance across the platform (where the conducting layer is least deep and the resistance highest). The changes in voltage/current/resistance across the tank from electrode to electrode, and the conditions within the gel, provide an interesting exercise in simple theoretical physics.

6. In comparison with the usual chromosomal DNA preparations that are strongly attached to a glass slide, comets are fragile, and not so strongly attached. Agarose gels can melt or detach under the standard stringent FISH conditions required for large DNA probes. All the steps, including slide preparations, posthybridization washes, and signal detection should be performed in the smallest possible volumes of solutions, to minimize the risk of agarose being dislodged. Recommended times in FISH protocols should be strictly adhered to, keeping processing times as short as possible.

7. As low melting point agarose normally used for making comet gels would melt at the temperature used for DNA denaturation in standard chromosomal preparations (70°C in 70% formamide/2× SSC), chemical denaturation with 0.5 N NaOH for 25 min is used. It is likely that unbroken DNA, in still-supercoiled loops, will undergo a considerable degree of renaturation on neutralization, while broken, relaxed loops will unwind completely and show little renaturation.

8. Another technical obstacle linked to the presence of agarose is that it takes more time for all the solutions to penetrate the gels. Additional washing rounds are therefore introduced in the protocol during the posthybridization and signal detection steps.

References

1. Collins, A.R. (2004) The comet assay for DNA damage and repair. *Mol Biotech* **26**, 249–261.
2. Pellestor, F., Paulasova, P., Macek, M., and Hamamah, S. (2004) The peptide nucleic acids: a new way for chromosomal investigation on isolated cells? *Hum Reprod* **19**, 1946–1951.
3. Larsson, C., Koch, J., Nygren, A., Janssen, G., Raap, A.K., Landegren, U., and Nilsson, M. (2004) In situ genotyping individual DNA molecules by target-primed rolling-circle amplification of padlock probes. *Nat Methods* **1**, 227–232.
4. Santos, S.J., Singh, N.P., and Natarajan, A.T. (1997) Fluorescence in situ hybridization with comets. *Exp Cell Res* **232**, 407–411.
5. Arutyunyan, R., Gebhart, E., Hovhannisyan, G., Greulich, K.O., and Rapp, A. (2004) Comet-FISH using peptide nucleic acid probes detects telomeric repeats in DNA damaged by bleomycin and mitomycin C proportional to general DNA damage. *Mutagenesis* **19**, 403–408.
6. Arutyunyan, R., Rapp, A., Greulich, K.O., Hovhannisyan, G., and Gebhart, E. (2005) Fragility of telomeres after bleomycin and cisplatin combined treatment measured in human leukocytes with the comet-FISH technique. *Exp Oncol* **27**, 38–42.
7. Shaposhnikov, S., Larsson, C., Henriksson, S., Collins, A., and Nilsson, M. (2006) Detection of Alu sequences and mtDNA in comets using padlock probes. *Mutagenesis* **21**, 243–247.
8. Shaposhnikov, S.A., Salenko, V.B., Brunborg, G., Nygren, J., and Collins, A.R. (2008) Single-cell gel electrophoresis (the comet assay): Loops or fragments? *Electrophoresis* **29**, 3005–3012.
9. McKenna, D.J., Rajab, N.F., McKeown, S.R., McKerr, G., and McKelvey-Martin, V.J. (2003) Use of the comet-FISH assay to demonstrate repair of the *TP53* gene region in two human bladder carcinoma cell lines. *Radiat Res* **159**, 49–56.
10. Kumaravel, T.S. and Bristow, R.G. (2005) Detection of genetic instability at HER-2/neu and p53 loci in breast cancer cells using comet-FISH. *Breast Cancer Res Treat* **91**, 89–93.
11. Horvathova, E., Dusinska, M., Shaposhnikov, S., and Collins, A.R. (2004) DNA damage and repair measured in different genomic regions using the comet assay with fluorescent in situ hybridization. *Mutagenesis* **19**, 269–276.
12. Bohr, V.A., Smith, C.A., Okumoto, D.S., and Hanawalt, P.C. (1985) DNA repair in an active gene: removal of pyrimidine dimers from the DHFR gene of CHO cells is much more efficient than in the genome overall. *Cell* **40**, 359–369.
13. Mellon, I., Spivak, G., and Hanawalt, P.C. (1987) Selective removal of transcription-blocking DNA damage from the transcribed strand of the mammalian DHFR gene. *Cell* **51**, 241–249.

Chapter 11

Simultaneous Labeling of Single- and Double-Strand DNA Breaks by DNA Breakage Detection-FISH (DBD-FISH)

José Luis Fernández, Dioleyda Cajigal, and Jaime Gosálvez

Abstract

DNA Breakage Detection-Fluorescence *In Situ* Hybridization (DBD-FISH) permits simultaneous and selective labeling of single- and double-strand DNA breaks in individual cells, either in the whole genome or within specific DNA sequences. In this technique, cells are embedded into agarose microgels, lysed and subjected to electrophoresis under nondenaturing conditions. Subsequently, the produced "comets" are exposed to a controlled denaturation step which transforms DNA breaks into single-stranded DNA regions, detected by hybridization with whole genome fluorescent probes or the probes to specific DNA sequences. This makes possible a targeted analysis of various chromatin areas for the presence of DNA breaks. The migration length of the DBD-FISH signal is proportional to the number of double strand breaks, whereas its fluorescence intensity depends on numbers of single-strand breaks.

The detailed protocol for detection of two types of DNA breaks produced by ionizing radiation is presented. The technique can be used to determine intragenomic and intercellular heterogeneity in the induction and repair of DNA damage.

Key words: DBD-FISH, FISH, DNA damage, DNA breaks, Comet assay

1. Introduction

DNA Breakage Detection-Fluorescence *In Situ* Hybridization (DBD-FISH) is a technique that permits determination of DNA breaks in the whole genome or within specific DNA sequence areas, cell by cell (1–3). The approach consists of embedding single cells in an inert agarose matrix on a slide, followed by incubation in one or two lysing solutions to remove proteins and membranes. This produces nucleoids, i.e. deproteinized nuclear DNA fiber loops that keep the general shape of the residual nucleus. Afterwards, the nucleoids are immersed in an alkaline unwinding solution that transforms all DNA breaks, i.e. single-strand DNA breaks (ssbs) and double-strand DNA breaks (dsbs) into restricted single-stranded DNA (ssDNA) motifs (4–6). Certain other DNA lesions, like abasic sites or some deoxyribose damage, behave as alkali-labile sites and may also be transformed by the alkaline solution into ssbs and subsequent ssDNA areas (7, 8).

These ssDNA motifs produced by alkaline treatment are detected by hybridization with fluorescent DNA probes. It was confirmed that after exposure to X-rays, as numbers of DNA breaks increase, more ssDNA is produced by the alkaline unwinding, thus more probes hybridize and more fluorescence is observed (9). This may be captured by a high sensitivity CCD camera coupled with a fluorescence microscope, and quantified using specific software for image analysis. DNA breaks are determined all over the genome when a whole genome probe is used in hybridization. When probes which hybridize to specific DNA regions are used, they permit detection and quantification of the local damage within these specific areas (9). Thus, DBD-FISH allows for simultaneous assessment of both the intercellular and intragenomic sensitivity to DNA damage induction and/or repair.

DBD-FISH is a versatile procedure that can be adapted for different purposes. For example, it can detect DNA regions hypersensitive to DNA denaturation. These constitutive alkali-labile sites vary depending on the cell type and the species and are likely related to native chromatin structure (10–14).

Ionizing radiation produces both ssbs and dsbs, the latter being induced 25-40 times less frequently (15, 16). More recently, DBD-FISH has allowed simultaneous detection and discrimination of ssbs and dsbs, induced by ionizing radiation, in the same cell and within specific DNA sequence areas, as well as in the whole genome (17–19). This was achieved by intercalating a neutral electrophoresis run before the unwinding step. Thus, migration produces a DBD-FISH signal with a head and a tail as in the comet assay, the length of which is related to the dsbs yield (16, 20). The DNA breaks, mainly ssbs spread all over the comet, are transformed into ssDNA by the adapted alkaline unwinding treatment, so that the intensity of fluorescence reflects the quantity of ssbs. The utility of the technique can be expanded by adding to it a bidimensional electrophoresis, which uses neutral pH in the first run and denaturing conditions in the second perpendicular run (21). This approach permits discrimination between ssbs and dsbs present in the same cell. Since the technique of DBD-FISH has been extensively detailed by us in the other volume of Methods in Molecular Biology (1), the protocol presented here describes the combination of DBD-FISH with electrophoresis.

2. Materials

2.1. Reagents and Technical Equipment

1. Normal melting point agarose.
2. Low melting point agarose.
3. Distilled water.
4. Pretreated slides (Chromacell SL, Madrid, Spain).
5. Electrophoresis power and chamber.
6. Epifluorescence microscope with appropriate filters and objectives.
7. High-sensitivity CCD camera coupled to a computer.
8. Image analysis software (Visilog 5.1, Noesis Vision Inc., Ville St-Laurent, Quebec, Canada).

2.2. Solutions

2.2.1. Slide Preparation

1. Lysing solution 1: 0.4 M Tris–HCl, 0.8 M DTT, 1% SDS, 0.05 M EDTA, pH 7.5. Store at 4°C and use at 42°C in a hood.
2. Lysing solution 2: 0.4 M Tris–HCl, 2 M NaCl, 1% SDS. Store at 4°C and use at 42°C.
3. TBE buffer: 0.09 M Tris–Borate, 0.002 M EDTA, pH 7.5. Store at room temperature.
4. Isotonic saline solution: 0.9% NaCl. Store at 4°C.
5. Alkaline unwinding solution: 0.03 M NaOH, pH 12.2. Make fresh and keep cold at 4°C. It is generally used at 7°C.
6. Neutralizing solution: 0.4 M Tris–HCl, pH 7.5. Store at 4°C and use at room temperature.
7. Ethanol baths, 70%, 90% and 100%.

2.2.2. FISH

1. DNA probes for FISH directly labeled with fluorochromes or with hapten (e.g. biotin or digoxigenin).

    An increasing variety of human DNA probes are available. DNA probes may be directed to specific families of repetitive DNA sequences, i.e. satellite DNA, like the classical satellite or alphoid satellite DNA probes, and also telomeric probes. There are "painting" probes to label most of DNA from a specific chromosome, cosmid or YAC probes and oligonucleotide probes for specific DNA sequences or chromosome regions. All are habitually supplied with their specific hybridization buffer. Suppliers provide information on preparation, incubation, washing, and detection conditions.

    A small number of examples of human DNA probes in hybridization buffer are presented:

    (a) Chromosome-specific satellite DNA probes (0.5 ng/μL final concentration) in 65% formamide/2× SSC, 10%

dextran sulfate, 100 mM calcium phosphate, pH 7.0 (1× SSC is 0.015 M NaCitrate, 0.15 M NaCl, pH 7.0).

(b) Whole genome (4.3 ng/μL final concentration), or Telomeric or Pancentromeric (all human alphoids) (0.5 ng/μL final concentration) DNA probes, in 50% formamide/2× SSC, 10% dextran sulfate, 100 mM calcium phosphate, pH 7.0.

Formamide is a potential carcinogen. It should be used in a deionized form.

2. DNA probe washing solutions. All should be freshly prepared and used at specific temperatures, depending on the stringency of the probe washings, as described in the methods (see Subheading 3.2.2, step 2).

3. Blocking solution: 5% BSA, 4× SSC, 0.1% Triton X-100. Store in aliquots at −20°C.

4. Detection reagent, e.g. streptavidin-Cy3 (Sigma, St. Louis, MO) for biotin-labeled probes, freshly diluted (1:200) in antibody detection buffer (1% BSA, 4× SSC, 0.1% Triton X-100), and (if both biotin and digoxigenin labeled probes were incubated in the same slide)/or anti-digoxigenin-FITC (Roche Molecular Biochemicals, Indianapolis, IN) (1:100) in the same buffer. The antibody detection buffer is stored in aliquots at −20°C.

5. Antibody washing solution (4× SSC, 0.1% Triton X-100, pH 7). Store at 4°C.

6. Counterstaining – Antifading solution: DAPI (2 μg/mL) in Vectashield (Vector Laboratories, Burlingame, CA). Store at 4°C in the dark. DAPI is a potential carcinogen.

3. Methods

3.1. Slide Preparation

3.1.1. Microgel Embedding of Cell Suspension

1. In a glass tube, dissolve 1% low melting point agarose in distilled water by microwave heating and keep the tube in a 37°C water bath at for at least 5 min to achieve this temperature.

2. Gently mix single cells in a suspension of culture medium, using a micropipette, with the low melting point agarose solution at 37°C to arrive at a final concentration of 0.7%. Cell concentration should be checked under phase-contrast microscopy and adjusted so that, after mixing with the agarose, cells do not overlap or are not excessively dispersed.

3. Pipette 50 μL of the cell-agarose suspension onto the coated surface of a pretreated slide (Chromacell SL, Madrid, Spain)

then cover with a 24×60 mm glass coverslip, taking care to avoid trapping any bubbles. The slide is deposited on a metallic or glass surface.

4. Place the metallic or glass plate in the freezer at 4°C for 5 min to solidify the agarose on the slide. The cell density of the microgel can be checked again under phase-contrast microscopy (see Notes 1 and 2).

3.1.2. Nucleoid Production

1. Remove coverslip and immediately immerse slide horizontally in a tray with abundant lysing solution 1, previously warmed to 42°C, and place in an oven at this temperature for 45 min.
2. Immerse slide horizontally in a tray with abundant lysing solution 2, also heated and kept at 42°C for 45 min (see Note 3).

3.1.3. Electrophoretic DNA Migration Under Neutral pH

1. Incubate slide horizontally in a box with abundant TBE buffer for 10 min.
2. Transfer slide to the electrophoresis chamber and electrophorese at 20 V (1 V/cm), 12 mA, for 10–15 min at room temperature (22°C) in TBE buffer.
3. Wash electrophoresed nucleoids in a tray with abundant isotonic saline solution for 2 min (see Note 4).

3.1.4. Transformation of DNA Breaks into ssDNA

1. Prepare fresh alkaline unwinding solution and store at 4°C. Transfer to a tray on ice until the temperature stabilizes at 7°C.
2. Remove the coverslip and immediately immerse the slide horizontally in the tray with alkaline unwinding solution for 2.5 min at 7°C. During this incubation, cover the tray with a sheet of aluminum foil to avoid light-induced DNA damage (see Notes 5–7).

3.1.5. Neutralization Washing and Dehydration

1. Horizontally immerse the slide in neutralizing solution at room temperature for 5 min.
2. Horizontally immerse the slide in TBE buffer at room temperature for 2 min.
3. Incubate slide sequentially with 70%, 90% and 100% ethanol baths, 2 min each, at room temperature.
4. Leave slide to dry horizontally on a sheet of filter paper, at room temperature or at 37°C, for 10–15 min. This causes the agarose layer to flatten, resulting in a very thin film with the nucleoids inside. After drying, the slides can be used immediately for hybridization or may be stored at room temperature in darkness for at least a month.

3.2. FISH

3.2.1. DNA Probe Hybridization

Denatured or single-stranded DNA probes, either directly labeled with fluorochromes or hapten-labeled, are incubated on dried slides. All types of probes can be hybridized.

1. Denature double-stranded DNA probe at 70°C for 8 min in an Eppendorf tube, with its specific hybridization buffer. Then immediately put on ice for 2–3 min.
2. Pipette 15 µL of probe solution onto the slide and cover with a coverslip (10×60 mm). At least two probes or probe combinations may be hybridized on the same slide, as long as their hybridization, incubation and washing conditions are the same. Then, encircle the slide with plastic parafilm, to avoid probe drying during incubation.
3. Incubate overnight in a dark moist chamber, lined with two sheets of wet filter paper. We incubate all probes at 37°C, except whole genome and telomeric probes, which are kept at room temperature (see Note 8).

3.2.2. DNA Probe Washings

1. Remove glass coverslip by gently immersing the slides vertically in a Coplin jar containing isotonic saline solution at room temperature.
2. Wash out unbound DNA probe by incubation in Coplin jars containing formamide/SSC solutions and then in SSC solutions, with gentle agitation, their stringency being that appropriate for the chosen probe.
   (a) Specific human alphoid satellite DNA probes: two washes in 60% formamide/2× SSC, pH 7, 42°C, 5 min each, followed by two washes in 2× SSC, pH 7, 37°C, 3 min each.
   (b) Specific human classical satellite DNA probes: two washes in 50% formamide/2× SSC, pH 7, 42°C, 5 min each, followed by two washes in 2× SSC, pH 7, 37°C, 3 min each.
   (c) Human pancentromeric probe: two washes in 50% formamide/2× SSC, pH 7, 37°C, 5 min each followed by two washes in 2× SSC, pH 7, 37°C, 3 min each.
   (d) Whole genome probe and telomeric probe: similar to the latter, but with all solutions at room temperature.
3. Transfer briefly to a Coplin jar containing antibody washing solution.
4. Remove slide from the antibody washing solution and blot excess fluid from the edge. Slides incubated with directly fluorescent-labeled probes are directly counterstained, whereas those with hapten-labeled probes are subjected to the detection steps.

3.2.3. DNA Probe Detection

1. Pipette 90 µL of blocking solution onto the slide and cover with a plastic coverslip. Incubate for 5 min at 37°C in a moist chamber.
2. Remove coverslip and tilt slide briefly to allow excess fluid to drain.

3. Apply 90 µL of detection reagent solution (streptavidin-Cy3 for biotin-labeled probes and/or anti-digoxigenin-FITC for digoxigenin-labeled probes) to the slide. Place a plastic coverslip on top of the solution, and incubate for 30 min in a humidified chamber at 37°C.

4. Remove coverslip and gently agitate in three volumes of antibody washing solution at room temperature, 2 min each.

5. Remove slide from the antibody washing solution and blot excess fluid from the edge (see Note 9).

3.2.4. Nucleoid Counterstaining and Microscopy

1. Pipette 20 µL of Counterstaining-Antifading solution onto slide and cover with a 24 × 60 mm glass coverslip, avoiding trapping air bubbles.

2. Visualize DAPI-stained nucleoids under an epifluorescence microscope using 10×, 40× and 100× objectives and a specific "blue" fluorescence filter. DBD-FISH signal is observed under 40× or 100× objectives, but using the appropriate fluorescence filter for the fluorochrome ("red" filter for Cy3, "green" filter for FITC).

3.3. Image Capture

Images are captured with a high-sensitivity CCD camera, such as the Ultrapix-1600 (Astrocam, Microphotonics Inc., Allentown, PA), which has a 1,536 × 1,024 pixel spatial resolution and distinguishes more than 16,000 gray level values.

1. Images are acquired using an intermediate resolution level (e.g. 3 × 3 binning factor).

2. The exposure time for image capture is determined by taking the highest and lowest signal intensities of the whole experiment into account. The highest signal intensity should be captured without gray level saturation within its field and the lowest signal intensity should be clearly discriminated. Always use the same objective (100×) and zoom magnification (1.5×) with the microscope. All subsequent images are taken at the same resolution level and exposure time.

3. Two calibration steps are necessary before fluorescent images can be captured:

   (a) A reference image from the current black level is obtained, corresponding to the electronic noise detected by the CCD camera under the specific conditions of image acquisition.

   (b) A flatfield image is captured from a dark area of the glass slide, without a FISH signal. This image must be acquired using the same filter, objective and magnification as are the DBD-FISH signals. It is a reference image of illumination variation within the visualization field in the sample due to the autofluorescence microscope optics, CCD sensitivity or external sources. Each glass slide should have its own flatfield image.

4. A sequence of 25–150 randomly selected images from DBD-FISH signals per experiment point is captured.

5. The black level and flatfield calibration images previously taken are subtracted from the images of the sequence, thus correcting them for background noise and illumination variation.

6. The sequence of images is saved in a file in the format of the camera (.apf).

7. A new sequence of images from a different experimental point is then captured, corrected for black level and flatfield images, saved, and so on.

8. Each sequence of images to be analyzed is exported from .apf into .img files (imager 2), in our case, since our digital image analysis system cannot operate with the file type of the camera. The transformation does not modify the gray levels, and each image appears separately and numbered.

3.4. Digital Image Analysis

The images captured with the CCD camera are analyzed using image analysis software. We have designed two semiautomatic routines that can run within an open system such as Visilog 5.1 (Noesis). One routine is dedicated to the analysis of the large extended DBD-FISH signals from the whole genome probe, while the other analyzes spot-like signals typical of those obtained with the rest of the probes. The main difference between the two routines is that the latter incorporates a "top-hat" transformation to highlight the spots.

The main steps in the digital image analysis are:

1. Selection of the area of interest within the whole image. This is usually a rectangle selected by the operator in the region containing the specific signal to be analyzed, and includes surrounding background fluorescence.

2. Thresholding to obtain a binary image. A fixed gray level of segmentation is chosen so the DBD-FISH signal is separated from the background. Higher gray levels correspond to DBD-FISH fluorescence, while lower gray levels correspond to background fluorescence.

3. The mean gray level from the background is automatically subtracted from the gray level corresponding to each pixel from the DBD-FISH signal.

4. Four main variables are measured for each DBD-FISH signal image, and are exported as .txt files to an Excel spreadsheet for statistical analysis.

(a) Surface area (A): number of pixels. DNA breaks relax the DNA loops of the nucleoids, thus increasing the surface of the signal.

(b) Length of migration (L): linear number of pixels, from one end to another end of the DBD-FISH signal, in the direction of the electrophoresic field.

(c) Mean Fluorescence Intensity (MFI): average pixel gray level.

(d) Whole Fluorescence Intensity (WFI = A × MFI): This is the sum of the gray levels from all pixels comprising the DBD-FISH signal. It should correspond to the entire quantity of hybridized probe.

5. When studying the dose-response effect within specific DNA targets to compare their possible differential sensitivity, the relative increase in the intensity (ssbs) and migration length (dsbs) of the DBD-FISH signal, with respect to its background signal, must be established for each target. This generates a dose-response curve for each target. The comparison of the coefficients of the dose-response curves provides information of the relative sensitivity among the different targets (see Notes 10 and 11).

6. When studying repair of ssbs within a target, the MFI of the control-untreated sample must be subtracted from that of the damaged samples. The remaining intensity should then be plotted as a function of repair time. When analyzing dsbs repair, the same operations must be performed, but on migration length.

4. Notes

1. Several microgels may be prepared on a single slide. This is habitually performed when analyzing DNA repair after exposure of a single dose of the DNA damaging agent. Four different repair times, as well as a zero dose and a damaged-unrepaired sample, can be analyzed on a single slide, each in a different microgel. Thus six microgels, three in the upper half and other three in the lower half of the slide, are simultaneously processed in the same slide, under the same technical conditions. The volume of the cell-agarose suspension to be pipetted must be proportionally adjusted to the correspondent microgel surface.

2. Cells can be exposed to the DNA-damaging agent either before mixing with the liquid agarose, in suspension or monolayer, or when included within the microgel monolayer.

For dose-response studies of ionizing radiation effect, it is preferable to expose the cells when they are in the gel on the microscope slide. To avoid DNA repair, a cold plate is placed under the slide, taking care not to freeze the agarose layer as this disrupts the gel and damages the cells. Only a fraction of each slide is exposed to the X-rays. An area is always left unexposed, either by protecting it or by keeping it out of the irradiated field, in order to act as an unirradiated control. At least three doses may be applied on a single slide, in addition to the unirradiated area. For DNA repair studies, cells must be irradiated in their culture flasks, previously cooled to avoid DNA repair. Then, the culture medium is changed to one that is heated to 37°C, and incubated at this temperature for the time of repair to be assayed.

3. The use of two lysing solutions assures effective protein removal. Lysing solution 1 is probably sufficient, but the appearance of nucleoid fibers is cleaner and tighter when using the lysing solution 2 afterwards, allowing a better discrimination of the "core" and "halo" of the nucleoid. These solutions can be used for shorter times, e.g. 20 min for lysis 1 and 10 min for lysis 2, at room temperature. This is enough for conventional DBD-FISH. Nevertheless, for successful electrophoresis DNA migration coupled to DBD-FISH, a very strong lysis is advisable to allow broken DNA duplexes to migrate as freely as possible (16). This is the reason to increase the incubation times to 45 min for each lysis, as well as to increase the lysing temperature to 42°C.

4. It is important to check for the length of electrophoresis DNA migration. After 8–10 min of electrophoresis, a control slide is stained with counterstaining-antifading solution and examined under the fluorescence microscope. This is to assure that the length of the comet is not higher than the visual field of the objective. If time is enough, the electrophoresis is stopped. If not, it may proceed for a few minutes more.

5. The ssDNA produced by the alkali may arise by ssbs, dsbs, alkali-labile sites (i.e. DNA lesions or modifications that turn into strand breaks by alkaline treatment) (4, 7, 8), and chromosome telomeric ends (9). Except in the case of telomeres, all of them can be induced by exposure to DNA-damaging agents or through internal cellular processes such as excision repair, apoptosis or the action of topoisomerases. In the case of ionizing radiation, most induced DNA lesions are ssbs and alkali-labile sites (15, 16). This is why the length of the comet, or the DBD-FISH signals visualized, is dependent on dsbs, whereas the intensity of the DBD-FISH signal reflects ssbs (17). Replicating structures behave as DNA strand breaks when denatured in alkaline assays (22). Thus, cells in S-phase could appear with strong labeling after DBD-FISH.

6. Unwinding begins on both sides of each break and proceeds along the DNA helix (5, 6). The rate of production or length of ssDNA produced from a break of origin will be dependent on the time, temperature, and ionic strength of the alkaline unwinding solution (5). These variables may be manipulated for each experimental situation. When analyzing the DNA breakage level exclusively induced within a specific DNA region, it is important to restrict the length of the ssDNA extended from the break as far as possible. This reduces the possibility of ssDNA being generated from DNA breaks relatively close to the probed sequence, as these could extend into the probed area with the consequence that DBD-FISH would not be absolutely restricted to the specific area. Thus, relatively big DNA targets like highly repetitive satellite DNA sequence regions are the best choice for determining internal DNA damage by DBD-FISH (Fig. 1). When assaying very short targets, they could be markers of their sensitivity and that of the surrounding DNA sequences. We suggest an incubation time of 2.5 min at 7°C since shorter times do not allow the alkali to diffuse homogeneously to all cells within the gel matrix (23). If less restrictive conditions are advisable, one may use from less to more denaturant NaOH 0.03 M at room temperature (22°C), NaOH 0.03 M 1 M NaCl at 7°C or the same solution at 22°C, for 2.5 min. Following this order of solutions, the production of ssDNA is progressively higher for the same incubation time.

7. It is estimated that 1 Gy of X rays induces 20–40 dsbs and 1,000 ssbs (15, 16). Since dsbs are much less frequently produced than ssbs, it is necessary to apply high doses to produce significant dsbs to be analyzed, i.e. in the range of 5–50 Gy. These high doses produce an enormous amount of ssbs, so the unwinding treatment may denature most of DNA with low doses, losing the linear increase of signal intensity with dose and the subsequent ability to relate ssbs with dose. In order to simultaneously discriminate ssbs and dsbs, the mean intensity should keep increasing linearly within the high dose range, as the migration length does. To this purpose, the unwinding treatment must be quite restrictive. This is other reason for using the alkaline solution without salt and at 7°C.

8. Several probes with different haptens or fluorochromes can be simultaneously hybridized, washed and detected, if their hybridization buffer and washing conditions are similar (Fig. 1a). If the hybridization buffer and/or the DNA probe washings are different, the probe with most stringent hybridization and washing conditions can be hybridized first. After the specific washings, the slide with the microgel film is immersed in increasing ethanol baths (70%, 90%, 100%, 2 min each) and then air-dried. The new DNA probe with less stringent

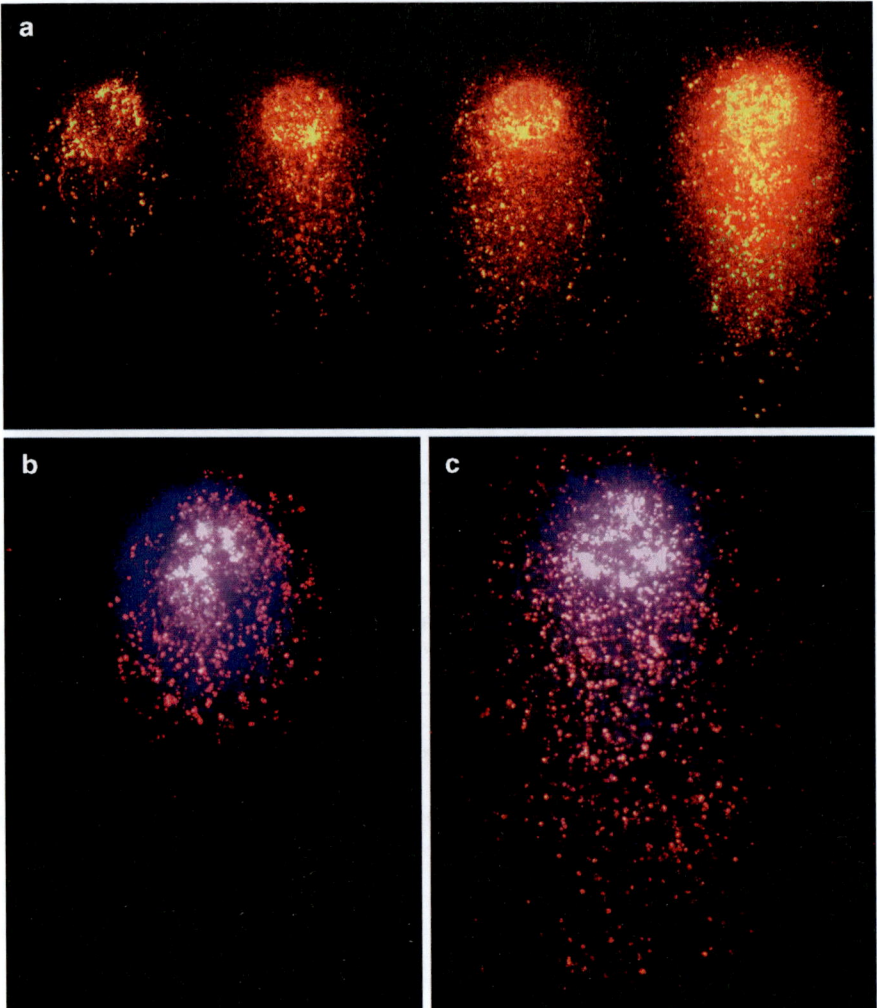

Fig. 1. Mammalian cells processed with the DBD-FISH to simultaneously determine dsbs and ssbs. (a) Chinese hamster cell line (CHO) exposed to increasing doses of X-rays, *from left to right*: 0, 12, 25 and 50 Gy. Most chromosomes from this cell line contain big centromeric blocks of arrays of interstitial telomeric repeat sequences (ITRS). After processing, the nucleoids were hybridized with a whole genome probe Cy3-labeled (*red*) mixed with a telomeric probe, FITC labeled (*green*). As the dose increases, there is an increase in the migration length (dsbs) and in the fluorescence intensity (ssbs), in both the whole genome and in the ITRS. (b, c) Mouse embryonic fibroblasts control (b) and exposed to 60 Gy of X-rays (c), processed, hybridized with a DNA probe specific for major satellite DNA Cy-3 labeled (*red*) and DAPI counterstained (*blue*). After irradiation, there is an increase in the migration length (dsbs) and in the fluorescence intensity (ssbs) in the pericentromeric major satellite DNA repeated sequences. Note that control unirradiated cells show a relatively high background signal from the ITRS (a, *left*) and from the major satellite DNA sequences (b). This is probably due to a specific DNA conformation of these repetitive DNA sequences that enhances the alkaline denaturation, i.e. constitutive alkali-labile sites.

hybridization and washing conditions is then hybridized and washed. Once the final wash is finished, the detection step may be initiated with a mixture of the different fluorochrome-conjugated antibodies to each different hapten-labeled DNA probe, if necessary.

9. Interslide and interexperimental variation is evident in DBD-FISH results, as in all DNA unwinding techniques. In each X-ray-irradiated slide, it is important to preserve an unirradiated area to provide a control signal to which that of the exposed area is compared (Fig. 1a *left*, b). It is also desirable to incubate two different probes on the same slide. This aids the more accurate evaluation of the relative differences in labeling between the different targets, in a similarly processed cell population. If the probes are labeled with the same fluorochrome, they may be hybridized in separate areas of the slide, including both unirradiated and irradiated regions. Moreover, if differentially labeled, they may be simultaneously or sequentially hybridized in the same area, and then the same cell measured with different filters to establish the fluorescence ratio between the signals (Fig. 1a).

10. Only relative increases in signal intensity and/or length with respect to background from control cells may be used for comparison of sensitivities in different targets. This is because each target has its own genomic size and background signal. Moreover, the different DNA probes for FISH could have different labeling efficiencies.

11. The alkaline unwinding treatment could produce a different length of ssDNA from a break, i.e. signal, depending on the sequence where it is located. If these possible differences in unwinding-renaturation rate are not corrected for, one could not be confident in the comparison of sensitivity among different targets. These differences could be estimated by irradiating deproteinized nucleoids, i.e. X-rays exposure after lysis (9, 18). Since the different chromatin structures disappear after lysis, ionizing radiation must induce DNA breaks practically at random. As a consequence, the DNA breaks per unit length of DNA should be similar in the different targets. If differences exist among different targets in relative increase of signal with respect to background after irradiating deproteinized nucleoids, these should be due to the different unwinding-denaturation rate. The slope of the linear dose-response curve obtained in the whole genome may be divided by that of the specific DNA sequence, thus obtaining its factor of sensitivity to unwinding-renaturation from DNA breaks, relative to that of the genome overall. In the dose-response studies in the cell with native chromatin, each intensity value of the signal from a specific DNA sequence is multiplied by its respective unwinding-renaturation factor, thus correcting for the differences in the unwinding-renaturation rate. Afterwards, comparisons of the sensitivity of different native targets can be more confidently performed. It must

be taken into account that the range of doses of X-rays applied to deproteinized nucleoids should be much lower than those applied to cells with organized chromatin (9, 18), since they are much more sensitive.

Acknowledgments

This work was supported by Fondo de Investigaciones Sanitarias (FIS PI070459), Xunta de Galicia (INCITE07PXI916201ES) and the Consejo de Seguridad Nuclear (CSN), Spain.

References

1. Fernández, J.L., and Gosálvez, J. (2002) Application of FISH to detect DNA damage: DNA breakage detection-FISH (DBD-FISH). *Methods Mol. Biol.* **203**, 203–216.
2. Fernández, J.L., Goyanes, V., and Gosálvez, J. (2002) DNA Breakage Detection-FISH (DBD-FISH). In: Rautenstrauss, B., and Liehr, T. (ed.), *FISH Technology – Springer Lab Manual.* Springer–Verlag, Heidelberg, pp. 282–290.
3. Fernández, J.L., Goyanes, V., and Gosálvez, J. (2005) DNA Breakage Detection-FISH (DBD-FISH). In: Fuchs, J., and Podda, M. (ed.), *Encyclopedia of Medical Genomics and Proteomics.* Marcel Dekker, Inc., New York, DOI: 10.1081/E-EDGP-12004027.
4. Ahnström, G., and Erixon, K. (1973) Radiation-induced single-strand breaks in DNA determined by rate of alkaline strand separation and hydroxylapatite chromatography: an alternative to velocity sedimentation. *Int. J. Radiat. Biol.* **36**, 197–199.
5. Rydberg, B. (1975) The rate of strand separation in alkali of DNA of irradiated mammalian cells. *Radiat. Res.* **61**, 274–287.
6. Ljungman, M. (1999) Repair of radiation-induced DNA strand breaks does not occur preferentially in transcriptionally active DNA. *Radiat. Res.* **152**, 444–449.
7. Téoule, R. (1987) Radiation-induced DNA damage and repair. *Int. J. Radiat. Biol.* **51**, 573–589.
8. Von Sonntag, C. (1987) *The Chemical Basis of Radiation Biology.* Taylor and Francis Ltd, London.
9. Vázquez-Gundín, F., Rivero, M.T., Gosalvez, J., and Fernández, J.L. (2002) Radiation-induced DNA breaks in different human satellite DNA sequence areas, analyzed by DNA breakage detection-fluorescence in situ hybridization. *Radiat. Res.* **157**, 711–720.
10. Fernández, J.L., Vázquez-Gundín, F., Rivero, M.T., Goyanes, V., and Gosálvez, J. (2001) Evidence of abundant constitutive alkali-labile sites in human 5 bp classical satellite DNA loci by DBD-FISH. *Mutat. Res.* **473**, 163–168.
11. Rivero, M.T., Vázquez-Gundín, F., Goyanes, V., Campos, A., Blasco, M., Gosálvez, J., and Fernández, J.L. (2001) High frequency of constitutive alkali-labile sites in mouse major satellite DNA, detected by DNA breakage detection-fluorescence in situ hybridization. *Mutat. Res.* **483**, 43–50.
12. López-Fernández C., Arroyo, F., Fernández, J.L., and Gosálvez, J. (2006) Interstitial telomeric sequence blocks in constitutive pericentromeric heterochromatin from *Pyrgomorpha conica (Orthoptera)* are enriched in constitutive alkali-labile sites. *Mutat. Res.* **599**, 36–44.
13. Muriel, L., Segrelles, E., Goyanes, V., Gosálvez, J., and Fernández, J.L. (2004) Structure of human sperm DNA and background damage, analyzed by in situ enzymatic treatment and digital image analysis. *Mol. Hum. Reprod.* **10**, 203–209.
14. Cortés-Gutierres, E.I., Dávila-Rodriguez, M.I., López-Fernández, C., Fernández, J.L., and Gosálvez, J. (2008) Alkali-labile sites in sperm cells from *Sus* and *Ovis* species. *Int. J. Androl.* **31**, 354–363.
15. Ward, J.F. (1988) DNA damage produced by ionizing radiation in mammalian cells: identities, mechanisms of formation, and reparability. *Prog. Nucleic Acid Res. Mol. Biol.* **35**, 95–125.
16. Olive, P.L. (1999) DNA damage and repair in individual cells: applications of the comet

17. Fernández, J.L., Vázquez-Gundín, F., Rivero, M.T., Genescá, A., Gosálvez, J., and Goyanes, V. (2001) DBD-FISH on neutral comets: simultaneous analysis of DNA single- and double- strand breaks in individual cells. *Exp. Cell Res.* **270**, 102–109.
18. Rivero, M.T., Mosquera, A., Goyanes, V., Slijepcevic, P., and Fernández, J.L. (2004) Differences in repair profiles of interstitial telomeric sites between normal and DNA double-strand break repair deficient Chinese hamster cells. *Exp. Cell Res.* **259**, 161–172.
19. Losada, R., Rivero, T., Goyanes, V., Slijepcevic, P., and Fernández, J.L. (2005) Effect of Wortmannin on the repair profiles of DNA double-strand breaks in the whole genome and in interstitial telomeric sequences of Chinese hamster cells. *Mutat. Res.* **570**, 119–128.
20. McKelvey-Martin, V.J., Green, M.H.L., Schmezer, P., Pool-Zobel, B.L., De Méo, M.P., and Collins, A. (1993) The single gel cell electrophoresis (comet assay): a European review. *Mutat. Res.* **288**, 47–63.
21. Rivero, M.T., Vázquez-Gundín, F., Muriel, L., Goyanes, V., Gosálvez, J., and Fernández, J.L. (2003) Patterns of DNA migration in two-dimensional single-cell gel electrophoresis analyzed by DNA breakage detection-fluorescence in situ hybridization. *Environ. Mol. Mutagen.* **42**, 223–227.
22. Olive, P.L., Chan, A.P.S., and Cu, C.S. (1988) Comparison between the DNA precipitation and alkali unwinding assays for detecting DNA strand breaks and cross-links. *Cancer Res.* **48**, 6444–6449.
23. Vázquez-Gundín, F., Gosálvez, J., de la Torre, J., and Fernández, J.L. (2000) DNA breakage detection-FISH (DBD-FISH): effect of unwinding time. *Mutat. Res.* **453**, 83–88.

(first entry continued from previous page:)
assay in radiobiology. *Int. J. Radiat. Biol.* **75**, 395–405.

Chapter 12

Co-localization of DNA Repair Proteins with UV-Induced DNA Damage in Locally Irradiated Cells

Jennifer Guerrero-Santoro, Arthur S. Levine, and Vesna Rapić-Otrin

Abstract

This chapter describes a technique in which indirect immunofluorescence is applied to visualize the process of nucleotide excision repair (NER) at the site of locally induced damage in DNA. UV-irradiation of cells through an isopore polycarbonate membrane filter generates cyclobutane pyrimidine dimers (CPD) and (6-4) photoproducts (6-4PP) on a subnuclear area, which corresponds to the size of a pore on the membrane. Specific antibodies to CPD and 6-4PP define the damaged spot. The NER components co-localize at the damaged-DNA subnuclear spot, where the proteins are stained with the appropriate fluorescent antibodies. This relatively simple and affordable method facilitates the examination of the sequential assembly of NER proteins in the chromatin-embedded DNA photoproducts. The method also enhances the identification of repair auxiliary proteins and complexes, such as ubiquitin E3 ligases, involved in the initiation of NER on non-transcribed DNA.

Key words: Nucleotide excision repair, Xeroderma pigmentosum, Cyclobutane pyrimidine dimers, (6-4) Photoproducts, Isopore polycarbonate filter, Localized UV-irradiation, Indirect immunofluorescence, Tagged-protein, CSK buffer, UV-damaged DNA binding protein 1, UV-damaged DNA binding protein 2, Cullin 4A, Cullin 4B, Ubiquitin E3 ligase

1. Introduction

1.1. Nucleotide Excision Repair

Cells use various molecular mechanisms for repairing damaged DNA to preserve the integrity of the genome. The major DNA repair process that removes helix-distorting lesions, including UV-induced cyclobutane pyrimidine dimers (CPD) and (6-4) photoproducts (6-4PP), is the nucleotide excision repair (NER) pathway (1). The general mechanism of NER in humans has been elucidated primarily through analyses of individuals with the inherited skin cancer-prone disorder xeroderma pigmentosum (XP), which is characterized by defects in NER (2, 3) and by reconstitution of the repair reaction *in vitro* with purified proteins (complementation groups XP-A through XP-G) (4, 5). In eukaryotic cells, NER operates on chromatin-embedded DNA substrates. However, the *in vivo* dynamics of NER in repair-proficient and -deficient cells remained poorly documented for many years due to the lack of a suitable methodologic approach.

Development of a method that utilizes an isopore polycarbonate membrane filter to produce UV-induced DNA lesions in a localized area of the cell nucleus (6), combined with indirect immunofluorescence, gave rise to a simple technique for studying the assembly and operation of the NER machinery *in vivo* (7). When a cell monolayer is covered with the filter membrane, UVC radiation exposure to the cells occurs only through pores of 3–8 μm diameter, depending on the filter (see Figs. 1a and 2) (6, 8). Specific antibodies to CPD and 6-4PP (9) define the damaged subnuclear spot. DNA repair proteins co-localize *in situ* at the damaged DNA, where the proteins and photolesions are stained with the appropriate fluorescent antibodies. A series of such experiments conducted in normal and XP cell lines support the theory based on *in vitro* experiments of sequential assembly of the repair proteins at the site of the damage (7, 10–12). The current model posits that the NER process results in active complex intermediates and that the NER reaction progresses from recognizing the DNA distortion around a lesion, to formation of an open structure and damage location, to recruiting the nucleases that excise the damaged DNA fragment, and finally to re-synthesizing DNA to fill the gap (1).

1.2. Application of Localized Nuclear Irradiation in Defining the Role of Repair Auxiliary Complexes, Ubiquitin E3 Ligases, in the Initiation of NER on Non-transcribed DNA

The NER system includes transcription-coupled repair, which removes lesions from the transcribed strand of active genes, and global genome repair, which processes lesions in the non-transcribed DNA. The fibroblasts from XP-E patients have a partial deficiency in global genome repair due to a defective UV-damaged DNA binding protein complex (UV-DDB) (13, 14). UV-DDB comprises two proteins: the 48-kDa UV-damaged DNA binding protein 2 (DDB2) and the 127-kDa UV-damaged DNA binding protein 1 (DDB1), encoded by the *DDB2* and *DBB1* genes, respectively (15, 16). Mutations in *DDB2* cause the XP-E phenotype (13, 17, 18). Although UV-DDB binds readily to fragments of DNA that contain various types of damage (19), it is an auxil-

Fig. 1. (continued) DNA (**b**). Merged images show that V5-CUL4A and Flag-DDB2 or HA-CUL4B and Flag-DDB2 co-localize in the UV-irradiated subnuclear spot (**c**, **d**). Anti Flag and anti-HA antibodies do not cross-react when used for indirect immunofluorescence on locally UV-irradiated cells (**d**) (see Note 11). A model of the DDB1-CUL4A^{DDB2} ubiquitin E3 ligase bound to a UV-damaged nucleosome (PDB ID: 1AOI) (**e**). This figure is modified from published figures (Copyright 2006, National Academy of Sciences, USA (26, 31)).

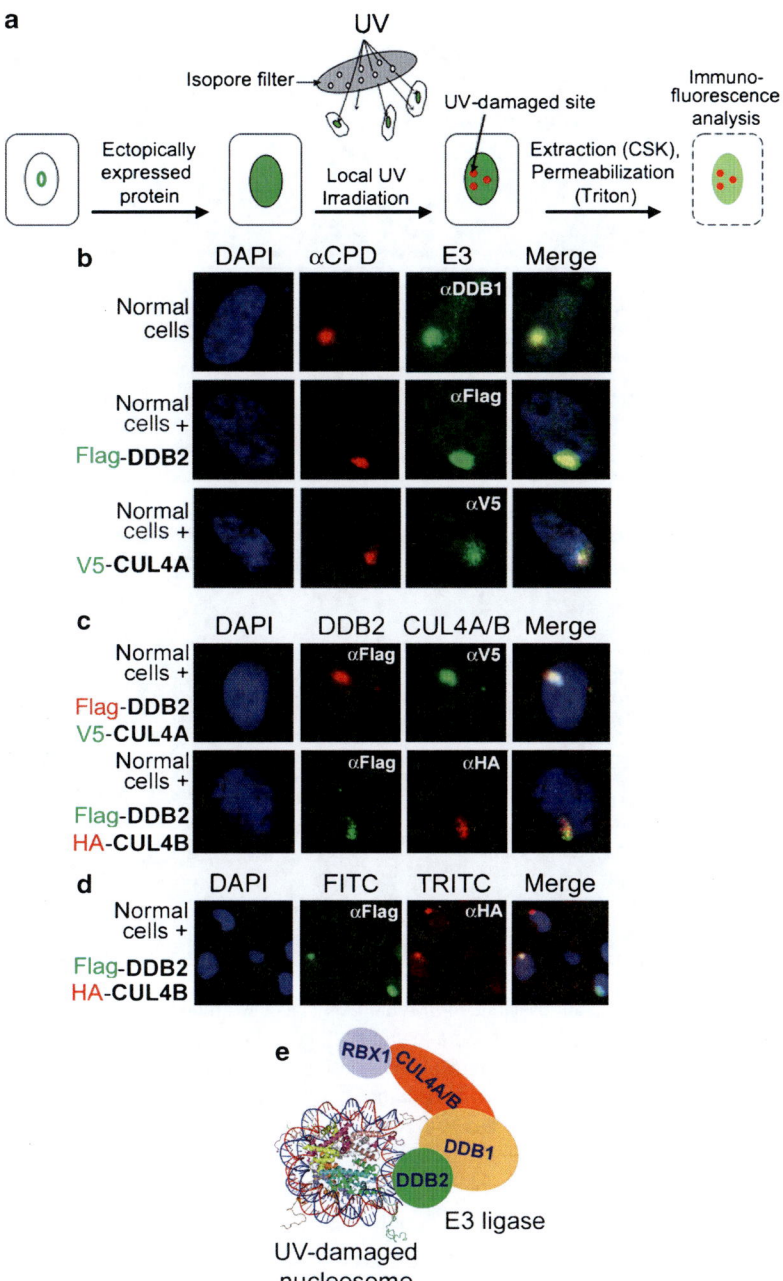

Fig. 1. DDB1-CUL4A^{DDB2} ubiquitin E3 ligase co-localizes at UV-damaged DNA sites. Transformed human fibroblasts (WI-38VA13 cells) were transiently transfected with full-length cDNA of the type indicated and treated as illustrated on *panel* (**a**). An 8-μm pore size filter and a dose of 60 J/m^2 were applied. UV-irradiated sites were visualized by fluorescent immunostaining using anti-CPD antibody (*red*), and endogenous DDB1 was visualized by an anti-chicken DDB1 antibody (*green*). Ectopically expressed proteins (DDB2 and CUL4A) were visualized by an antibody to the epitope present on the transfected protein (*green*). The cells were counterstained with DAPI. Merged images show that components of the E3 ligase (DDB1, DDB2, and CUL4A) co-localize with damaged.

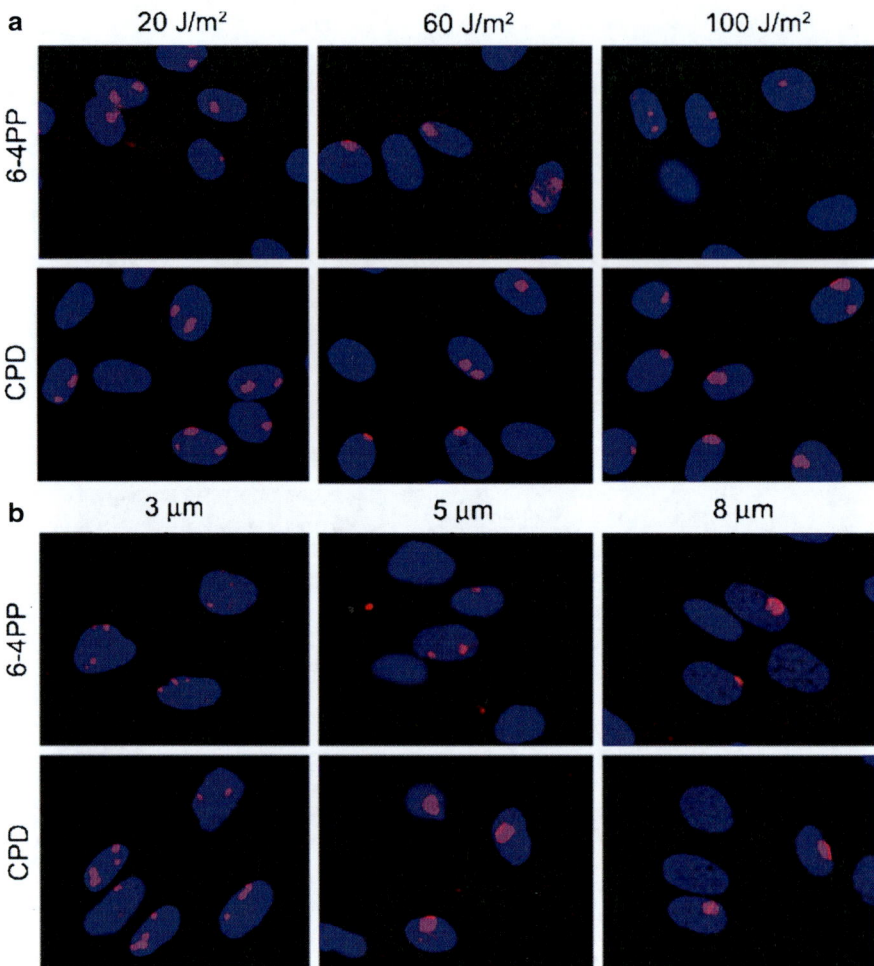

Fig. 2. *In situ* detection of UV-induced DNA damage in locally irradiated human fibroblasts. The effect of UVC dose (**a**) or pore size of the isopore polycarbonate membrane filter (**b**) on indirect immunofluorescent visualization of the CPD and 6-4PP foci. The 5 μm pore size filters (**a**) and a dose of 60 J/m^2 (**b**) were applied. The cells were fixed and permeabilized immediately after irradiation. After denaturation of DNA, CPDs and 6-4PPs were detected with TKM53 and 64M-2 antibodies, respectively, and visualized with Alexa Fluor 594-conjugated antibody (*red*). Nuclear DNA (*blue*) was counterstained with DAPI.

iary factor for the NER process *in vitro* (4, 20), and its involvement in NER was not defined until recently. In contrast to the NER deficiency in intact XP-E cells, XP-E cell extracts can repair naked DNA substrates *in vitro*, strongly suggesting that UV-DDB plays a role in the repair of DNA in the chromatin environment (21). Moreover, DDB2 is the first molecule in the activity sequence of NER protein to co-localize with the DNA lesion (22, 23), and facilitates the subsequent recruitment of XPC (24, 25). These observations led us to hypothesize that UV-DDB plays a role in initiating repair in the non-transcribed DNA by modifying chromatin in the vicinity of DNA lesions, which in turn may facilitate the assembly of the repair machinery on DNA (21, 26, 27). UV-DDB has been identified as a component of the Cullin 4A

(CUL4A)-RBX1-based ubiquitin E3 ligase (DDB1-CUL4A^{DDB2}), which further establishes its role in NER (28). Ubiquitin E3 ligase is a multiprotein complex that targets a substrate for posttranslational modification (i.e., ubiquitination) (29, 30). We recently demonstrated that DDB1-CUL4A^{DDB2} ubiquitin ligase co-localizes with UV-damaged DNA *in vivo* and targets histone H2A for monoubiquitination at the site of the photolesion (26). Further, we confirmed that Cullin 4B (CUL4B) forms another E3 ligase with UV-DDB (DDB1-CUL4B^{DDB2}) and binds to damaged subnuclear spots (31). Applying a localized nuclear irradiation technique (Fig. 1) was very instrumental in defining *in vivo* the role of XP-E factor (DDB2) and DDB1-CUL4^{DDB2} ubiquitin ligases in the initiation of NER on non-transcribed DNA.

This chapter describes a protocol in which indirect immunofluorescence is applied to visualize the co-localization of NER proteins at the site of local damage when cells are UV-irradiated through an isopore polycarbonate membrane filter. We emphasize the co-localization of ectopically expressed tagged proteins with the CPD photolesion.

2. Materials

2.1. Cell Culture

1. SV40 transformed human fibroblasts WI-38 VA13 (obtained from ATCC). This cell line required handling at Biosafety Level 2 containment.
2. 6-Well cell culture clusters (Costar) (see Note 1).
3. Microscope glass coverslips (cover glass #2, circle, 22 mm, Fisherbrand), sterilized by autoclaving or clean with methanol/ethanol mixture (1:1), rinse and store in 75% (v/v) ethanol. Watchmaker's forceps, curved.
4. Phosphate-buffered saline (PBS).
5. DMEM culture media supplemented with 10% fetal bovine serum, 0.1 mM essential and nonessential amino acids, and 2 mM L-glutamin.
6. Trypsin–EDTA solution.
7. Cell culture accessories: sterile hood (Biosafety Level 2) and 5% CO_2 incubator at 37°C.
8. FuGENE 6 Transfection Reagent (Roche) for the transient transfection of cells with the constructs of tagged protein.

2.2. Local UV-Irradiation

1. UV bench lamp, short wave 254 nm (Spectroline X-15G).
2. Radiometer DSE-2000A with interchangeable sensors (Spectroline).
3. UV safety eyeglasses and face shield.

4. Isopore polycarbonate membrane filters (Millipore), diameter 47 mm, pore size: 3, 5, and 10 μm (see Note 2).

5. Cell culture plate 100 mm.

2.3. In Situ CSK Buffer and Detergent Extractions

1. Cytoskeleton (CSK) buffer: 100 mM NaCl, 300 mM sucrose, 10 mM Pipes, pH 6.8, 3 mM $MgCl_2$, store in aliquots at −20°C. Before use, add the following: 0.2% Triton X-100 and Complete EDTA-free protease inhibitor cocktail tablet (Roche), and keep on ice after reconstitution.

2. 2% Paraformaldehyde in PBS: prepare fresh from 37% paraformaldehyde (Sigma).

3. Triton buffer: 0.2% Triton X-100 in PBS.

4. Washing buffer (WB): PBS containing 0.5% bovine serum albumin (BSA, Sigma) and 0.15% glycine.

5. Always prepare fresh and keep on ice.

6. Blocking buffer: 5% BSA in PBS, keep at 4°C.

7. 0.4 M NaOH.

2.4. Visualization of the UV-Damaged DNA Sites and Co-localized Proteins

1. Primary antibodies against the UV-induced DNA damage (see Note 3):
   (a) Anti-6-4PP monoclonal antibody, clone 64M-2 (Cosmo Bio). Dilution 1:400.
   (b) Anti-CPD monoclonal antibody, clone KTM53 (Kamiya Biochemical). Dilution 1:400.

2. Primary antibodies against the tagged epitope present in the proteins (see Note 4):
   (a) Anti-Flag polyclonal antibody developed in rabbit (Sigma). Dilution 1:200.
   (b) Anti-HA.11 monoclonal antibody, clone 16B12 (Covance Research Product Inc). Dilution 1:400.
   (c) Anti-V5 FITC conjugated polyclonal antibody developed in goat (Bethyl Laboratories Inc). Dilution 1:400.

3. Secondary antibodies conjugated to fluorophore: Alexa Fluor 594 (goat anti-mouse and goat anti-rabbit) or Alexa Fluor 488 (goat anti-mouse and goat anti-rabbit) (Invitrogen). All secondary antibodies were used as 1:1,000 dilutions.

4. Vectashield mounting medium with DAPI (blue) (Vector Laboratories). DAPI (4′,6-diamidino-2-phenylindole) produces a blue fluorescence with excitation at about 360 nm and emission at about 460 nm when bound to DNA.

5. Glass slides for mounting coverslips.

6. Kimwipes, general purpose laboratory wiper.

7. Fluorescence microscope with 60× and 100× oil-immersion lenses.

3. Methods

3.1. Cell Culture

1. The coverslip can be autoclaved in a box, picked with a sterile aspirator pipette or washed in 75% ethanol while holding with forceps, and flamed briefly before placing in a 6-well culture cluster.
2. One or two days prior to experiment, plate $2–5 \times 10^5$ cells, so that the cells are 70–80% confluent at the day of irradiation.
3. For immunodetection of transiently expressed proteins, perform transfection of the cells 30–48 h before the irradiation (see Note 5).

3.2. Local UV-Irradiation (See Note 6 and Fig. 3)

1. Prior to irradiation, aspirate the medium and keep at 37°C. Subsequently, rinse the cells twice with PBS at room temperature. Carefully lift a coverslip with the cells from a well using a forceps and place onto the surface of plastic support (i.e., a cell culture plate) (Fig. 3, Step 1). To prevent the cells from drying, add a few drops of PBS to form a thin layer of buffer on top of the cover glass.
2. Presoak an isopore polycarbonate membrane filter in PBS and place gently over the cell monolayer (Fig. 3, Step 2).
3. To irradiate locally, place the filter-shielded cells immediately under UVC lamp and irradiate at a dose rate of 1–5 $J/m^2/s$. Apply a dose between 20 and 100 J/m^2 by varying the time of irradiation (Fig. 3, Step 3) (see Note 7 and Fig. 2).
4. Remove the filter and either proceed with extraction of the cells immediately or return the irradiated cells to the original medium at 37°C for post-irradiation incubation (Fig. 3, Step 4) (see Note 8).

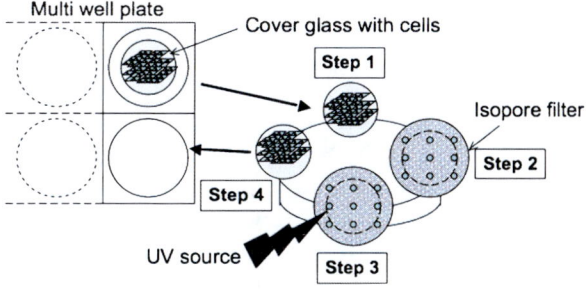

Fig. 3. Schematic illustration of the UV-irradiation of the cells, grown on a coverslip, through an isopore polycarbonate membrane filter (see Subheading 3.2 and Note 6).

3.3. In Situ CSK Buffer and Detergent Extractions (See Note 9)

1. Rinse the cells twice with PBS at 4°C (simply place the culture cluster on ice).
2. Rinse the cells with CSK buffer. Gently add 2 mL of CSK buffer containing 0.2% Triton X-100 and incubate for 5 min at 4°C. Rinse once with CSK buffer, remove the buffer completely, and transfer the coverslip to a new well.
3. Fix the cells by adding 2 mL of 2% paraformaldehyde and incubate for 15 min at room temperature.
4. Permeabilize the cells by rinsing with 0.2% Triton, and incubating in 2 mL twice, 10 min each at 4°C.
5. Rinse the cells with WB at 4°C.
6. Add 2 mL of 5% BSA and block the cells for 20 min at 4°C. Repeat steps 4 and 5.

3.4. Visualization of the UV-Damaged DNA Sites and Co-localized Proteins

1. Add 2 mL of 0.4 M NaOH and denature DNA for 4 min at room temperature (see Note 10).
2. Rinse the cells five times with 2 mL of PBS at room temperature.
3. Dilute primary antibody in WB (see Note 11). Rinse the cells with WB and remove buffer completely from a well. Add 0.8–1 mL of diluted antibody to immerse coverslip completely. Alternatively, place a piece of Parafilm in the bottom of a cover plate of the 6-well cell culture cluster. In a grid pattern that replicates the six wells, label the appropriate place for each coverslip with a marker. Carefully remove each coverslip from a well of the culture cluster with forceps, blot the excess buffer by touching the edge of the coverslip to a Kimwipe, and place on the Parafilm. Add diluted antibody as a drop (~250 µL) onto coverglass. Wet some Kimwipes in PBS and add to the plate to maintain humidity and prevent the coverslips from drying. Cover the plate with Parafilm.
4. Incubate the cells with primary antibody for 90 min at room temperature. Protect the plate from light with aluminum foil when primary antibody is FITC or TRITC conjugated.
5. Rinse the cells with 0.2% Triton, and incubate in 2 mL twice, for 10 min each at 4°C. Detergent treatment makes the cells permeable again, so any residue of primary antibody is removed efficiently.
6. Rinse the cells with WB at room temperature.
7. Dilute fluorophore-conjugated secondary antibody in WB and repeat the procedure for primary antibody as in step 3. Cover the plate and protect from light with aluminum foil. Incubate for 60 min at room temperature.
8. Repeat step 4.

9. Rinse the cells with PBS before mounting the coverslip onto slide.
10. Add 10 µL of DAPI/Vectashield medium onto slides. Lift the coverslip from a well and blot excess PBS with a Kimwipe. Invert the coverslip, making sure that the side with the cells is facing down onto slide. Blot mounted coverslip with Kimwipes and seal the glass with a rim of nail polish.
11. Analyze immediately or store in dark at room temperature (for a week) or at −20°C for a longer period.
12. Obtain and capture fluorescent images with the fluorescence microscope equipped with appropriate filters for fluorophores: DAPI, FITC, or Alexa Fluor 594 and Alexa Fluor 488 (see Note 12).

4. Notes

1. The 35-mm cell culture dishes can be used as well, but the 6-well cell culture clusters are more convenient for processing multiple samples.
2. The isopore polycarbonate membrane filters (Millipore) are available in various diameters and pore sizes. We routinely used the 8-µm pore size filter (Fig. 1) that has been discontinued. Millipore offers 10 µm filters now.
3. The anti-CPD (clone TDM-2) antibody established by Mori et al. (9) and distributed by Cosmo Bio is used primarily in studying the DNA repair process. We also routinely used clone KTM53 because it was the only commercially available anti-CPD antibody when we initiated the localized nuclear UV-irradiation study (Figs. 1 and 2).
4. The anti-CPD and anti-6-4PP monoclonal antibodies (listed in Subheading 2.4) were developed in mice. This could impose an obstacle for successful co-localization with the damaged DNA if the specific antibody against the cellular protein of interest is of mouse origin. Ectopic expression of the tagged protein of interest overcomes this problem. A wide spectrum of primary and conjugated anti-tag antibodies is commercially available now.
5. Optimize the transfection time and method for the protein of interest, and confirm that the ectopically expressed protein has retained the biological properties of the endogenous counterpart.
6. The original method for local UV-irradiation through isopore filters involved placing a filter over a monolayer of cultured cells in 35-mm glass bottom dishes before irradiating them

from above (6). We found handling a filter over the cells, placed in a well, awkward and time consuming. We adapted the following modified protocol (Fig. 3, Steps 1–4). A coverslip with the cells is taken from a well, placed onto a cell culture plate, then covered with a wet filter, and the whole plate with shielded cells is placed under the UVC source and irradiated. This modification enables the replicates (coverslips with the cells) of the same experimental condition to be irradiated simultaneously; one 10-cm cell culture plate can hold up to four "sandwiches" of a 22-mm coverslip shielded with a 47-mm filter. If more than one replica ("sandwich") is irradiated, ensure that each "sandwich" is exposed to the same dose of UVC.

7. Always protect your face with UV safety eyeglasses and face shield when working with any UV source. Power on the UVC lamp at least 10 min before irradiation. Monitor the dose rate using radiometer with a 254-nm sensor before and after irradiation. An equivalent amount of CPD and 6-4PP products is produced per cell with local doses of 5 J/m^2 and 100 J/m^2 (32). However, a common practice is to expose the cells to a higher UVC dose when local irradiation is applied than when uniform (without filter) irradiation is used (6, 22, 25). The size of the CPD and 6-4PP foci increases with the size of filter pores (Fig. 2) (6).

8. The DDB2/XPE factor and the DDB2-based complexes (UV-DDB and E3 ligases) co-localize with UV-damaged DNA immediately after irradiation (Fig. 1). Detection of the other NER proteins with CPD and 6-4PP foci occurs later and requires post-irradiation incubation of the cells (7, 11).

9. *In situ* extraction with CSK buffer eliminates an excessive amount of cellular proteins unbound either to the scaffold or UV-damaged DNA. The level of ectopically expressed NER protein(s) often exceeds the number of CPDs and 6-4PPs, which results in an increased background of unbound protein (25). The extraction is very efficient in eliminating the abundant nuclear fraction of endogenous DDB1, which is not a part of the ubiquitin E3 ligase bound to CPD (Fig. 1a). Test whether a particular protein of interest would benefit from CSK extraction, and adjust the numbers of washes and incubation time.

10. The antibodies against CPD or 6-4PP (TDM-2 and 64M-2) react only with photoproducts in single-stranded DNA; consequently, the cellular DNA has to be denatured before immunostaining. Even though Kamiya Biochemical's data sheet for KTM53 claims that the antibody reacts specifically with thymine dimers, induced by UV-irradiation, either in double- or single-stranded DNA, we have always used this

antibody to probe denatured (single-stranded) DNA (Figs. 1 and 2).

11. As mentioned in Note 4, co-localization of DNA repair protein with UV-induced DNA damage imposes certain challenge in selecting a specific primary antibody that can be incubated together with anti-CPD (anti-6-4PP) and without cross-reacting with the secondary antibody to anti-CPD. The anti-Flag and anti-V5 antibodies (listed in Subheading 2.4) meet this requirement (Fig. 1b). We used the anti-Flag together with anti-V5 or anti-HA antibodies to detect the corresponding tagged proteins, which co-localized within a subnuclear spot (Fig. 1c). Figure 1d illustrates that anti-Flag and anti-HA antibodies do not cross-react when used for indirect immunofluorescence on locally UV-irradiated cells. Flag-DDB2 and HA-CUL4B were visualized with the antibodies against the Flag (green) or HA (red) epitope present on the proteins, respectively. The cells were counterstained with DAPI (blue). Merged image shows proteins co-localized within the irradiated area. In some cases, the irradiated area is stained for DDB2 and not CUL4B and vice versa, indicating that there are cells that express only one of the two transfected cDNAs. Any cross-reactivity between the Flag and HA antibodies in this setting would result in double staining in all transfected cells, which is not the case.

12. We obtained the fluorescent images (Figs. 1 and 2) with a Olympus Provis AX70 fluorescence microscope equipped with appropriate filters for fluorophores. The digital images were then captured with a cooled CCD camera, and processed and superimposed using SPOT software (Diagnostic Instruments, Sterling Heights, MI).

Acknowledgments

This work was supported by a University of Pittsburgh Medical School startup fund and a grant from The Pittsburgh Foundation (to V. R.-O.).

References

1. Friedberg, E. C., Walker, G. C., Siede, W., Wood, R. D., Schultz, R. A., and Ellenberger, T. (2005) DNA Repair and Mutagenesis, 2nd Edition, ASM Press: Washington, DC
2. Wood, R. D., Mitchell, M., Sgouros, J., and Lindahl, T. (2001) Human DNA repair genes. *Science* **291**, 1284–9.
3. Bootsma, D., Kraemer, K. H., Cleaver, J. E., and Hoeijmakers, J. H. J. (2002) Nucleotide excision repair syndromes: xeroderma pigmentosum, cockayne syndrome, and trichothiodystrophy. In: The Genetic Basis of Human Cancer (Vogelstein, B., Kinzler, K. W., eds.). 2nd Edition, McGraw-Hill: New York; 211–37.

4. Aboussekhra, A., Biggerstaff, M., Shivji, M. K., Vilpo, J. A., Moncollin, V., Podust, V. N., Protic, M., Hubscher, U., Egly, J. M., and Wood, R. D. (1995) Mammalian DNA nucleotide excision repair reconstituted with purified protein components. *Cell* **80**, 859–68.

5. Riedl, T., Hanaoka, F., and Egly, J. M. (2003) The comings and goings of nucleotide excision repair factors on damaged DNA. *EMBO J* **22**, 5293–303.

6. Katsumi, S., Kobayashi, N., Imoto, K., Nakagawa, A., Yamashina, Y., Muramatsu, T., Shirai, T., Miyagawa, S., Sugiura, S., Hanaoka, F., Matsunaga, T., Nikaido, O., and Mori, T. (2001) In situ visualization of ultraviolet-light-induced DNA damage repair in locally irradiated human fibroblasts. *J Invest Dermatol* **117**, 1156–61.

7. Volker, M., Mone, M. J., Karmakar, P., van Hoffen, A., Schul, W., Vermeulen, W., Hoeijmakers, J. H., van Driel, R., van Zeeland, A. A., and Mullenders, L. H. (2001) Sequential assembly of the nucleotide excision repair factors in vivo. *Mol Cell* **8**, 213–24.

8. Wang, Q. E., Zhu, Q., Wanim, M. A., Wani, G., Chen, J., and Wani, A. A. (2003) Tumor suppressor p53 dependent recruitment of nucleotide excision repair factors XPC and TFIIH to DNA damage. *DNA Repair* **2**, 483–99.

9. Mori, T., Nakane, M., Hattori, T., Matsunaga, T., Ihara, M., and Nikaido, O. (1991) Simultaneous establishment of monoclonal antibodies specific for either cyclobutane pyrimidine dimer or (6-4) photoproduct from the same mouse immunized with ultraviolet-irradiated DNA. *Photochem Photobiol* **54**, 225–32.

10. Mone, M. J., Bernas, T., Dinant, C., Goedvree, F. A., Manders, E. M., Volker, M., Houtsmuller, A. B., Hoeijmakers, J. H., Vermeulen, W., and van Driel, R. (2004) In vivo dynamics of chromatin-associated complex formation in mammalian nucleotide excision repair. *Proc Natl Acad Sci U S A* **101**, 15933–7.

11. Green, C. M., and Almouzni, G. (2003) Local action of the chromatin assembly factor CAF-1 at sites of nucleotide excision repair in vivo. *EMBO J* **22**, 5163–74.

12. Dunand-Sauthier, I., Hohl, M., Thorel, F., Jaquier-Gubler, P., Clarkson, S. G., and Scharer, O. D. (2005) The spacer region of XPG mediates recruitment to nucleotide excision repair complexes and determines substrate specificity. *J Biol Chem* **280**, 7030–7.

13. Rapic-Otrin, V., Navazza, V., Nardo, T., Botta, E., McLenigan, M., Bisi, D. C., Levine, A. S., and Stefanini, M. (2003) True XP group E patients have a defective UV-damaged DNA binding protein complex and mutations in DDB2 which reveal the functional domains of its p48 product. *Hum Mol Genet* **12**, 1507–22.

14. Chu, G., and Chang, E. (1988) Xeroderma pigmentosum group E cells lack a nuclear factor that binds to damaged DNA. *Science* **242**, 564–7.

15. Keeney, S., Chang, G. J., and Linn, S. (1993) Characterization of a human DNA damage binding protein implicated in xeroderma pigmentosum E. *J Biol Chem* **268**, 21293–300.

16. Dualan, R., Brody, T., Keeneym, S., Nichols, A. F., Admon, A., and Linn, S. (1995) Chromosomal localization and cDNA cloning of the genes (DDB1 and DDB2) for the p127 and p48 subunits of a human damage-specific DNA binding protein. *Genomics* **29**, 62–9.

17. Nichols, A. F., Ong, P., and Linn, S. (1996) Mutations specific to the xeroderma pigmentosum group E Ddb- phenotype. *J Biol Chem* **271**, 24317–20.

18. Itoh, T., Mori, T., Ohkubo, H., and Yamaizumi, M. A. (1999) Newly identified patient with clinical xeroderma pigmentosum phenotype has a non-sense mutation in the DDB2 gene and incomplete repair in (6-4) photoproducts. *J Invest Dermatol* **113**, 251–7.

19. Wittschieben, B. O., Iwai, S., and Wood, R. D. (2005) DDB1-DDB2 (xeroderma pigmentosum group E) protein complex recognizes a cyclobutane pyrimidine dimer, mismatches, apurinic/apyrimidinic sites, and compound lesions in DNA. *J Biol Chem* **280**, 39982–9.

20. Kulaksiz, G., Reardon, J. T., and Sancar, A. (2005) Xeroderma pigmentosum complementation group E protein (XPE/DDB2): purification of various complexes of XPE and analyses of their damaged DNA binding and putative DNA repair properties. *Mol Cell Biol* **25**, 9784–92.

21. Rapic Otrin, V., Kuraoka, I., Nardo, T., McLenigan, M., Eker, A. P. M., Stefanini, M., Levine, A. S., and Wood, R. D. (1998) Relationship of the xeroderma pigmentosum group E DNA repair defect to the chromatin and DNA binding proteins UV-DDB and replication protein A. *Mol Cell Biol* **18**, 3182–90.

22. Wakasugi, M., Kawashima, A., Morioka, H., Linn, S., Sancar, A., Mori, T., Nikaido, O., and Matsunaga, T. (2002) DDB accumulates at DNA damage sites immediately after UV

irradiation and directly stimulates nucleotide excision repair. *J Biol Chem* **277**, 1637–40.

23. Luijsterburg, M. S., Goedhart, J., Moser, J., Kool, H., Geverts, B., Houtsmuller, A. B., Mullenders, L. H., Vermeulen, W., and van Driel, R. (2007) Dynamic in vivo interaction of DDB2 E3 ubiquitin ligase with UV-damaged DNA is independent of damage-recognition protein XPC. *J Cell Sci* **120**, 2706–16.

24. El-Mahdy, M. A., Zhu, Q., Wang, Q. E., Wani, G., Praetorius-Ibba, M., and Wani, A. A. (2006) Cullin 4A-mediated proteolysis of DDB2 protein at DNA damage sites regulates in vivo lesion recognition by XPC. *J Biol Chem* **281**, 13404–11.

25. Fitch, M. E., Nakajima, S., Yasui, A., and Ford, J. M. (2003) In vivo recruitment of XPC to UV-induced cyclobutane pyrimidine dimers by the DDB2 gene product. *J Biol Chem* **278**, 46906–10.

26. Kapetanaki, M. G., Guerrero-Santoro, J., Bisi, D. C., Hsieh, C. L., Rapic-Otrin, V., and Levine, A. S. (2006) The DDB1-CUL4A^{DDB2} ubiquitin ligase is deficient in xeroderma pigmentosum group E and targets histone H2A at UV-damaged DNA sites. *Proc Natl Acad Sci U S A* **103**, 2588–93.

27. Rapic-Otrin, V., McLenigan, M. P., Bisi, D. C., Gonzalez, M., and Levine, A. S. (2002) Sequential binding of UV DNA damage binding factor and degradation of the p48 subunit as early events after UV irradiation. *Nucleic Acids Res* **30**, 2588–98.

28. Groisman, R., Polanowska, J., Kuraoka, I., Sawada, J., Saijo, M., Drapkin, R., Kisselev, A. F., Tanaka, K., and Nakatani, Y. (2003) The ubiquitin ligase activity in the DDB2 and CSA complexes is differentially regulated by the COP9 signalosome in response to DNA damage. *Cell* **113**, 357–67.

29. Petroski, M. D., and Deshaies, R. J. (2005) Function and regulation of cullin-RING ubiquitin ligases. *Nat Rev Mol Cell Biol* **6**, 9–20.

30. Hannah, J., and Zhou, P. (2009) Regulation of DNA damage response pathways by the cullin-RING ubiquitin ligases. *DNA Repair* **8**, 536–43.

31. Guerrero-Santoro, J., Kapetanaki, M. G., Hsieh, C. L., Gorbachinsky, I., Levine, A. S., and Rapic-Otrin, V. (2008) The cullin 4B-based UV-damaged DNA-binding protein ligase binds to UV-damaged chromatin and ubiquitinates histone H2A. *Cancer Res* **68**, 5014–22.

32. Imoto, K., Kobayashi, N., Katsumi, S., Nishiwaki, Y., Iwamoto, T. A., Yamamoto, A., Yamashina, Y., Shirai, T., Miyagawa, S., Dohi, Y., Sugiura. S., and Mori, T. (2002) The total amount of DNA damage determines ultraviolet-radiation-induced cytotoxicity after uniformor localized irradiation of human cells. *J Invest Dermatol* **119**, 1177–82.

Part III

Detection in Live Tissues, Blood, Urine, Sperm

Chapter 13

Ultrasound Imaging of Apoptosis: Spectroscopic Detection of DNA-Damage Effects at High and Low Frequencies

Roxana M. Vlad, Michael C. Kolios, and Gregory J. Czarnota

Abstract

A new noninvasive method for the detection of DNA damage using mid-to high-frequency ultrasound (10–60 MHz) has been developed. Ultrasound imaging and quantitative analysis methods are used to detect cell death occurring in response to anticancer therapies in cell samples *in vitro*, in rat brain tissue *ex vivo*, and in cancer mouse models *in vivo*. Experimental evidence indicates that the mechanism behind this ultrasonic detection is linked to changes in the size and acoustic properties of the cell nucleus occurring with forms of cell death, and in particular apoptosis. Nuclear changes associated with cell death can result in up to 16-fold increase in ultrasound backscatter intensity and changes in spectral slope that are consistent with theoretical predictions. Furthermore, color-coded images can be generated based on specific ultrasound parameters in order to identify the regions of cell death in tumor ultrasound images with treatments. These results provide a foundation for future investigations regarding the use of ultrasound in preclinical and clinical settings to noninvasively monitor tumor responses to specific interventions that induce cell death.

Key words: Apoptosis, Cell death, Ultrasound, Spectroscopy, Radiation, Photodynamic therapy, High-frequency ultrasound, Low-frequency ultrasound

1. Introduction

1.1. High-Frequency Ultrasound Imaging of DNA damage

This chapter describes a novel noninvasive method for the detection of DNA damage. Depending on the cell type, the type of DNA damage and cellular environment, cells may die by different mechanisms (i.e., apoptosis, oncosis, mitotic arrest/catastrophe) when exposed to different type of injuries. Each of these mechanisms is characterized by distinct biochemical processes leading to distinct cellular and nuclear structural changes (1–3). For example, nuclei condense and fragment in apoptotic cell death, whereas in mitotic arrest the nuclei increase in size. The diversity of the processes involved in cell death makes it difficult to find a mechanism common to all different forms of cell death that can be potentially exploited as an image-based biomarker of cell death. Therefore, current standard clinical routines for detecting cell death still require tissue excision (e.g., biopsy) and specialized tissue staining. These specialized types of staining may detect one specific form of cell death (e.g., apoptosis) but may not detect other modes of cell death (e.g., mitotic arrest/catastrophe, oncosis).

Our research has demonstrated the capability of high-frequency ultrasound to detect apoptosis *in vitro* and *in vivo* (4–8), mitotic arrest/catastrophe *in vitro* (9), and oncosis/necrosis *ex vivo*, the latter in livers exposed to ischemic injury (10). This type of noninvasive detection of cell death has applications to numerous clinical scenarios. For instance, it could be used to monitor tumor responses to anticancer therapies (8), guide tissue biopsies to areas of response and assess organ viability for transplantation (10).

Furthermore, apoptosis or programed cell death has become essential to the understanding of the development of normal tissues, the evolution of the carcinogenic process and the response of tumors to anticancer therapies (11, 12). This chapter describes the techniques of using mid- to high-frequency ultrasound to detect apoptosis and other forms of cell death *in vitro*, *ex vivo*, and *in vivo* in cell samples and animal systems exposed to different anticancer therapies.

The first section of this chapter introduces the basic principles of ultrasound imaging and quantitative ultrasound methods relevant to the process of detecting cell death. The next two sections describe the experimental systems, the materials, and the protocols for monitoring cell death using high-frequency ultrasound. In Subheading 4, we discuss the methods and present representative results. Lastly, we conclude with a discussion about the limits and limitations of the method and future developments.

1.2. Background

1.2.1. Ultrasound Imaging

A variety of information can be extracted from verbal communication, music, and other sources of sound. Sound is formed by pressure waves that range from 10 Hz to 20 kHz, whereas ultrasound is formed from pressure waves at frequencies above the human audible range. In order to understand ultrasound interactions with biological tissues, a tissue may be regarded as a collection of small structures with different acoustic properties (e.g., cells, connective tissue, and microvasculature). An ultrasound transducer sends pulses of sound into the tissue. Where the sound wave encounters structures with different mechanical properties, part of the sound is reflected and part of it detected by the transducer. The strength of the reflected signal and its frequency content depend on the properties of these structures (e.g., size, density, and compressibility). Therefore, ultrasound is highly sensitive to the mechanical properties of tissue structures that often change during diseases or with cell death. Medical diagnostic ultrasound is a modality that uses ultrasound energy to map the variations in the acoustic properties of the body and display these as general grayscale images.

Diagnostic imaging is commonly performed using ultrasound in the range of 1–10 MHz because these frequencies can ensure the visualization of deep body structures of interest with a reasonable resolution (approximately 0.15–1.5 mm). Frequencies greater than 10 MHz have been used to increase the resolution of ultrasound images in diverse applications, including intravascular, skin, ocular and, recently, small animal imaging (13). Within the range of frequencies of 10–60 MHz, high-frequency ultrasound offers a high spatial resolution up to 25 µm and can enable longitudinal studies in mice with applications in developmental biology, cardiovascular disease (13–15), tumor growth (16, 17) angiogenesis and assessing antiangiogenic drug effects (18).

Experimentally, ultrasound imaging and spectrum analysis techniques have been applied by our group (4, 5, 19) to detect cell death in cell samples and tissues exposed to cancer therapies. Specifically, within the frequency range of 10–60 MHz, ultrasound imaging and tissue characterization techniques have been used to detect cell death, *in vitro* with cell samples exposed to the chemotherapeutic drug cisplatinum (4, 5, 7), *ex vivo* (4, 10), and *in vivo* using tumor mouse models exposed to cancer therapies (4, 8).

The resolution of ultrasound imaging, even at the frequencies of 10–60 MHz is below the resolution offered by optical microscopy. However, ultrasound confers the advantage of a relatively low attenuation permitting deeper penetration of ultrasound waves into the tissue (i.e., 1–5 cm deep) compared to the micron depths obtained using optical microscopy. Rather than resolving individual structures (e.g., cells and nuclei) the parameters computed from the analysis of ultrasound signals give information about the statistical properties of tissue structure at the scale of a cell. In the next section, we provide a brief description of ultrasound spectrum analysis methods used in this chapter to characterize cellular death in cell samples and biological tissues.

1.2.2. Quantitative Ultrasound Methods for the Characterization of Cell Death

Large-scale structures (larger than the ultrasound wavelength) can be clearly imaged with ultrasound because simple tissue interfaces, such as organ boundaries, produce well-defined backscatter signals. However, most structures of interest, including organ parenchyma and tumors contain a complex spatial distribution of small scatterers (smaller than the ultrasound wavelength) that result in a complex distribution of tissue mechanical properties. Ultrasound images reflect this complexity, exhibiting an average grayscale level (indicative of average scattering strength) with a speckle pattern that depends on tissue microstructure characteristics.

This differing pattern is useful in the location of lesions, such as fibrotic tissue and detecting cysts, but cannot give information on its own about the microscopic structure of tissue scatterers.

The microscopic properties of tissue scatterers can, however, be quantified using signal processing methods to analyze radio-frequency data (before applying an envelope detection used to construct conventional B-mode ultrasound images), (Fig. 1a). Characterization of tissue microstructure by examining the frequency-dependent information is termed quantitative ultrasound. Ultrasound parameters that further quantify tissue properties can be computed by comparing ultrasound measurements obtained from experiments with theory.

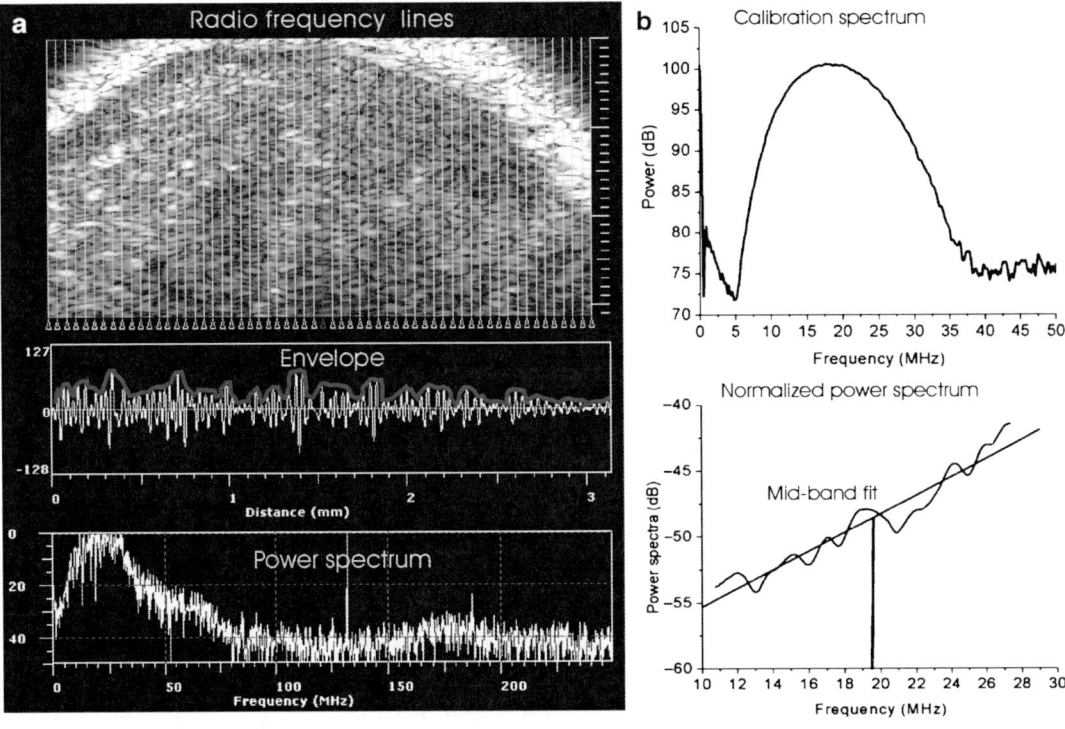

Fig. 1. (**a**) This *panel* describes the acquisition of radio-frequency data. The *upper panel* represents a selection from an ultrasound image with the scanning radio-frequency lines (pre-image data) detected by the ultrasound transducer; the *middle panel* represents the radio-frequency signal with the outline of the signal (envelope) corresponding to one of the scan lines. In order to create the ultrasound images typically displayed by the conventional ultrasound machines, these raw radio-frequency data undergo different signal processing steps, e.g., envelope detection. The *bottom panel* represents the power spectrum corresponding to the same scan line from the *upper panel*. (**b**) This *panel* describes the method of ultrasound spectrum analysis. The *upper panel* represents the calibration spectrum which is a power spectrum measured from a flat quartz immersed in water and placed at the transducer focus as described in Subheading 3.4. The normalized power spectrum (*lower panel*) is calculated by dividing the power spectrum calculated from a region of interest by the calibration spectrum. Linear regression analysis is then applied to the resulting spectrum. The spectral slope (dB/MHz) and the spectral intercept (dB) are the slope and the intercept of this line, respectively. The ultrasound-integrated backscatter (dB) is calculated by integrating the normalized power spectrum over the transducer's −6 dB bandwidth. This is typically similar to the mid-band fit that is the value of the regression line calculated at a frequency corresponding to the middle of the bandwidth.

In order to characterize cell death in cell samples and tissues, ultrasound spectrum analysis method developed by Lizzi et al (20–22) have been used. This method is briefly presented in Fig. 1b. Averaged power spectra measured from various tissues exhibit a quasilinear shape when expressed in dB (23–29). Therefore, one approach to analyzing such data is to apply linear regression analysis to the measured average power spectra (21). The linear fit to the averaged power spectra provides three parameters that can be used to characterize tissue properties, i.e., slope, mid-band fit and intercept. These parameters can be related to different histological properties of tissue scatterers as detailed in Subheading 3.4. The theoretical framework of ultrasound spectrum analysis and the relationship of ultrasound estimates with tissue structures have been described in various books and publications (20–22, 30, 31).

Experimental evidence from our previous work has indicated that the main scattering structure at 10–60 MHz is the cell nucleus (6, 8–10, 32). Assuming the nucleus the dominant scatterer, changes in ultrasound spectral parameters are a direct consequence of changes in nuclear size and acoustic properties. For example, nuclei change their size and properties during cell death, i.e., condense and fragment during apoptosis and increase in size during mitotic cell death. Therefore, it should be possible to detect and differentiate such different forms of cell death.

2. Materials

2.1. High-Frequency Ultrasound Imaging

A VisualSonics VS40B high-frequency ultrasound device (VisualSonics Inc., Toronto, Ontario, Canada, http://www.visualsonics.com) was used to collect ultrasound images and radio-frequency data from cell samples and tumor xenografts (6). A newer version of the VS40B device, the Vevo 770, has seen wide-spread adoption by many small animal imaging facilities (Fig. 2a). We use such instruments configured with single element transducers and a "digital RF" module capable of acquiring, digitizing, and exporting raw radio-frequency data for spectral analysis. New high-frequency linear array transducers (frequencies from 9 to 70 MHz) optimized for specific research applications have been recently released commercially (Vevo 2100). Such instruments are termed "micro-ultrasound" imaging systems and are dedicated to *in vivo* longitudinal imaging studies for small animal phenotyping. The transducer choice on this range of instruments depends on system capabilities, application and the specimen to be imaged. For most of the applications described here, one can use two or three transducers to image a single specimen in order to cover a larger range of frequencies.

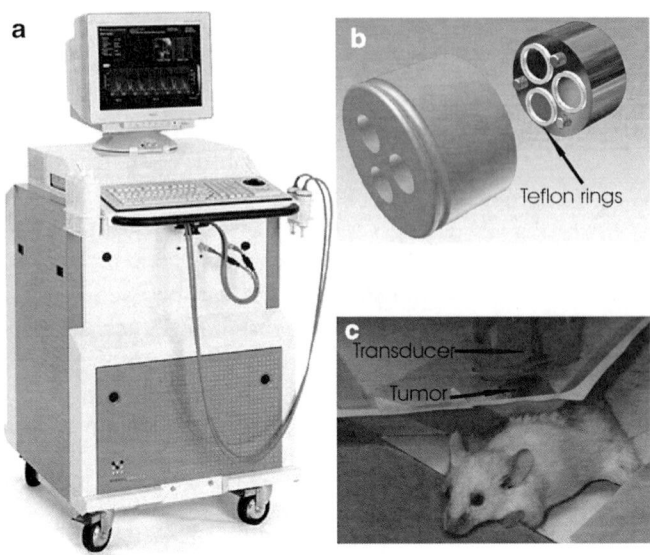

Fig. 2. (**a**) The Vevo 770 high-resolution microimaging system for real time small animal imaging. Similarly to its VS40B predecessor, the VEVO 770 has the ability to be used visualize and quantify small animal anatomical, hemodynamic and therapeutic targets and intervention effects with resolution of up to ~30 μm. Using these instruments, the region of therapeutic intervention can be monitored in the same animal over time (i.e., before, during, and after therapy). The Vevo 770 has an automated 3D acquisition capability. The data presented in this chapter are collected with a VS40B that features analog to digital conversion with 8-bits at 500-MHz sampling frequency. (**b**) Photograph of the sample holder. The cell sample holder is made of stainless steel and has two parts: a flat bottom very finely polished to have the surface roughness much smaller than the ultrasound wavelength and a stainless steel disk with three cylindrical holes cut through it, each 8×7 mm (diameter × height). The flat bottom is attached with screws to the stainless steel disk, and the bottoms of the wells are cushioned with Teflon rings. These Teflon rings ensure that the content of the wells is not spilled between compartments during the sample centrifugation. Two samples can thus be prepared once by simultaneous centrifugation of each in a sample well. The other well left empty serves as calibration reference. (**c**) The experimental setup for ultrasound data collection from a mouse tumor. The leg bearing the tumor is immersed in the coupling liquid (distilled and degassed water at room temperature) and data are collected from the entire tumor. Similarly, the sample holder containing the cell samples is immersed in phosphate-buffer saline and ultrasound data are collected from the cell samples.

2.2. DNA Damage

The model systems described in this chapter are:

1. Acute myeloid leukemia cells (AML-5) in which apoptosis is induced by treating cells with the chemotherapeutic drug cisplatinum.
2. Rat brain in which apoptosis is induced by treatment with photodynamic therapy.
3. Human tumor xenografts grown in mice in which apoptosis is induced by treatment with radiotherapy.
4. Hep-2 (epidermoid carcinoma of the larynx) cells in which mitotic arrest/catastrophe is induced by treating cells with radiotherapy.

2.2.1. Apoptosis in Cell Samples

1. AML-5 cells: available from the Ontario Cancer Institute, 610 University Avenue, Toronto, ON, Canada M5G 2M9.
2. Cisplatinum: a stock solution available at a concentration of 1 mg/mL should be stored in dark (Mayne Pharma).
3. Alpha-minimal essential medium for growing cells (Invitrogen Inc.).
4. Antibiotics: penicillin and for growing cells (Novapharm Biotech Inc).
5. Fetal bovine serum for growing cells (Cansera International Inc.).
6. Flat-bottom cryotubes (1 mL) or a custom designed sample cell holder (Fig. 2).

2.2.2. Apoptosis in Tissues

2.2.2.1. Photodynamic Therapy

1. Male Fischer rats (Charles River Laboratories Inc).
2. Photofrin II, a porphrin-derivative which confers sensitivity of cells to light (QLT, Canada).
3. Laser (632 nm) with an optical power irradiance of 100 mW/cm^2.
4. For anesthesia purposes: 100 mg/kg Ketamine, 5 mg/kg, and 1 mg/kg Acepromazine.

2.2.2.2. Radiotherapy

1. For animal experiments male severe combined immunodeficient (SB-17, SCID) mice (Charles River Laboratories Inc).
2. For anesthesia purposes: 100 mg/kg Ketamine, 5 mg/kg, and 1 mg/kg Acepromazine.
3. Depilatory cream Nair (Nair).
4. For cell culture radiation experiments: C666-1 cell line provided by Dr. Fei-Fei Liu's laboratory, (Princess Margaret Hospital, 610 University Avenue, Toronto, ON, Canada M5G 2M9), and RPMI 1640 cell culture media (Invitrogen Inc.) with antibiotics and sera as above.

2.3. Other Modalities of Cell Death, Mitotic Arrest/Catastrophe

1. Hep-2 cell culture (American Type Culture Collection, Manassas, USA).
2. Antibiotics as above, trypsin (Invitrogen).

2.4. Delivery of Radiotherapy

A small animal and cell irradiator Faxitron Cabinet X-ray System (Faxitron X-ray Corporation, Wheeling, IL, USA) is used to deliver radiotherapy. The instrument was used to deliver 160 keV, X-rays at a rate of 200 cGy/min and should be calibrated before use.

2.5. Flow-Cytometry, Measurement of DNA Content

1. Cell lysis buffer for cytometry (0.2% Triton X-100 in PBS-citrate, 0.1 mg/mL RNase A, 0.05 mg/mL propidium iodide).

2.6. Analysis of Ultrasound Backscatter Signals

For the analysis of backscatter signals, we developed customized programs in Matlab (The Mathworks Inc., Natick, MA, USA). Analysis of backscatter signals can be carried out using a variety of standard numerical analysis packages on the computing platform of one's choice. These are left to the individual researcher. Our preference is to use Matlab running on a 2.5 GHz Windows XP (Microsoft, Washington) or UNIX-based computer system. See Subheading 3 for explicit details.

3. Methods

DNA Damage (leading to apoptosis, mitotic arrest, or mitotic catastrophe) may be induced by exposing cell samples, rat brain, and tumor xenografts to different anticancer therapies. Users are free to use other materials and reagents to induce apoptosis and mitotic arrest/catastrophe. For example, in some of our previous experiments, AML cells were treated with colchicine that arrests cells in mitosis (5, 6) and Hep-2 cells were exposed to camptothecin that induces mainly apoptosis in that cell line (33).

For *in vitro* imaging of cell samples, adequate numbers of cells required to provide a sample of packed cells of at least 150 mm^3 in volume. Potentially, any cell line may be used. To date, we have worked with leukemia, melanoma, breast, cervix, prostate, and head and neck cancer cell cultures. Cell lines that do not grow in liquid suspension culture, once confluent, need to be trypsinized in order to free the cells from the growth surface of the flask. In the experiments described in this chapter, we have used AML and Hep-2 cell lines to demonstrate apoptosis and mitotic arrest/catastrophe. These protocols can be adapted for other cell culture systems. For any experimental time or condition, experiments are typically completed in triplicate.

3.1. Apoptosis in Cell Samples

1. For each experiment, AML cells obtained from frozen stock samples are cultured at 37°C in 150 mL alpha minimum essential media and antibiotics (100 mg/L penicillin and 100 mg/L streptomycin) with 5% fetal bovine serum. Frozen stock is used as this cell line is unstable.
2. To induce apoptosis, AML-5 cells are exposed to cisplatinum at 10 µg/mL for 0, 6, 12, 24, and 48 h. This drug is a DNA intercalator that causes a p53-dependent apoptosis in this cell line (34). To confirm and quantify the presence of apoptosis, we use different methods: the examination of cell cultured using phase-contrast microscopy, the examination of hematoxylin and eosin-stained samples or cell cycle analysis using flow-cytometry.
3. Wash viable and treated cells in phosphate-buffered saline and subsequently centrifuged in three steps at $800 \times g$, $1,500 \times g$,

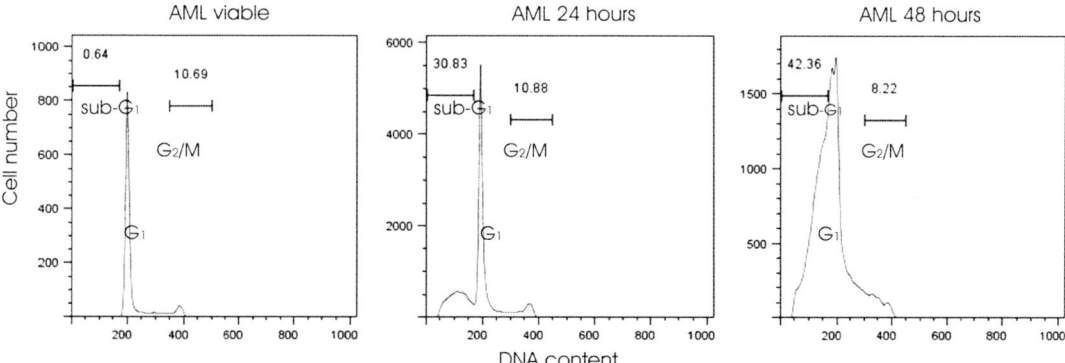

Fig 3. Cell cycle analysis corresponding to AML cells treated with cisplatin for 0, 24, and 48 h. The sub-G_1 fractions are identified as apoptotic cells because the nucleus becomes fragmented during apoptosis. The sub-G_1 peak can represent, in addition to apoptotic cells, mechanically damaged cells and cell fragments resulting from advanced stages of cell death. The sub-G_1 peak increased from 0.6% in viable AML cell samples to 30.8 and 42.4% after 24 and 48 h, respectively, exposure to cisplatin. The G_1 peak represents phenotypic normal cells and the G_2/M peak is identified as cells in mitosis and mitotic arrest. No significant changes are observed in the G_2/M peak with the treatment (10.7% at 0 h; 10.9% at 24 h; and 8.2% at 48 h).

and $1,900 \times g$. After each centrifugation, the cell culture media is aspirated and cells are washed in phosphate-buffer saline. The last centrifugation takes place in cryotubes or in a custom sample holder (Fig. 2b) resulting in a small solid sample of packed cells that emulates a segment of tissue.

4. Immerse the sample holder or the cryotube in phosphate-buffered saline, which acts as a coupling medium through which ultrasonic waves propagate. Each of the wells or tubes containing the viable and treated cells is imaged with high-frequency ultrasound (see Notes 1 and 2).

Representative results are presented in Figs. 3 and 4 and discussed in Notes 3–8.

3.2. Other Modalities of Cell Death

1. We use Hep-2 cells obtained from frozen stock samples in 15 mL alpha-minimum essential media with 0.1% gentamycin and 10% fetal bovine serum and grown in a humidified atmosphere at 37°C, containing 5% CO_2 for experiments as they can be induced to undergo different forms of cell death.

2. Irradiate flasks containing cell samples with 8 Gy radiation dose using a Faxitron Cabinet X-ray System.

3. Structural changes that are characteristic of apoptotic and mitotic response (i.e., increase in cellular and nuclear size, membrane ruffling, cytoplasmic vacuolization, nuclear fragmentation, and condensation and formation of apoptotic and mitotic bodies) are used as an indication of response to radiotherapy. These structural changes are observed in the Hep-2 cell line typically 48 h after exposure to radiotherapy.

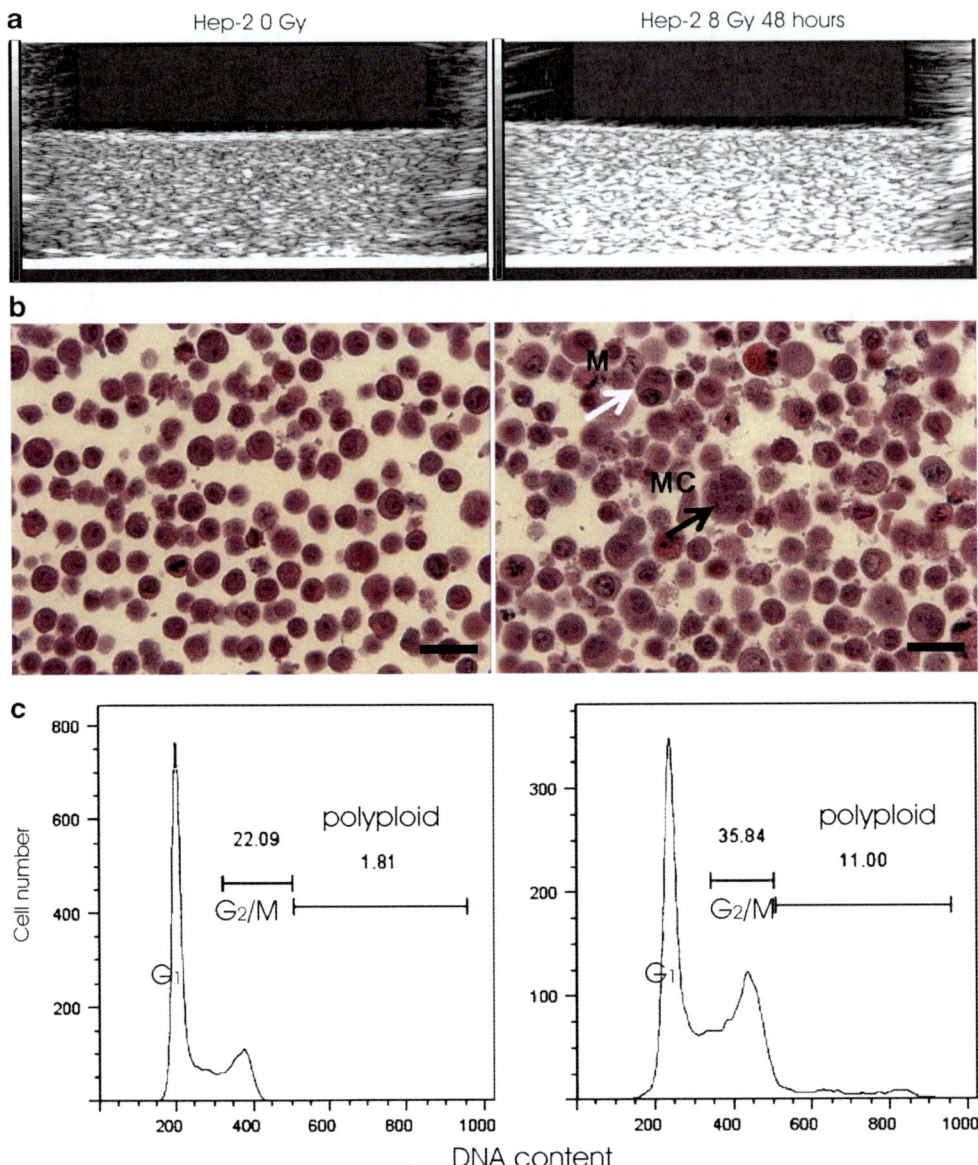

Fig. 4. Ultrasound imaging of mitotic arrest/catastrophe, *in vitro*, corresponding histology and cell cycle analysis. (**a**) The ultrasound images indicate an increase in ultrasound backscatter of the cell sample treated with radiotherapy. (**b**) The corresponding hematoxylin and eosin images demonstrate cells and nuclei with larger size consistent with cell death by mitotic arrest (*white arrow*) and mitotic catastrophe (*black arrow*). The scale bar represents 40 μm. (**c**) The corresponding cell cycle analysis indicates an increase in the mitotic cell fraction quantified by the G_2/M fraction (22.1% at 0 Gy and 35.8% at 8 Gy) and a sixfold increase in the polyploid cell fraction (1.8% at 0 Gy and 11.9% at 8 Gy) after radiotherapy. No sub-G_1 fractions are detectable.

4. Harvest viable and control cell samples by trypsinization and subsequent centrifugation at $800 \times g$. To prepare the cells for high-frequency imaging, cells are subsequently centrifuged and washed in phosphate-buffer saline as described at 0 from steps 3 to 5.

Representative results are presented in Fig. 4 and discussed in Notes 9 and 10.

3.3. Apoptosis in Tissues

Apoptosis is induced *ex vivo* in a rat brain system model treated with photodynamic therapy and *in vivo* in mouse tumor xenografts treated with radiotherapy.

3.3.1. Photodynamic Therapy in Rat Brain

1. Treat male Fischer rats with 12.5 mg/kg of Photofrin II injected intraperitoneally to photosensitize and keep animals in dark for 24 h prior to drug activation.
2. Anesthetize animals using ketamine, xylazine, and acepromazine injected intraperitoneally. Anesthesia consists of 100 mg/kg ketamine, 5 mg/kg xylazine, and 1 mg/kg acepromazine, typically, in 0.1 mL saline injected intravenously. This mixture can sedate the mice for approximately 1 h, sufficient time for the entire imaging and irradiation procedure.
3. Generate a 5.5 mm craniotomy in each side of the rat's skull with a mechanical surgical drill, taking care to avoid significant mechanical stress to the underlying cortex.
4. Treat exposed brain for 30 s using a red laser light with a wavelength of 632 nm and a spot size of 3 mm in diameter. This spot size allows simultaneous visualization of the treated region next to an untreated region in the 4-mm field of view of the high-frequency ultrasound scanner.
5. Several treatment irradiance powers of 1, 3, 5, and 17 J/cm^2 may be used. In order to minimize cerebral swelling post-therapy, the irradiance power of 3 J/cm^2 was selected in the past for further study. The optical irradiance power at the dural surface is 100 mW/cm^2.
6. To study the treatment effects at different time points, the animals are sacrificed at 1.5, 3, and 24 h after the application of photodynamic therapy.

Representative results are discussed in Notes 11 and 12.

3.3.2. Radiotherapy in Mice

1. Use C666-1 cells grown in RPMI 1640 cell culture media supplemented with 10% fetal bovine serum and antibiotics (100 mg/L penicillin and 100 mg/L streptomycin) to generate tumor xenografts from (~1.0×10^6 cells) cells injected intradermally into the left hind leg of each mouse. Primary tumors are measured weekly with a caliper and are allowed to develop for approximately 4 weeks until they reach a diameter of 6–10 mm.
2. Collect ultrasound data over 2 consecutive days, before exposure to radiotherapy and 24 h after exposure to radiotherapy. Other times may be used at the experimenter's discretion.

3. Prior to ultrasound imaging, anesthetize mice (as above) and depilate the tumor and surrounding area using Nair.

4. Immerse the leg bearing the tumor in a coupling liquid (distilled and degassed water at room temperature) and collect ultrasound data typically from the entire tumor (Fig. 2c). Ultrasound images are collected from 10 to 20 different scan planes with a distance between planes of 0.5 mm, by scanning the whole tumor sequentially from one side to another in order to sample the entire tumor. The distance between the radio-frequency lines and scanning planes should be at least one ultrasound beam-width to ensure statistical independence of the data collected for analysis.

5. After ultrasound imaging place the mouse in a standard mouse Lucite restraining box and shield the whole mouse body with 3 mm lead sheets except the tumor. Tumors are then typically irradiated using the Faxitron canet X-ray system with doses of 0, 2, 4, and 8 Gy.

6. After 24 h following exposure to radiation, mouse tumor xenografts are imaged again using ultrasound following the same sequence applied the day prior.

Representative results are shown in Fig. 5 and discussed in Notes 11 and 12.

3.4. Ultrasound Data Collection and Spectrum Analysis

1. In the experiments presented in this chapter, one may use a VS40B high frequency ultrasound instrument with 20-MHz focused transducer (20-mm focal length, 8-mm aperture diameter, –6 dB bandwidth of 11–28 MHz) and a 40 MHz-focused transducer (9 mm focal length, 6-mm aperture diameter, –6 dB bandwidth of 25–55 MHz). This instrument and a newer version are described previously in Subheading 2.

2. Collect radio-frequency data as needed. Radio-frequency data are typically collected from five different scan planes from cell

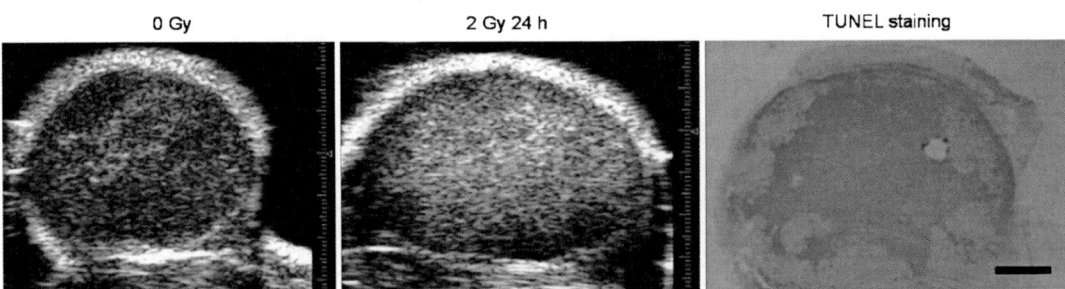

Fig. 5. Ultrasound images of apoptosis, *in vivo*, in response to radiotherapy and corresponding TUNEL staining. Representative ultrasound images of a C666-1 tumor, before and at 24 h after treatment with radiotherapy, presenting a large hyperechoic region after treatment. This region corresponds to the area of cell death in the TUNEL staining. The scale bar represents 1 mm.

samples and from five to ten different scan planes when using larger samples such as biological tissue. Each plane contains 15–40 8-bit radio-frequency lines sampled at 500 MHz and separated by a distance equal to the beam-width of the transducer used in the respective application.

3. Select an appropriate region of interest for analysis. The region of interest chosen to calculate the average ultrasound parameters is typically 4–6 mm wide and 1 mm in height centered at the transducer focus. These regions of interest are selected where the images appear to be homogeneous with no interfaces or large echoes.

4. Ultrasound scan lines from each bracketed line segment are multiplied by a Hamming weighting function to suppress spectral lobes and the Fourier transform is computed. The squared magnitudes of the resultant spectra from all regions of interest are averaged and divided by the power spectrum computed from a flat-quartz calibration target in order to calculate the normalized power spectra. This procedure removes system and transducer transfer function (instrumentation response) and provides a common reference for data collected with various transducers and systems. Linear regression analysis is then applied to the normalized backscatter power spectra in order to calculate the spectral parameters as presented in Fig. 1b. The ultrasound integrated backscatter is similar to the mid-band fit described by the spectrum analysis framework developed by Lizzi et al. (20, 22) and is determined by the effective scatterer size, concentration and difference in acoustic impedance between the scatterers and surrounding medium. The spectral slope can be related to the effective scatterer size (24, 27) (i.e., an increase in the spectral slope corresponds to a decrease in the average scatterer size) and spectral intercept depends on effective scatterer size, concentration, and relative acoustic impedance (Fig. 1). Further details on the theoretical and signal analysis considerations and how spectral parameters are related to tissue microstructure can be found elsewhere (20, 22).

A representative example of spectral analysis and how this can be used to differentiate apoptosis in the regions of the tumor that responded to therapy is presented in Fig. 6 (see Notes 13–21).

3.5. Flow-Cytometry, Cytology and Histology Analysis

3.5.1. Flow-Cytometry Cell-Cycle Analysis

1. Harvest cells, and then wash twice in fluorescent-activated cell sorting buffer (phosphate-buffer saline/0.5% bovine serum albumin). Resuspend cells in 1 mL of FACS buffer and fix for 1 h on ice in 3 mL of ice-cold 70% ethanol. Wash cells once before resuspending them in 500 µL of fluorescent-activated cell sorting buffer supplemented with 40 µg/mL

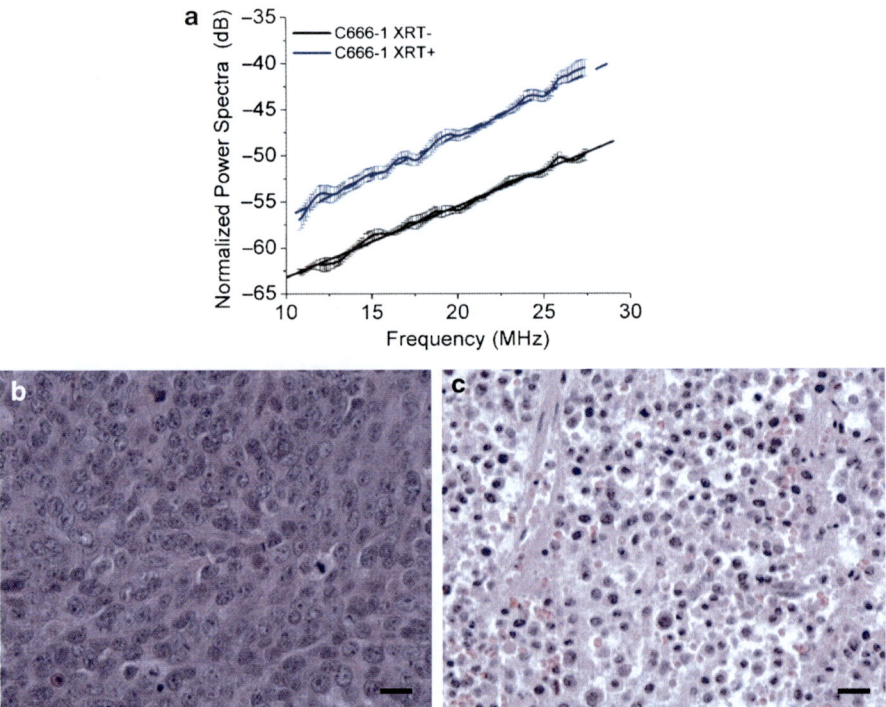

Fig. 6. Ultrasound spectrum analysis and histology demonstrating tumor response to radiotherapy. (**a**) Averaged normalized power spectra increased by 8.2 ± 0.8 dB and averaged spectral slopes increased from 0.77 ± 0.03 to 0.90 ± 0.05 dB/MHz for six C666-1 tumors before (XRT−) and at 24 h (XRT+) after treatment with radiotherapy. The spectra are corrected for attenuation as described in Notes 4–6. Error bars represent the standard error for $n = 6$ tumors. (**b**) Hematoxylin and eosin staining of a tumor tissue before radiotherapy and (**c**) after radiotherapy exhibiting the morphological changes characteristics to apoptosis with smaller condensed and fragmented nuclei. The scale bars represent 20 μm.

RNAse A and 50 μg/mL propidium iodine. Incubate cells at room temperature for 30 min in the dark before being analyzed in a fluorescent-activated cell sorting device using FL-2A and FL-2W channels.

2. Obtained results are typically represented as percentage of cells found in different phases of the cell cycle. This classification is dependent on DNA content and helps to identify the forms of cell death following each of the applied therapies (Figs. 3 and 4).

3.5.2. Cytology and Histology Analysis

1. Fix cell samples, rat brains, and tumor xenografts in 10% neutral-buffered formalin for 24–48 h and embed in paraffin.

2. Excise tumor xenografts and typically section in four locations in the same nominal orientation to best match ultrasound scanning planes. Hematoxylin and eosin for routine histological analysis and terminal deoxynucleotidyl transferase dUTP nick end labeling (TUNEL) staining to assess cell

death are then performed on tumor xenografts and cell samples. In this type of stain, the free 3′-OH DNA ends that result from chromatin fragmentation in apoptotic nuclei are typically labeled with a green fluorescent stain and cytoplasm is marked red with a propidium iodide counterstain.

4. Notes

General Notes

1. All reagents are stored and prepared according to manufacturer's instructions.
2. We used distilled, deionized, and degassed water in all of our animal experiments (resistivity >18.0 MΩ/cm and total organic content of less than ten parts per billion). This standard is referred to as "water" in the text.

Apoptosis in Cells

3. In order to minimize the work required for cell culture and preparation of the experiments, cells that grow in liquid suspension and have a short doubling time are preferred to adherent cells. For example, four to six flasks of adherent cells (175 cm^2 – the area of one flask with 75% cell confluence) are needed to provide a suitable volume of cellular material for a cell sample. We have preferred AML cells for our proof-of-principle experiments because they grow in suspension, have a relative short doubling time of 12 h, and do not need trypsinization. Furthermore, they are a good experimental system for apoptosis studies because the cells have a round regular shape with a large nucleus (~8.5 μm vs. cellular size ~10 μm). This makes it relatively easy to follow nuclear morphological changes under light microscopy.
4. AML cells kept a long time in cell culture (>4 weeks) may change their phenotype due to genetic instability and increase in size (32). This can affect their response to cytotoxic agents (e.g., cisplatin) and affect their ultrasound spectra.
5. Cells need to be carefully washed with phosphate-buffered saline and cell culture medium has to be removed before ultrasound imaging. Cell culture media has tensioactive properties (tends to form bubbles) that can result in significant artifacts when imaging with ultrasound.
6. The volume of a cell sample has to be reduced from 600 mL (four flasks of AML cells) to ~0.45–1.00 mL (the volume of a well in the custom sample holder or cryotube) during the three steps of cell centrifugation. At this stage, the cell material

that is very viscous can be difficult to pipette. The last centrifugation should result in a 0.10–0.20-mL solid cell sample. The recommended height of a sample is at least ~3 mm to facilitate artifact free ultrasound data collection. However, the height of the sample depends on the transducer specifications and needs to be optimized accordingly.

7. To avoid trapping small bubbles at the surface of the cell sample the well or cryotube containing the cell sample has to be gently overfilled with phosphate-buffer saline before placing it into a larger volume of the same solution for ultrasound imaging.

8. For imaging cell samples *in vitro*, we have designed a custom sample holder (Fig. 13.2b). This optimizes the quantity and geometry of cells available for an imaging session and allows data collection in a certain manner for the calculation of other ultrasounds parameters (e.g., speed of sound and attenuation coefficient) (32). For rigorous calculations of cell pellet properties, this (or a similar) setup is required for the appropriate corrections. An alternative technique is to prepare the samples in flat bottom cryotubes. However, these may be too tall (height of the tube > transducer focal length) to access cell samples with certain transducers and may have to be height adjusted.

Other Forms of Cell Death

9. To induce other forms of cell death, we have used head and neck cancer lines that die predominantly by mitotic arrest/catastrophe after exposure to radiotherapy (11). This modality of cell death is characterized by enlarged cells and nuclei, in contrast to nuclear condensation and fragmentation and cellular shrinking characteristic of apoptotic cell death. Minor evidence of cell undergoing apoptosis is observed in the Hep-2 cell culture examined under light microscopy, but no significant apoptotic fraction is measured by cytometry (Fig. 13.4).

10. The Hep-2 cell line has a cell cycle of up to 48 h and up to 50% of the cells express mitotic arrest/catastrophe around this time after treatment. Light microscopy images of cell culture do not reveal significant amount of damage at earlier time points. Based on these observations, we consider that 48 h a good time to observe a significant effect in ultrasound backscatter since keeping the treated cells longer than 2 days in the cell culture would allow the surviving cells to further divide decreasing the chance to effectively image early cell death.

Apoptosis in Tissues

High frequency ultrasound imaging may be used to detect apoptosis that occurred *ex vivo* in a rat model treated with photodynamic therapy. Similar results have been obtained *in vivo* on skin rat treated with photodynamic therapy (4).

11. Any type of treatment able to induce apoptosis in tissue may result in changes in ultrasound images and ultrasound parameters in the treated tissue compared with the same region before treatment. However, it is recommended that the tissue chosen for investigation to have a relative homogeneous structure because ultrasound backscatter is also sensitive to the degree of randomness in scatterer arrangement. In preliminary experiments, we have measured no increase in ultrasound backscatter from cell samples with large degree of randomness in the positions of nuclei although flow-cytometric measurements indicated that a large majority of cells (>50%) underwent death in these cell samples (35).

12. We have found that tumors larger than 10 mm (the maximum dimension) may exhibit large hyperechoic patches in ultrasound images before any treatment. Histological examination confirms that these patches are spontaneous regions of cell death. We consider that such tumors with large necrotic core cannot be used in the experiments as they may bias results. However, we have been able to utilize tumors that exhibited smaller hyperechoic patches (0.5 mm^2). This pattern of small necrotic/apoptotic regions inside the tumor prior to radiation exposure (or any other type of treatment) mimics well some human tumors, and hence we have considered that it is a valid approach to evaluate those tumors in analyses. After radiotherapy, the size of these patches typically increases covering larger tumor regions.

Ultrasound Data Collection and Spectrum Analysis

13. If the goal of the study is to observe the tumor response to therapy, we suggest scanning the tumor in the same orientation from one day to another and then match the tumor histology with the orientation of ultrasound scans.

14. Some readers may find it more convenient to use ultrasound gel instead of degassed and distilled water to ensure the coupling between the transducer and tumor. This is also a good alternative if care is taken to minimize bubble formation in the gel. Bubbles may be removed by centrifugation of gel.

15. An example of ultrasound spectrum analysis characterizing responses to radiotherapy in cancer mouse models is presented in Fig. 6. Provided that several assumptions about the nature of scattering hold (20, 28) the spectral slope can be used to compute an estimated effective scatterer size that is related to the sizes of the scatterers in the analyzed tissue. To calculate the effective scatterer size, we used theory derived by Lizzi et al. (21). The scatterer size estimates, calculated from spectral slopes measured from the same tumor before and after exposure to radiotherapy, yielded values of 9.5 and 5.2 μm, respectively. These values are close to the average size

of nuclei of 10.9 and 6.3 μm estimated from the tumor histology before and after exposure to radiotherapy (Fig. 6).

16. Spatial maps of local estimates of ultrasound spectral parameters (parametric images) can be computed if ultrasound spectrum analysis is applied on much smaller region of interest (i.e., a sliding window with a size of 500 μm and a 90% overlap that progressively analyzes radio-frequency data along the individual scan lines).

17. These parametric images have been able to distinguish areas of the tumor that respond to radiotherapy from the areas of the tumor that do not respond (Fig. 7). This type of imaging may have diverse applications, e.g., noninvasive assessment of cell death in tumors, the detection of tumor regions that respond to therapies, or determining the location of tissue biopsies.

18. To accurately estimate effective scatterer size, it is necessary to account for the attenuation losses due to the propagation of ultrasound waves through cell samples or tissues. To calculate the effective scatterer size in C666-1 tumors, we compensate the normalized power spectra for the attenuation losses in the skin and tumor tissue. The thickness of the tumor skin before and after exposure to radiotherapy is measured from the ultrasound images, yielding values of 0.30 ± 0.06 and 0.45 ± 0.15 mm, respectively. An attenuation coefficient of 0.2 dB/mm/MHz is used for the tumor skin (36). The attenuation coefficient considered for the tumor tissue is 0.06 dB/mm/MHz.

Limitations and Future Developments

19. In ultrasound imaging, the ultrasound frequencies determine the scale of structures that significantly scatter ultrasound,

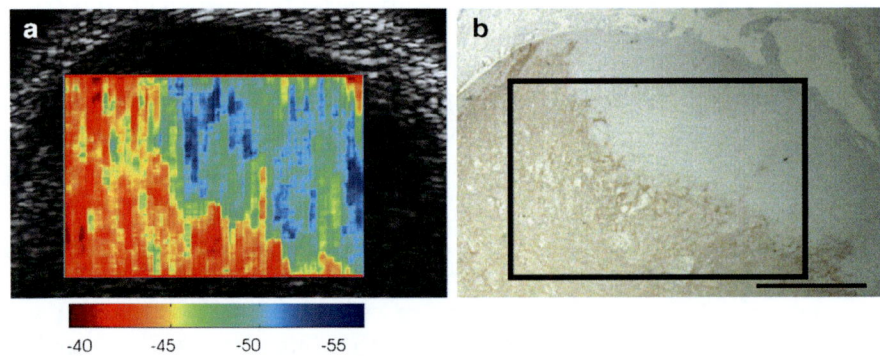

Fig. 7. Spatial map computed from the local estimates of ultrasound-integrated backscatter superimposed on the ultrasound image and corresponding TUNEL staining. The TUNEL-stained image presents an area of cell death of similar shape as the parametric image computed from the local estimates of the ultrasound-integrated backscatter. The frame in the image of the histology corresponds approximately to the spatial map superimposed on the ultrasound image. The color-bar indicates the range of the corresponding estimates. The scale bars represent 1 mm.

thus higher ultrasound frequencies are more sensitive to smaller structures. For the range of the ultrasound frequencies used in this study of (10–60 MHz), the corresponding ultrasound wavelengths of 150–25 μm approach the size of cells and nuclei (10–20 μm), and hence are more sensitive to changes in cellular and nuclear structure than conventional ultrasound. However, at these high ultrasound frequencies, the ultrasound signal is highly attenuated, and hence the penetration depth is typically limited to between 1 and 5 cm. This penetration depth allows the technique to be applicable to skin cancers, certain cancers of the breast and cancer that can be reached with endoscopic probes, such as nasopharyngeal and gastrointestinal cancers. The latest generation of ultrasound imaging systems dedicated to small animal imaging (VEVO 2100) uses pulse encoding techniques that provide better penetration depth at a given frequency. Ongoing studies in our laboratory have investigated the potential of detecting similar effects with lower frequency ultrasound (down to 5 MHz) that may expand the range of applications (37). Figure 8 presents data from corresponding low-frequency and high-frequency ultrasound imaging experiments conducted on mouse tumor xenografts, indicating an increase in the ultrasound backscatter that correlates with the presence of cell death in the histology.

20. Although some of the changes in spectral parameter estimates can be interpreted in terms of changes in nuclear size and acoustic impedance, recent theoretical and experimental evidence indicates that increases in nuclear randomization may in addition influence significantly the magnitude of ultrasound backscatter (35). In these circumstances, an understanding of the relative contribution of each of these nuclear changes (acoustic impendence vs. randomization) to ultrasound scattering is essential in order to accurately quantify cell death. Ongoing studies in our laboratory aim to precisely and reliably measure acoustic properties as a function of treatment (38).

21. In conclusion, ultrasound imaging and quantitative ultrasound methods can be used as a method of detecting noninvasively cell structural changes resulting from DNA damage in a variety of settings. In this chapter, we demonstrate the detection of such DNA damage *in vitro*, *ex vivo*, and *in vivo* following cell death after exposure to various anticancer therapies. High-frequency ultrasound imaging and quantitative ultrasound methods are able to provide very specific information about changes in cellular structure (i.e., differentiate between apoptotic and mitotic cell death). The technique and its extension to the lower frequency range has the potential to noninvasively detect cell death in numerous clinically relevant scenarios, including detecting tumor responses to specific

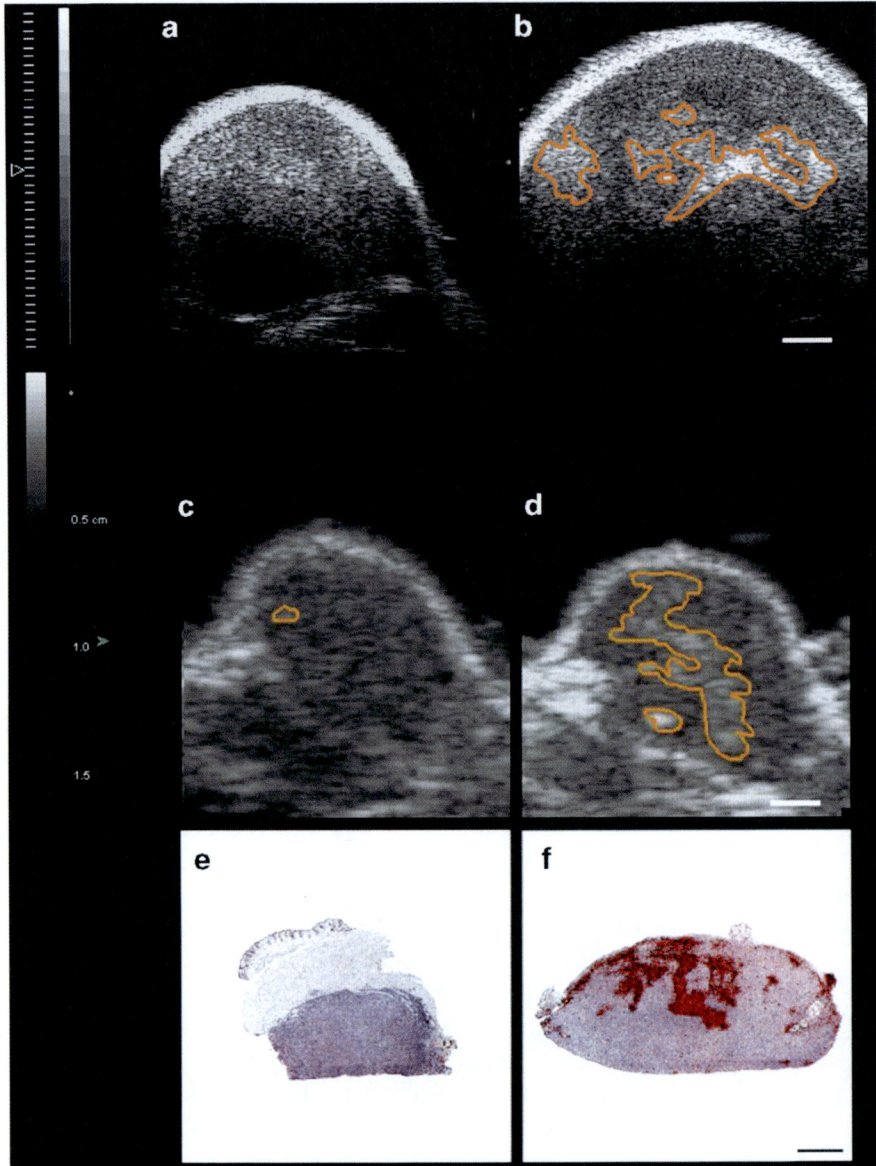

Fig. 8. Ultrasound images of a mouse tumor before and after treatment with an 8 Gy dose of radiation. The tumor is imaged using a transducer with a center frequency of 30 MHz. An increase of 11 dB in the ultrasound-integrated backscatter is found for the area of the tumor outlined in orange compared with the same tumor before treatment. Panels (**a**) and (**b**) are high frequency images of the same tumor before and after radiotherapy obtained with a Visual Sonics instrument operating at 30 MHz. Panels (**c**) and (**d**) are the ultrasound images of the same tumor in a slightly different orientation, before and after radiotherapy, imaged with an Ultrasonix instrument (Ultrasonix Medical Corporation, http://www.ultrasonix.com) using a transducer with a center frequency of 7 MHz. This frequency is close to the ultrasound frequency used in the conventional clinical ultrasound machines. An increase of 6 dB in the ultrasound-integrated backscatter is found for the area of the tumor outlined in orange compared with the same tumor before exposure to the treatment. Panels (**e**) and (**f**) present TUNEL-stained histology images from the same tumors. The regions outlined in orange indicate cell death. The scale bar represents 1 mm.

therapies and guiding tissue biopsies and thus help in developing tools for the clinical management of cancer treatment. Since monitoring therapy response is becoming an essential component of drug development, the ultrasound methods described here can be used in preclinical studies to assess responses to new experimental anticancer therapies in animal cancer models. Imaging of therapy responses in animal cancer models permits repeated assessments of the tumor and may provide spatial and temporal information regarding target organs and the heterogeneity of the response. This information can be correlated with tumor biopsies and histopathological methods to further provide correlative therapy response endpoints related to ultrasound methods.

Acknowledgments

We gratefully thank Drs. Fei-Fei Liu and Deborah Foster for providing some of the cell lines used in this work. We thank Drs. Nehad Alajez, Sebastian Brand, and Anoja Giles for providing technical support. We thank Tennyson and Bradbury D. Bear for continued indefatigable scientific discourse. This research was supported by the American Institute of Ultrasound in Medicine's Endowment for Education and Research Grant, Canadian Institutes of Health Research Strategic Training Fellowship Excellence in Radiation Research for the twenty-first century to RMV; Natural Sciences Engineering Research Council of Canada, Canada Research Chair Programme, Canadian Institutes of Health Research, Canadian Foundation of Innovation/Ontario Innovation Trust, Ryerson University to MCK; Sunnybrook Health Sciences Centre, Natural Sciences Engineering Research Council of Canada, Cancer Care Ontario Cancer Imaging Network of Ontario grants to GJC. Support from a Tier II Canada Research Chair Award and GJC from a Ontario Clinician Scientist Award from the Ontario Ministry of Health and Long-Term Care.

References

1. Van Cruchten S., Van den Broeck W. (2002) Morphological and biochemical aspects of apoptosis, oncosis and necrosis. *Anat Histol Embryol* **31**, 214–223.
2. Darzynkiewicz Z, Juan G, Li X, Gorczyca W, Murakami T, Traganos F. (1997) Cytometry in cell necrobiology: analysis of apoptosis and accidental cell death (necrosis). *Cytometry* **27**, 1–20.
3. Darzynkiewicz Z, Juan G, Bedner E. (2001) Determining cell cycle stages by flow cytometry. *Curr Protoc Cell Biol* Chapter 8:Unit 8.4.
4. Czarnota GJ, Kolios MC, Abraham J, et al. (1999) Ultrasound imaging of apoptosis: high-resolution non-invasive monitoring of programmed cell death in vitro, in situ and in vivo. *Br J Cancer* **81**, 520–527.

5. Kolios MC, Czarnota GJ, Lee M, Hunt JW, Sherar MD. (2002) Ultrasonic spectral parameter characterization of apoptosis. *Ultrasound Med Biol* **28**, 589–597.

6. Czarnota GJ, Kolios MC, Hunt JW, Sherar MD. (2002) Ultrasound imaging of apoptosis. DNA-damage effects visualized. *Methods Mol Biol* **203**, 257–277.

7. Tunis AS, Czarnota GJ, Giles A, Sherar MD, Hunt JW, Kolios MC. (2005) Monitoring structural changes in cells with high-frequency ultrasound signal statistics. *Ultrasound Med Biol* **31**, 1041–1049.

8. Banihashemi B, Vlad RM, Giles A, Kolios MC, Czarnota GJ. (2008) Ultrasound imaging of apoptosis in tumour response: novel monitoring of photodynamic therapy effects. *Cancer Res* **68**, 8590–8596.

9. Vlad RM, Alajez NM, Giles A, Kolios MC, Czarnota GJ. (2008) Quantitative ultrasound characterization of cancer radiotherapy effects in vitro *Int J Radiat Oncol Biol Phys* **72**, 1236–1243.

10. Vlad RM, Czarnota GJ, Giles A, Sherar MD, Hunt JW, Kolios MC. (2005) High-frequency ultrasound for monitoring changes in liver tissue during preservation. *Phys Med Biol* **50**, 197–213.

11. Tannock IF, Hill RP, Bristow RG, Harrington L. The basic science of oncology. New York: McGraw-Hill, 2005.

12. Lockshin RA, Tilly JL, Zakeri Z. When cells die: a comprehensive evaluation of apoptosis and programmed cell death. New York: Wiley-Liss, 1998.

13. Foster FS, Pavlin CJ, Harasiewicz KA, Christopher DA, Turnbull DH. (2000) Advances in ultrasound biomicroscopy. *Ultrasound Med Biol* **26**, 1–27.

14. Foster FS, Zhang MY, Zhou YQ, et al. (2002) A new ultrasound instrument for in vivo microimaging of mice. *Ultrasound Med Biol* **28**, 1165–1172.

15. Foster FS, Zhang M, Duckett AS, Cucevic V, Pavlin CJ. (2003) In vivo imaging of embryonic development in the mouse eye by ultrasound biomicroscopy. *Invest Ophthalmol Vis Sci* **44**, 2361–2366.

16. Graham KC, Wirtzfeld LA, MacKenzie LT, et al. (2005) Three-dimensional high-frequency ultrasound imaging for longitudinal evaluation of liver metastases in preclinical models. *Cancer Res* **65**, 5231–5237.

17. Cheung AM, Brown AS, Hastie LA, et al. (2005) Three-dimensional ultrasound biomicroscopy for xenograft growth analysis. *Ultrasound Med Biol* **31**, 865–870.

18. Goertz DE, Yu JL, Kerbel RS, Burns PN, Foster FS. (2002) High-frequency Doppler ultrasound monitors the effects of antivascular therapy on tumor blood flow. *Cancer Res* **62**, 6371–6375.

19. Czarnota GJ, Kolios MC, Vaziri H, et al. (1997) Ultrasonic biomicroscopy of viable, dead and apoptotic cells. *Ultrasound Med Biol* **23**, 961–965.

20. Lizzi FL, Greenebaum M, Feleppa EJ, Elbaum M, Coleman DJ. (1983) Theoretical framework for spectrum analysis in ultrasonic tissue characterization. *J Acoust Soc Am* **73**, 1366–1373.

21. Lizzi FL. (1997) Ultrasonic scatterer-property images of the eye and prostate. *Proc IEEE Ultrason Symp* **2**, 1109–1118.

22. Lizzi FL, Astor M, Liu T, Deng C, Coleman DJ, Silverman RH. (1997) Ultrasonic spectrum analysis for tissue assays and therapy evaluation. *Int J Imaging Syst Technol* **8**, 3–10.

23. Lizzi FL, Ostromogilsky M, Feleppa EJ, Rorke MC, Yaremko MM. (1987) Relationship of ultrasonic spectral parameters to features of tissue microstructure. *IEEE Trans Ultrason Ferroelectr Freq Control* **34**, 319–329.

24. Lizzi FL1, Feleppa EJ, Kaisar AS, Deng CX. (2003) Ultrasonic spectrum analysis for tissue evaluation. *Pattern Recognit Lett* **24**, 637–658.

25. Ursea R, Coleman DJ, Silverman RH, Lizzi FL, Daly SM, Harrison W. (1998) Correlation of high-frequency ultrasound backscatter with tumor microstructure in iris melanoma. *Ophthalmology* **105**, 906–912.

26. Lizzi FL, Astor M, Feleppa EJ, Shao M, Kalisz A. (1997) Statistical framework for ultrasonic spectral parameter imaging. *Ultrasound Med Biol* **23**, 1371–1382.

27. Feleppa EJ, Kalisz A, Melgar S, et al. (1996) Typing of prostate tissue by ultrasonic spectrum analysis. *IEEE Trans Ultrason Ferroelectr Freq Control* **43**, 609–619.

28. Lizzi FL, Astor M, Kalisz A, et al. (1996) Ultrasonic spectrum analysis for assays of different scatterer morphologies: theory and very-high frequency clinical results. *Proc IEEE 1996 Ultrason Symp* **2**, 1155–1159.

29. Lizzi FL, King DL, Rorke MC, et al. (1988) Comparison of theoretical scattering results and ultrasonic data from clinical liver examinations. *Ultrasound Med Biol* **14**, 377–385.

30. Shung KK, Thieme GA, Ultrasonic Scattering in Biological Tissues, Insana MF, Brown DG, Chapter 4. Boca Raton, FL: CRC Press, 1993.

31. Insana MF, Wagner RF, Brown DG, Hall TJ. (1990) Describing small-scale structure in random media using pulse-echo ultrasound. *J Acoust Soc Am* **87**, 179–192.

32. Taggart LR, Baddour RE, Giles A, Czarnota GJ, Kolios MC. (2007) Ultrasonic characterization of whole cells and isolated nuclei. *Ultrasound Med Biol* **33**, 389–401.
33. Brand S, Solanki B, Foster D, Czarnota GJ, Kolios MC. (2009) Monitoring of cell death in epithelial cells using high frequency ultrasound spectroscopy. *Ultrasound Med Biol* **35**, 482–93.
34. Zamble DB, Lippard SJ. (1995) Cisplatin and DNA repair in cancer chemotherapy. *Trends Biochem Sci* **20**, 435–439.
35. Vlad RM, Orlova V, Hunt JW, Kolios MC, Czarnota GJ. (2008) Changes measured in the backscatter ultrasound signals during cell death can be potentially explained by an increase in cell size variance. *Ultrasonic Imaging* **29**, 256.
36. Shung KK. Diagnostic ultrasound: Imaging and blood flow measurements. Boca Raton, FL: CRC Press, 2005, p. 155.
37. Czarnota GJ, Papanicolau N, Lee J, Karshafian R, Giles A, Kolios MC. (2008) Novel low-frequency ultrasound detection of apoptosis in vitro and in vivo. . *Ultrasonic Imaging* **29**, 237–238.
38. Brand S, Weiss EC, Lemor RM, Kolios MC. (2008) High frequency ultrasound tissue characterization and acoustic microscopy of intracellular changes. *Ultrasound Med Biol* **34**, 1396–1407.

Chapter 14

Quantifying Etheno–DNA Adducts in Human Tissues, White Blood Cells, and Urine by Ultrasensitive ^{32}P-Postlabeling and Immunohistochemistry

Jagadeesan Nair, Urmila J. Nair, Xin Sun, Ying Wang, Khelifa Arab, and Helmut Bartsch

Abstract

Exocyclic etheno–DNA adducts are formed by the reaction of lipid peroxidation products, such as 4-hydroxy-2-nonenal (HNE) with DNA bases to yield 1,N^6-etheno-2'-deoxyadenosine (εdA), 3,N^4-etheno-2'-deoxycytidine (εdC), and etheno-2'-deoxyguanosine. These adducts act as a driving force for many human malignancies and are elevated in the organs of cancer-prone patients suffering from chronic inflammation and infections. Here, we describe the ultrasensitive and specific techniques for the detection of εdA and εdC in tissue and white blood cell (WBC) DNA. This approach is based on combined immunopurification by monoclonal antibodies and ^{32}P-postlabeling analysis. The detection limit is about five adducts per 10^{10} parent nucleotides, requiring 5–10 μg of DNA. In addition, we describe techniques for immunohistochemical detection of εdA and εdC in tissue biopsies, and the approaches for the analysis of εdA and εdC excreted in urine. The utility of these detection methods for human studies is based on: (1) high sensitivity and specificity, (2) low amounts of DNA required, (3) capability to detect "background" levels of etheno–DNA adducts in biopsies, WBC, and urine samples of healthy subjects, and (4) reliable monitoring of the disease-related increase of these substances in patients.

The described methods are useful in diagnosis and monitoring of chronic degenerative diseases, including cancer, atherosclerosis, and neurodegenerative disorders.

Key words: Etheno–DNA adducts, Ultrasensitive detection, Oxidative stress, Lipid peroxidation, Carcinogenesis, Chronic degenerative diseases, Human biomonitoring

1. Introduction

Persistent oxidative stress is currently implicated in over 100 human and animal chronic degenerative diseases (CDD), including cancer, inflammatory, infectious, cardiovascular, and neurological disorders. Oxidative stress enhances lipid peroxidation

(LPO), and by-products are implicated in the initiation and progression stages of CDD, in particular under conditions of chronic inflammation and infections (1). Exocyclic etheno–DNA adducts are formed by the reaction of LPO end products, such as 4-hydroxy-2-nonenal (HNE) with DNA bases dA, dC, and dG and are strong promutagenic DNA lesions causing point mutations (2). Based on immunopurification by monoclonal antibodies combined with ^{32}P-postlabeling, we have developed ultrasensitive and specific detection methods for 1,N^6-etheno-2′-deoxyadenosine (εdA) and 3,N^4-etheno-2′-deoxycytidine (εdC) in tissue and white blood cell (WBC) DNA (3). In addition, we describe methods for the immunohistochemical detection of εdA and εdC in tissue biopsies (4, 5) and for the analysis of human urine samples where εdA and εdC are excreted (6, 7).

Etheno adducts were first detected in target tissues of rodents treated with the carcinogens vinyl chloride or urethane where these promutagenic adducts, if not repaired, initiate carcinogenesis. These adducts are also elevated in cancer (or CDD)-prone organs of patients suffering from chronic inflammation and infections that act as a driving force to many human malignancies.

Levels of etheno–DNA base adducts, often elevated by up to two orders of magnitude, were detected in (1) colon tissue of ulcerative colitis and Crohn's disease patients and in the pancreas of patients with chronic pancreatitis (8), (2) in urine of viral- and alcohol-associated chronic hepatitis liver cirrhosis and hepatocellular carcinoma patients (4), (3) in atherosclerotic plaques of the abdominal aorta from patients with cardiovascular disease (1), and (4) in breast ductal epithelial cells from breast cancer patients (9). High dietary intake of ω-6 polyunsaturated fatty acids (a cancer risk factor) markedly increased etheno adduct levels in WBC of female volunteers (10) and dietary vitamin E and vegetable intake was protective against such damage (11). In alcoholic liver disease patients, ethanol-induced cytochrome P4502E1 expression strongly correlated with hepatic etheno-DNA lesions (12).

Together these studies indicate that human biomonitoring of etheno–DNA adducts, by methods we describe here, holds great promise (1) to diagnose causes and progression of human CDD, including cancer, atherosclerosis, and neurodegenerative disorders and (2) to verify the efficacy of therapeutic and chemopreventive interventions.

The utility of our detection methods for the application in human translational studies is based on: (1) high sensitivity and specificity, (2) only low amounts of DNA required, (3) capable to detect "background" values of etheno–DNA adducts in biopsies, WBC and urine samples of healthy subjects, and (4) reliable monitoring of any disease-related increase in at-risk patients.

2. Materials

2.1. DNA Isolation from Buffy Coat Using a Modified QIAGEN Midi Kit Protocol

1. Isopropanol.
2. Ethanol.
3. 0.22 µm Filters (Millipore).
4. QIAGEN DNA extraction Midi Kit (Qiagen).
5. Desferroxamine mesylate (Sigma-Aldrich).
6. Proteinase K (Roche).
7. RNase A and RNase T1 (Roche).

2.2. Enzymatic Hydrolysis of DNA by MN-SPD (DNA Digestion)

1. Filter Microcon YM 50, (Millipore).
2. Micrococcal nuclease (MN) (Merck).
3. Spleen phosphodiesterase (SPD) (Sigma-Aldrich).
4. HPLC, with C18 column reverse-phase 5 µm (250 × 4.0 mm) column.
5. Tris base (Sigma).
6. Tris–HCl (Sigma).
7. Affi-Gel Hz Immunoaffinity Kit (Biorad).

2.3. Preparation of Immunoaffinity Column

1. CNBr-Sepharose powder.
2. 1 mM HCl.
3. Coupling buffer: (10×) diluted to (1×) in 0.1 M $NaHCO_3$, pH 8.3, containing 0.5 M NaCl.
4. Washing buffer:
   A: 0.1 M acetate buffer, pH 4.0 containing 5.0 M NaCl.
   B: 0.1 M Tris–HCl, pH 8.0 containing 0.5 M NaCl.
5. Polystyrene minicolumn: i.d. 8 mm, length 102 mm, 2 mL bed volume (Pierce).
6. *Commercial Mabs 1G4 for etheno-dA (Santa Cruz)*: Mabs EM-A-1(etheno-dA) and EM-C-1(etheno-dC) from: Prof. Manfred F. Rajewsky and Ms. Kerstin Heise, Institute of Cell Biology (Cancer Research), University of Duisburg-Essen Medical School and West German Cancer Center Essen.

2.4. Immunoaffinity Clean Up

1. Phosphate-buffered saline, pH 7 containing 0.02% NaN_3.
2. Methanol–water 1:1 (v/v).
3. Speedvac.

2.5. TLC Plate Preparation (for Three Plates)

1. Plastic plate 120 × 20 cm.
2. Cellulose powder MN 301 for TLC; (Macherey & Nagel, 816250).

3. 5% polyethyleneimine in double distilled water (ddH$_2$O), pH 6.
4. Mixer.
5. Cellulose spreader.
6. Rubbing tool.

2.6. ^{32}P-Labeling, TLC and Quantitation of Etheno Adducts (εdA, εdC)

1. [γ-^{32}P] ATP 5,000 Ci/mmol.
2. Saturated ammonium sulfate.
3. 1 M Acetic acid, pH 3.5 adjusted with ammonia.
4. TLC boxes with radio-protection.
5. Liquid scintillation counter.
6. TLC plates kept at 4°C for a minimum of 2–3 weeks.
7. T4-polynucleotide kinase.
8. 100 mM Tris–HCl, 20 mM CaCl$_2$ pH 6.8.
9. Standard solutions of normal nucleotide-3′-monophosphates in ddH$_2$O (20 pmol/μL); the concentration in stock solutions is determined from the molar absorption at λ_{260} nm which is for dA = 14.75; for dC = 7.35; for dG = 11.75; and for $T = 8.75 \times 10^3$ L/M/cm.
10. 100 mM ammonium formate, pH 7.5 adjusted with ammonia.

2.7. Detection εdA in Urine by Immunoaffinity-HPLC-Fluorescence

1. HPLC buffer: 25 mM ammonium acetate, pH 6.8, containing 14.2% methanol.
2. Commercial Mabs 1G4 for etheno-dA (Santa Cruz); Mabs EM-A-1(etheno-dA) and EM-C-1(etheno-dC) from: Prof. Manfred F. Rajewsky and Ms. Kerstin Heise, Institute of Cell Biology (Cancer Research), University of Duisburg-Essen Medical School and West German Cancer Center Essen.
3. Rabbit IgG (Sigma-Aldrich, Schnelldorf).
4. Tris–HCl buffer: 10 mM Tris–HCl, pH 7.5, 140 mM NaCl, 3 mM NaN$_3$, containing 1% bovine albumin, and 0.1% rabbit IgG.
5. Methanol–water 1:1 (v/v).
6. HPLC with C18 column reverse-phase 5 μm (250 × 4.0 mm).
7. Creatinine kit (Sigma-Aldrich).
8. εdA standard (Sigma)

2.8. Quantification of εdC in Urine by ^{32}P-Postlabeling/TLC

1. 5′-Bromo-2′-deoxyuridine (BrdU) (Sigma-Aldrich); stock BrdU solution contains 50 fmol/μL (10 μL), add 490 μL ddH$_2$O: final concentration is 1 fmol/μL, add 2 μL to the reaction mixture (see 3.2.2.4).
2. LC-18 solid-phase silica column (500 mg, 3 mL) (Supelco Park, Bellefonta).

3. The multisubstrate deoxyribonucleoside kinase cloned from *Drosophila melanogaster* (provided by Dr. A. Karlsson, Karolinska Institute, Stockholm, Sweden).

4. Methanol–water (1:1 v/v).

5. HPLC with a Hypersil ODS column 5 μm (250×8 mm) connected with an automatic fraction collector.

6. Stock εdC solution contains 5 fmol/μL (5 μL) add 20 μL ddH$_2$O: Final concentration is 1 fmol/μL, add 2 μL to the reaction mixture.

7. X-ray films (Fuji medical X-ray film, 100 NIF, 28×20 cm).

2.9. Immunohistochemical Detection of εdA and εdC in Liver Tissue

1. Tween-20 (polyoxyethylene sorbitan monolaurate).
2. Triton X-100 (*t*-octylphenoxypoly-ethoxyethanol).
3. APES (3-aminopropyl-triethoxysilan).
4. Bovine serum albumin (BSA).
5. Tris–HCl (Sigma-Aldrich).
6. Tris base (Sigma-Aldrich).
7. Hematoxylin.
8. Kaiser's glycerin-gelatin mounting media.
9. Perhydrol (H$_2$O$_2$).
10. Methanol p.A.
11. KH$_2$PO$_4$ p.a. (Merck).
12. Na$_2$HPO$_4$ p.a. (Merck).
13. Biotinylated goat antimouse IgG and avidin–biotin–peroxidase (ABC) complex (Vectastain elite ABC kit).
14. 3,3′-Diaminobenzidine (DAB) substrate kit (Linaris).
15. 4,6-Diamidino-2-phenyl-indole diacetate (DAPI).
16. ImmunoPure (A/G)IgG purification kit (Pierce).
17. Microcon microconcentrators (Millipore).
18. Proteinase K (Roche).
19. RNase (Roche).
20. Commercial Mabs 1G4 for etheno-dA (Santa Cruz); Mabs EM-A-1(etheno-dA) and EM-C-1(etheno-dC) from: Prof. Manfred F. Rajewsky and Ms. Kerstin Heise, Institute of Cell Biology (Cancer Research), University of Duisburg-Essen Medical School and West German Cancer Center Essen, Hufeland-Strasse 55, D-45122 Essen Germany.
21. Blocking-solution: 10% BSA, 2% horse serum, 0.05% Tween-20, 0.05% Triton X-100 in PBS.
22. RNase A solution: Preheat RNase A at 80°C for 10 min, and prepare a solution containing 100 μg/mL in 50 mM Tris base pH 7.4.

3. Methods

3.1. 1,N^6-Ethenodeoxyadenosine (εdA) and 3,N^4-Ethenodeoxycytine (εdC) in WBC-DNA Quantified by Immunoaffinity/^{32}P-Postlabeling (3) (Fig. 1)

3.1.1. DNA Isolation from WBCs Using a Modified QIAGEN Midi Kit

1. Extract DNA from buffy coats corresponding to 5–6 mL of blood using the Qiagen Midi kit and protocol from Qiagen (see Note 1).
2. Prepare digestion buffer (G2) by adding desferroxamine at a concentration of 5 mmol/L.
3. Suspend buffy coat in 1 mL G2 buffer, vortex and mix thoroughly for 5 min.
4. Adjust the total volume to 10 mL with G2 buffer; add 100 µL proteinase K (25 µg/µL in water).
5. Incubate at 37°C for 4 h, and then leave overnight at 4°C.
6. Inactivate DNase activity of RNase A by heating for 10 min at 80°C.
7. Add 25 µL of RNase A (10 µg/µL heated before to inactivate the DNase, see step 6) and 20 µL of RNase T1 (5 U/µL) to the samples. Incubate for 60 min at 37°C.
8. Equilibrate the Midi column with 4 mL equilibrating buffer (QBT buffer and allow the buffer to drain off).
9. Vortex the samples thoroughly for 5 min and load on the equilibrated column.
10. Wash the column with 2 × 7.5 mL QC buffer.

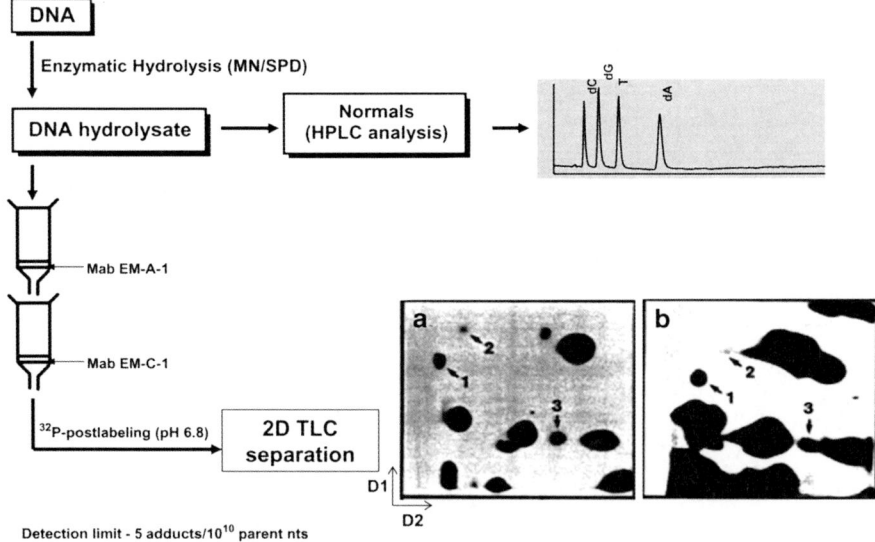

Fig. 1. Scheme depicting the immunoaffinity/^{32}P-postlabeling-TLC method for the analysis of 1,N^6-ethenodeoxyadenosine (εdA) and 3,N^4-ethenodeoxycytine (εdC) in tissue and WBC-DNA. Typical autoradiograms of εdA and εdC on TLC (**a**): Standard εdA (spot no. 1), εdC (no. 2), and internal standard dU (no. 3); (**b**): Corresponding spots from human WBC DNA.

11. Add 1.6 mol/L of NaCl to elution buffer (QF) and adjust the pH with HCl to 7.4 by adding 1.14 g/50 mL buffer and few drops of HCl; before use, heat the QF buffer to 50°C.
12. Elute the DNA with 1 × 5 mL preheated QF buffer.
13. Precipitate the DNA at room temperature by adding 3.5 mL of cooled isopropanol (see Note 2).
14. Centrifuge the tube at 7,500 × g for 10 min, discard the supernatant carefully (do not disturb the pellet).
15. Wash the pellet with 1 mL 70% cold ethanol (kept at 4°C) and air dry the samples.
16. Redissolve the pellet in 0.5 mL water keep at 4°C overnight.
17. Dilute (1–20) a small aliquot of DNA as needed (in duplicate samples) in ddH$_2$O.
18. Measure the OD at 260 nm and record the ratios 230/260 and 260/280 nm; expected yield of DNA per sample is 100 µg (see Note 3).

3.1.2. Enzymatic Hydrolysis of DNA by MN-SPD (DNA Digestion)

3.1.2.1. Reaction

1. Dissolve 5–25 µg of DNA in 92 µL of ddH$_2$O.
2. Add 100 µL of 100 mM Tris–HCl, 20 mM CaCl$_2$, pH 6.8.
3. Add 4 µL (0.2 U/µL) of MN (final concentration 0.004 U/µL).
4. Add 4 µL (0.025 U/µL) of SPE (final concentration 0.0005 U/µL); total volume is 200 µL.
5. Incubate the samples at 37°C with shaking for 4 h (reduced to 3 h in case of low amount of DNA).

3.1.2.2. Filtration

1. Prewash: 100 µL is added to microcon tubes, centrifuge at 10,000 × g for 10 min at 4°C.
2. Discard tube containing washing ddH$_2$O.
3. Load the total solution (200 µL) from the DNA digestion reaction in the microcon tubes, centrifuge at 10,000 × g for 10 min at 4°C.
4. Rinse the microcon tubes with 100 µL ddH$_2$O at 10,000 × g for 10 min at 4°C; after filtration, the total volume of the solution is 300 µL.

3.1.2.3. Separation and Quantification of the Normal Nucleotides

1. Run HPLC in an isocratic elution mode with ammonium formate 100 mM, pH 7.5 (1.0 mL/min), like a Hewlett Packard liquid chromatography system equipped with a UV-monitor at $\lambda = 254$ nm and a C18 LC column.
2. Inject 20 µL of the standards and digested DNA solution in the HPLC system.
3. Quantify the different nucleotides using the standards chromatograms.

3.1.3. Preparation of Immunoaffinity Column

1. Suspend CNBr-Sepharose 4B gel (powder) 0.8 g in 5 mL of 1 mM HCl (minisorp polyethylene immunotube (Nunc)).
2. Wash with 1 mM HCl, 10× with 5 mL each; centrifuge at 700 ×g for 10 min at 4°C.
3. Resuspend in 5 mL of coupling buffer.
4. Wash twice with coupling buffer; centrifuge at 2,000 ×g for 10 min at 4°C.
5. Add an equivalent of 0.5 µg of monoclonal antibody (EM-A-1 for εdA, EM-C-1 for εdC).
6. Stir at 4°C overnight.
7. Centrifuge at 2,000 ×g for 10 min at 4°C.
8. Wash with coupling buffer 5× with 5 mL each.
9. Centrifuge at 2,000 ×g for 10 min at 4°C.
10. Resuspend gel into 5 mL of 0.1 M Tris–HCl buffer, pH 8.
11. Stir end-over-end at 4°C overnight.
12. Centrifuge at 2,000 ×g for 10 min at 4°C.
13. Wash with washing buffer A and B, 5× with 5 mL each.
14. Centrifuge at 2,000 ×g for 10 min at 4°C.
15. Wash twice with 5 mL of PBS buffer containing 0.02% NaN_3.
16. Pour 1/6 aliquots of gel into polystyrene minicolumn.
17. Before use, wash column with PBS buffer containing 0.02% NaN_3 for further use.
18. Fill the immunoaffinity columns with PBS and store them at 4°C.

3.1.4. Immunoaffinity Clean Up

1. Wash the columns sequentially at room temperature with 10 mL water, 15 mL methanol–water, 5 mL water, and 5 mL PBS.
2. When the PBS is almost drained out, take the columns to the cold room and load 280 µL of DNA hydrolysate to each column made of monoclonal antibodies (EM-A-1 for εdA and EM-C-1 for εdC).
3. Allow to drain out and wash the columns with 8 × 5 mL PBS followed by 2 × 5 mL water; discard the washings.
4. Bring the columns to room temperature and elute the etheno adducts (εdA, εdC) from each pair of columns with 2 × 2.5 mL methanol–water.
5. Stir gently the elute and aliquot into 2 × 5 mL and dry in speedvac overnight.
6. Resuspend the sample in 400 µL methanol–water.
7. Transfer to 0.5 mL Eppendorf tube.
8. Dry the tubes in a speedvac and store at −20°C for 3 months maximum.

3.1.5. TLC Plate Preparation (for Three Plates)

3.1.5.1. Preparation of PEI Cellulose

1. Dissolve 67.5 g cellulose powder in 285 mL of ddH$_2$O.
2. Add 60 mL of 5% PEI, pH 6 and mix well.
3. Remove bubbles under vacuum for 1 h.

3.1.5.2. Preparation of Plastic Plates

1. Rub one side of plastic by rubbing tool (till getting equally rough surface).
2. Put on glass plate.
3. Clean twice with ddH$_2$O and dry.
4. Clean once with detergent, without surface modifying substance, clean again with ddH$_2$O twice and dry.
5. Clean with 100% ethanol.

3.1.5.3. Plating

1. Spread PEI cellulose solution with a rubbing tool (thickness 0.3 mm).
2. Let drying overnight.
3. Cut 20 cm long (5–6 plates/1 long plate).
4. Dry in an oven at 50°C for 2 h.
5. Pack in aluminum foil and keep at 4°C for a minimum of 2–3 weeks before use.

3.1.6. ^{32}P-Labeling, TLC and Quantitation of Etheno Adducts (εdA, εdC)

1. Dissolve the dried samples obtained from the immunoaffinity in a 10 μL mixture containing 50 mM Tris–HCI, pH 6.8, 10 mM MgCl$_2$, 10 mM dithiothreitol, 100 amol 3′-dUMP as internal standard (IS), [γ-^{32}P]ATP (10 nCi, spec. act. 5,000 Ci/mmol), and 10 U T4-polynucleotide kinase; keep tubes in ice.
2. Incubate the tubes at 37°C for 90 min, centrifuge and spot onto the left side of prewashed TLC plates at 1.5 cm from the margins (see Note 4).
3. Run the first direction (D1) with 1 M acetic acid, pH 3.5, for 3 h.
4. Dry the plates and clip at 2 cm above the origin to remove the excess of radioactive ATP and develop in the second direction (D2, perpendicular to the first) with saturated ammonium sulfate at 20±2°C, pH 3.5 (adjusted with H$_2$SO$_4$) overnight.
5. Run in parallel plates containing 100 amol each of the standard 3′-εdAMP, 3′-εdCMP, and 3′-dUMP; label and separate in a similar way as described above.
6. Trim at 1.5 cm from the top of the second direction to remove radioactive inorganic phosphate and expose to X-ray films in cassettes supported with intensifying screens at −80°C for 6–12 h.
7. Cut out the sections of the TLC plates corresponding to the spots of the etheno nucleotides (εdA and εdC as 5′-monophosphates) and of 5′-dUMP; determine radioactivity by liquid scintillation counting.

8. Measure the background counts from the appropriate blank areas on the TLC plates and subtract.
9. Calculate the amount of εdA or of εdC (in amol) in the spots by (F2/F1) × 100, whereby F is equal to c.p.m. of εdAMP (or of εdCMP) divided by c.p.m. of dUMP in standard; F2 is equal to c.p.m. of εdAMP (or of εdCMP) divided by c.p.m. of dUMP in sample.
10. Calculate the numbers of εdA or εdC adducts per parent nucleotide in DNA from the ratio: Quantity of etheno adducts measured on TLC over quantity of parent nucleotide in the sample derived from the HPLC-analysis.

3.2. Urinalysis of εdA and εdC

3.2.1. Detection εdA in Urine by Immunoaffinity-HPLC-Fluorescence (6) (Fig. 2)

3.2.1.1. Preparation of Urine Samples

1. Take 2 mL of urine sample and filter through 0.22 μm filter.
2. Spike with [2,8-^{3}H]-1,N^6-ethenoadenine (^{3}H-εA, IS).
3. Add 1 mL of cold ethanol and keep at −20°C for 30 min.
4. Vortex thoroughly for 30 s.
5. Centrifuge at 2,000 × g for 10 min to precipitate proteins.
6. Transfer supernatant to a new 7 mL tube.
7. Dry under vacuum.

3.2.1.2. C18 Column Enrichment

1. Dissolve in 0.5 mL of ddH$_2$O.
2. Load onto a C18 column (500 mg, 3 mL).
3. Prewash with 10 mL of methanol following 10 mL of ddH$_2$O.
4. After loading the sample, wash with 10 mL of NaH$_2$PO$_4$, pH 6.8, followed by 10 mL of ddH$_2$O and 10 mL of 10% methanol–water to remove the bulk of normal nucleosides at 4°C.

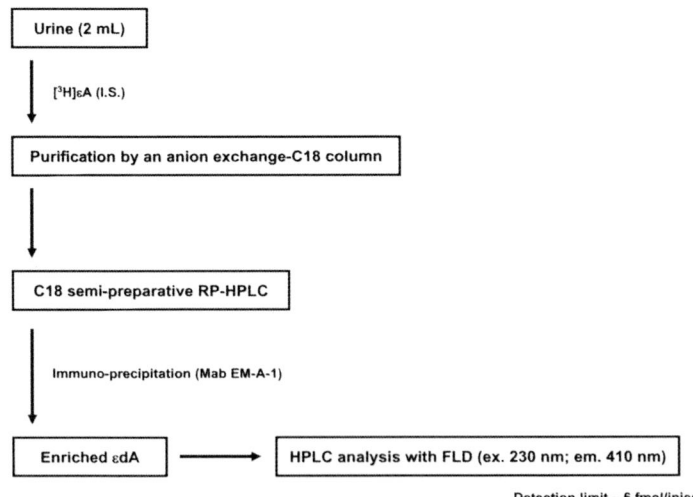

Fig. 2. Scheme depicting the immunoaffinity-HPLC-fluorescence method for the detection of εdA in human urine.

5. Elute with 5 mL of 50% methanol at room temperature.
6. Dry under vacuum.

3.2.1.3. Separation by Preparative HPLC

1. Dissolve the dried sample in 100 μL methanol, centrifuge, carefully remove the supernatant and dry it.
2. Dissolve the dried sample in HPLC buffer and resolve onto a semipreparative C18 column (260 × 8 mm i.d.).
3. Select the fractions containing the IS [^{3}H]-εA (fraction 1) and εdA (fraction 2).
4. Count radioactivity of fraction 1 containing [^{3}H]-εA for recovery correction.
5. Concentrate fraction 2 containing εdA by vacuum centrifugation followed by immunopurification for HPLC-FD analysis.

3.2.1.4. εdA Determination in Urine

1. Redissolve the dried sample in Tris–HCl buffer and monoclonal antibody EM-A-1.
2. Precipitate the antigen–antibody complex with saturated ammonium sulfate.
3. Wash the precipitate with methanol:water (1:1 v/v).
4. Dry under vacuum.
5. Analyze the residue by HPLC consisting of an HP1100 pump with an HP1046A fluorescence detector (Hewlett Packard).
6. Detect εdA by $\lambda_{excitation}$ 230 nm/$\lambda_{emission}$ 410 nm and quantify from a standard curve using standard εdA.
7. Measure urinary creatinine levels by picric acid-based method using a kit according to supplier's protocol; normalize εdA concentration using urinary creatinine levels and express as fmol εdA/μmol creatinine.

3.2.2. Quantification of εdC in Urine by ^{32}P-Postlabeling/TLC (7) (Fig. 3)

3.2.2.1. Preparation of Urine Samples

1. Take 1 mL of urine sample and filter through 0.22 μm filter.
2. Spike with [2.8-^{3}H]-1, N^6-ethenoadenine (^{3}H-εA), as an IS to check the HPLC purification yield.
3. Add 1 mL of cold ethanol and keep at −20°C for 30 min.
4. Vortex thoroughly for 30 s.
5. Centrifuge at 2,000 × g for 10 min to precipitate proteins.
6. Transfer supernatant to a new 7 mL tube.
7. Dry under vacuum.

3.2.2.2. C18 Column Enrichment

1. Dissolve in 0.5 mL of ddH$_2$O.
2. Load onto a C18 column (500 mg, 3 mL).
3. Prewash with 10 mL of methanol followed by 10 mL of ddH$_2$O.
4. Load sample, wash with 10 mL of NaH$_2$PO$_4$, pH 6.8, followed by 10 mL of ddH$_2$O and 10 mL of methanol:water (1:10 v/v) to remove the bulk of normal nucleosides at 4°C.

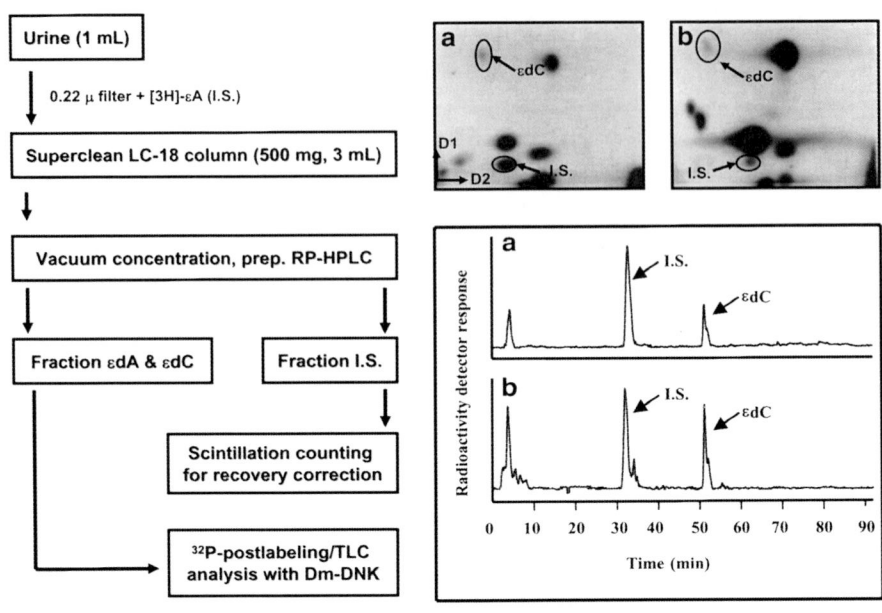

Fig. 3. Scheme depicting the ^{32}P-postlabeling-TLC method for the detection of εdC in human urine. Typical autoradiograms of εdC on TLC (*top panels*) and HPLC profiles (*bottom panels*): (**a**) Standard of εdC and internal standard (IS); (**b**) Human urine sample showing εdC and IS.

5. Elute with 5 mL of methanol:water (1:1 v/v) at room temperature.
6. Dry under vacuum.

3.2.2.3. Separation by Preparative HPLC

1. Dissolve the dried sample in 100 µL ddH$_2$O.
2. Inject into HPLC equipped with 10×250 mm, 5 µm, ODS column.
3. Select fractions: fraction 1 IS ([^{3}H]-εA); fraction 2, εdC.
4. Count radioactivity of fraction 1 containing [^{3}H]-εA for recovery correction.
5. Concentrate fraction 2 containing the εdC by vacuum centrifugation followed by ^{32}P-postlabeling/TLC analysis.

3.2.2.4. ^{32}P-Labeling, TLC and Quantitation of εdC

1. Carry out the ^{32}P-Postlabelling of the urinary εdC as described in Subheading 3.1.6 with minor modifications (3).
2. Add 1 fmol of 5′-bromo-2′-deoxyuridine (BrdU) to the dried samples obtained from pre-HPLC and adjust to 10 µL with a buffer containing 50 mM Tris–HCl, pH 6.8, 10 mM MgCl$_2$, 10 mM dithiothreitol, [γ-^{32}P]ATP (10 µCi, spec. act. 6,000 Ci/mmol), and 0.5 µg deoxyribonucleoside kinase from *Drosophila melanogaster* (Dm-DNK).
3. In parallel, label 2 fmol of εdC standard in separate tubes, in a similar way.

4. Incubate the samples at 37°C for 2 h, centrifuge and spot on the left side of prewashed PEI-cellulose plates.
5. Develop the plates by two-dimensional TLC and expose to X-ray films.
6. Cut out the sections of the TLC plates corresponding to the spots of [^{32}P]-εdC-5′-monophosphate and of [^{32}P]-BrdU-5′-monophosphate and determine the radioactivity by liquid scintillation counting.
7. Determine the background counts using appropriate blank areas on the TLC plates and subtract them.
8. Calculate the fmol of εdC in the spots by the term: (F2/F1)×100, whereby F1 = c.p.m. of εdC/c.p.m. of BrdU (IS) in standard and F2 = c.p.m. of εdC/c.p.m. of BrdU (IS) in sample.
9. Normalize the εdC concentration using urinary creatinine levels and express as fmol εdC/μmol creatinine.

3.3. Immunohistochemical Detection of εdA and εdC in Human and Rat Livers

3.3.1. Immunostaining for εdA and εdC in Human Liver Tissue (4, 12)

1. Using a cryotome cut frozen liver into sections of 6 μm thickness, immediately fix with cold acetone at 4°C for 10 min and air dry.
2. Dip in PBS for 10 min.
3. Dip in 0.3% H_2O_2 in absolute methanol at room temperature for 10 min to quench endogenous peroxidase activity.
4. Wash in PBS for 5 min.
5. Treat with proteinase K (10 μg/mL in ddH_2O) at room temperature for 10 min to remove histone and nonhistone proteins from DNA, to increase antibody accessibility.
6. Wash in PBS for 5 min.
7. Treat the slides with RNase A solution for 1 h at 37°C.
8. Wash in PBS for 5 min.
9. Treat with 4 N HCl for 5 min at RT to denature DNA; rinse with PBS followed by ddH_2O; neutralize the pH by incubating for 5 min at room temperature with 50 mM Tris base buffer, pH 7.4.
10. Block nonspecific binding sites by incubation with a blocking solution at 37°C for 20 min.
11. Incubate with primary antibody (EM-A-1 for εdA detection or EM-C-1 for εdC detection, dilution 1:20) containing 2% horse serum in PBS at 4°C overnight.
12. Wash in PBS for 5 min 3×.
13. Incubate with horse antimouse IgG (Vectastain, dilution 1:400) containing 2% horse serum in PBS at room temperature for 1 h.
14. Wash in PBS for 5 min.
15. Incubate with ABC solution for 30 min at room temperature.

16. Wash in PBS for 5 min.
17. Stain with fresh working solution of DAB for 2 min (see Note 5).
18. Wash in water.
19. Stain with DAPI (2 µg/mL = 6 µL from 1 mg/mL DAPI into 3 mL PBS) for 30 s.
20. Wash in PBS.
21. Cover with Kayser's gelatine.

3.3.2. Immunohistochemical Detection of εdA in Rat Liver Tissue (5)

1. Cut frozen liver using a cryotome (Fig. 4) into 6 µm sections and immediately fix with cold acetone at 4°C for 10 min and air dry.
2. Dip in PBS for 10 min.
3. Dip in 0.3% H_2O_2 in absolute methanol at room temperature for 20 min to quench endogenous peroxidase activity.
4. Wash in PBS for 5 min.
5. Treat with proteinase K (20 µg/mL in ddH_2O) at room temperature for 10 min to remove histone and nonhistone proteins from DNA to increase antibody accessibility.
6. Wash in PBS for 5 min.
7. Treat the slides with RNase A solution for 1 h at 37°C.
8. Wash in PBS for 5 min.
9. Incubate in blocking solution at 37°C for 10 min.
10. Wash in PBS.
11. Denature with 4 N HCl 5 min at room temperature.
12. Dip in ddH_2O then PBS.
13. Neutralize pH with 50 mM Tris base pH 7.4 for 5 min at room temperature.
14. Dip in ddH_2O then PBS.
15. Incubate at 37°C for 20 min, in a solution containing four drops of avidin solution per mL of blocking solution.
16. Rinse in PBS.
17. Incubate in a solution containing four drops of biotin solution per 1 mL of primary antibody solution at 4°C overnight (primary antibody solution: EM-A-1 1:20, and 2% horse serum in PBS).
18. Wash in PBS for 5 min 3×.
19. Incubate with horse antimouse IgG 1:400 and 2% horse serum in PBS at room temperature for 1 h.
20. Wash in PBS for 5 min.
21. Incubate with ABC solution for 30 min at room temperature.
22. Wash in PBS for 5 min.

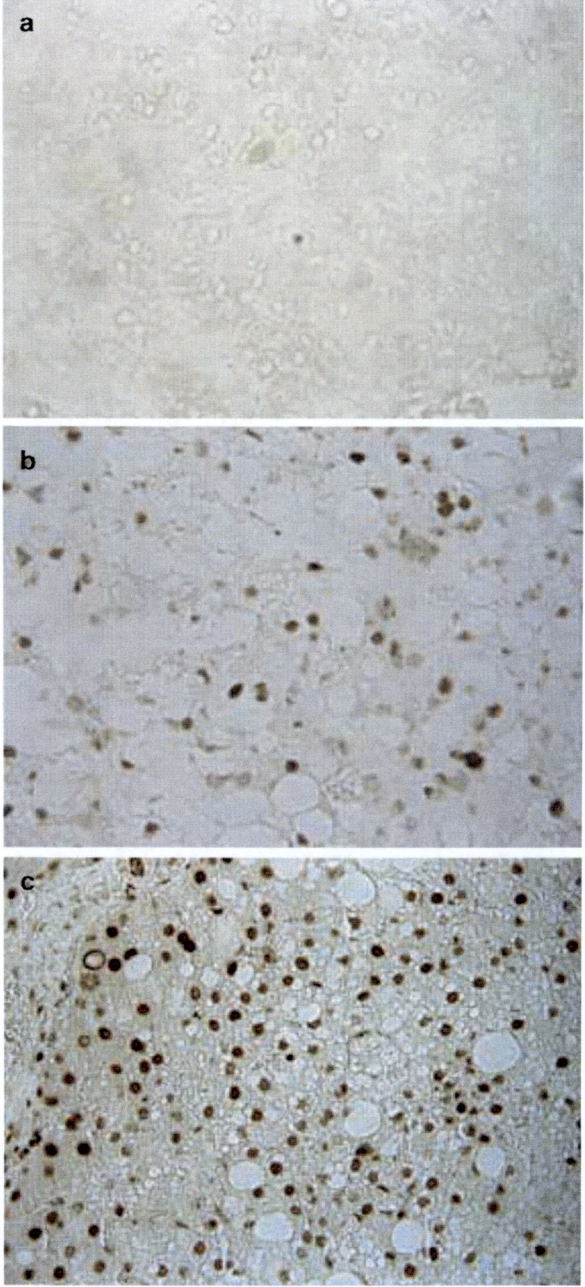

Fig. 4. Immunohistochemical staining of human liver sections showing the formation of εdA adducts in nuclear DNA: (**a**) negative control, (**b**) patient with alcoholic liver disease, and (**c**) patient with chronic hepatitis C viral infection. In these patients, who are prone to liver cancer, the levels of etheno–DNA adducts are often elevated by up to two orders of magnitude.

23. Stain for 2 min with freshly prepared DAB-Ni (see Notes 6 and 7).
24. Wash in water.
25. Stain with DAPI (2 μg/mL = 6 μL from 1 mg/mL DAPI into 3 mL PBS) for 30 s.

26. Wash in PBS.
27. Cover with Kayser's gelatine.
28. Take images and do semiquantitative analysis and statistical evaluation of εdA-prevalence.

3.3.3. Imaging and Semiquantitative Analysis of εdA

Make images with a microscope equipped with a camera Axiocam (Carl Zeiss) for color pictures and the imaging software Axiovision. Measure the relative mean pixel intensity of randomly selected nuclei using the imaging software (Lucia G, Nikon). Count the positively stained cell nuclei that have an arbitrary mean pixel density ≥0.5. Use the ratio of positively stained nuclei over the total number of cells × 100 to express the percentage of relative εdA-prevalence. When the needle biopsies are small, count a minimum of 25 cells per slide.

3.3.4. Statistical Analysis

Test the statistical significance of a difference using the Kruskal–Wallis test; analyze the rank correlations by Kendall's *tau* test. Use a program such as "ADAM" (DKFZ Heidelberg) for statistical evaluation.

4. Notes

1. This kit can be used effectively for a maximum of 15 mL of blood.
2. Sometimes you may not see the floating DNA. In such a case, store the mixture at 4°C overnight.
3. A solution of 200 μg DNA/mL when diluted tenfold will give approx. $OD_{260} = 0.4$.
4. Spot the total volume at one shot, otherwise double spots are observed.
   This is critical, to avoid confusion with εA and εC spots from RNA contamination which run close to εdA and εdC spots.
5. Prepare the working DAB solution using the kit immediately before use. To 5 mL of distilled water, add two drops of buffer stock, four drops DAB stock solution, two drops of Hydrogen peroxide solution, and mix well for 2 min.
6. For preparing the working DAB-Ni add two drops of Nickel solution to the DAB solution (see Note 5) and mix well. *Caution*: Both DAB and nickel chloride are suspected carcinogens. Take appropriate care when using these reagents.
7. DAPI staining is the same for both human and rat liver tissue. DAB staining is slightly different for the two tissues. In the case of rat liver tissue, DAB-Ni is used because nickel salt is necessary to reduce the background and increase the sensitivity.

Acknowledgments

The authors greatly acknowledge the earlier contributions to method development/standardization by A. Barbin, M. Hollstein, and Y. Guichard (IARC, Lyon, France), A. Frank, Y. Yang, S. Dechakhamphu, and M. Meerang (DKFZ, Heidelberg). We thank S. Fuladdjusch for excellent secretarial help and Prof. Manfred F. Rajewsky and coworkers, Institute of Cell Biology (Cancer Research), University of Duisburg-Essen Medical School and West German Cancer Center Essen, Hufeland-Strasse 55, D-45122 Essen Germany for agreeing to provide the antibodies upon request. This article is dedicated to Jagadeesan Nair who passed away prematurely in August 2007. Without his perseverance, these achievements would have not been possible (see obituary: Dr Jagadeesan Nair, Senior Scientist at the German Cancer Research Center (DKFZ). *Carcinogenesis*. 2008; **29**, 887–888).

References

1. Nair J, De Flora S, Izzotti A, Bartsch H. (2007) Lipid peroxidation-derived etheno–DNA adducts in human atherosclerotic lesions. *Mutat Res.* **62**, 95–105.
2. IARC Scientific Publications. (1999) Exocyclic DNA adducts in mutagenesis and carcinogenesis. Proceedings of the 2nd international conference. Heidelberg, Germany, September 1998. *IARC Sci Publ.* **150**, 1–361. Eds. Singer, B. and Bartsch, H.
3. Nair J, Barbin A, Guichard Y, Bartsch H. (1995) 1,N^6-Ethenodeoxyadenosine and 3,N^4-ethenodeoxycytine in liver DNA from humans and untreated rodents detected by immunoaffinity/^{32}P-postlabeling. *Carcinogenesis* **16**, 613–617.
4. Frank A, Seitz HK, Bartsch H, Frank N, Nair J. (2004) Immunohistochemical detection of 1,N^6-ethenodeoxyadenosine in nuclei of human liver affected by diseases predisposing to hepato-carcinogenesis. *Carcinogenesis* **25**, 1027–1031.
5. Yang Y, Nair J, Barbin A, Bartsch H. (2000) Immunohistochemical detection of 1,N^6-ethenodeoxyadenosine, a promutagenic DNA adduct, in liver of rats exposed to vinyl chloride or an iron overload. *Carcinogenesis* **21**, 777–781.
6. Nair J. (1999) Lipid peroxidation-induced etheno–DNA adducts in humans. *IARC Sci Publ.* **150**, 55–61.
7. Sun X, Karlsson A, Bartsch H, Nair J. (2006) New ultrasensitive ^{32}P-postlabelling method for the analysis of 3,N^4-etheno-2'-deoxycytidine in human urine. *Biomarkers* **11**, 329–340.
8. Nair J, Gansauge F, Beger H, Dolara P, Winde G, Bartsch H. (2006) Increased etheno–DNA adducts in affected tissues of patients suffering from Crohn's disease, ulcerative colitis, and chronic pancreatitis. *Antioxid Redox Signal.* **8**, 1003–1010.
9. Nair J, Sun X, Adzersen K, Bartsch H. (2007) Target organ vs surrogate levels of oxidative stress and lipid peroxidation-induced DNA adducts in women with benign and malignant breast diseases. In: Proceedings of the 98th Annual Meeting of the American Association for Cancer Research; Los Angeles, CA: AACR 2007: abstract No. 1719.
10. Nair J, Vaca CE, Velic I, Mutanen M, Valsta LM, Bartsch H. (1997) High dietary omega-6 polyunsaturated fatty acids drastically increase the formation of etheno–DNA base adducts in white blood cells of female subjects. *Cancer Epidemiol Biomarkers Prev.* **6**, 597–601.
11. Hagenlocher T, Nair J, Becker N, Korfmann A, Bartsch H. (2001) Influence of dietary fatty acid, vegetable, and vitamin intake on etheno–DNA adducts in white blood cells of healthy female volunteers: a pilot study. *Cancer Epidemiol Biomarkers Prev.* **10**, 1187–1191.
12. Wang Y, Millonig G, Nair J, Patsenker E, Stickel F, Mueller S, Bartsch H, Seitz HK. (2009) Ethanol-induced cytochrome P4502E1 causes carcinogenic etheno-DNA lesions in alcoholic liver disease. *Hepatology* **50**, 453–461

Chapter 15

ELISpot Assay as a Tool to Study Oxidative Stress in Peripheral Blood Mononuclear Cells

Jodi Hagen, Jeffrey P. Houchins, and Alexander E. Kalyuzhny

Abstract

Enzyme-Linked Immuno Spot (ELISpot) assay is widely used for vaccine development, cancer and AIDS research, and autoimmune disease studies. The output of an ELISpot assay is a formation of colored spots which appear at the sites of cells releasing cytokines, with each individual spot representing a single cytokine-releasing cell.

We have shown that hydrogen peroxide-induced oxidative stress was causing ~twofold decrease in the number of lymphocytes secreting the TH1 cytokines IFN-gamma and IL-2, as well as chemokine IL-8 and cytokine TNF alpha. However, the number of cells secreting TH2 cytokines IL-4 and IL-5 in hydrogen peroxide-treated group did not change. Our ELISpot data indicate that oxidative stress may affect TH1-TH2 cytokine secretion balance which, in turn, may underlie developments of various pathological conditions. We adopted ELISpot assay for studying oxidative stress in human peripheral blood lymphocytes by analyzing the acute effect of hydrogen peroxide treatment on the frequency of cells secreting IFN-gamma, IL-2, IL-4, IL-5, IL-8, and TNF-alpha.

Key words: ELISpot, Oxidative stress, Peripheral blood mononuclear cells, PBMCs, IFN gamma, TNF alpha, IL-2, IL-8, TH1 and TH2 cytokines, Reactive oxygen species, ROS

1. Introduction

Reactive oxygen species (ROS), such as hydrogen peroxide (H_2O_2), can be generated in live cells and tissues either via normal oxidative intracellular metabolism or induced by extracellular toxins which inhibit the activity of antioxidant enzymes. Reduced endogenous antioxidant capacity causes natural overproduction of ROS which, in turn, leads to oxidative stress. H_2O_2 can affect different biochemical reactions in human lymphocytes, including alterations in enzymatic activities, lipid peroxidation, and damage to DNA, and it was reported that short-term exposure of lymphocytes to H_2O_2

suppresses NFκB, AP-1, and NFAT transcription factors which play important roles in regulating the production of cytokines (1).

The objective of this study was to examine the effect of oxidative stress induced by H_2O_2 in unstimulated peripheral blood mononuclear cells (PBMCs) on their capacity to secret IFN-gamma, IL-2, IL-4, IL-5, IL-8, and TNF-alpha. Our oxidative stress model utilized short-term treatment of lymphocytes cultured *in vitro* with H_2O_2 in a concentration that does not impair lymphocyte viability (2). We employed ELISpot assays (3–5) which are more sensitive than ELISA (6) and permit the determination of frequency of cytokine-secreting cells.

ELISpot allows the detection of just a few cytokine-releasing cells out of tens of thousands making this technique a method of choice for vaccine development (7–9), cancer research (10), AIDS research (11–13), allergy research (14), and autoimmune disease studies (15–17).

Our results indicate that H_2O_2-induced oxidative stress significantly inhibits the secretion of the TH1 cytokines IFN-gamma and IL-2, as well as proinflammatory chemokine IL-8 and inflammatory cytokine TNF-alpha (Table 1, Fig. 1). It appeared that acute oxidative stress induced by H_2O_2 did not affect the frequency of cell secreting TH2 cytokines: the regulator of adaptive and humoral immunity IL-4 and B-cell activator IL-5. Interestingly, there was no linear correlation between the number of cultured cells and the number of spots for each cytokine tested (Table 1). The lack of such a correlation is not completely understood and additional studies are required to address this issue.

We find that ELISpot assay can be easily adapted for studying oxidative stress in human lymphocytes and, if combined with multi-donor testing (18), ELISpot can be used for high-throughput screening of multiple oxidative stress substances.

2. Materials

2.1. Isolation of Human PBMCs

1. Ficoll-Paque PLUS (GE Biosciences, St. Giles, UK).
2. 50 mM Phosphate-buffered saline (PBS), pH 7.2.
3. Red Blood Cells lysing solution: 155 mM NH_4Cl cell-lysing solution, 10 mM $NaHCO_3$, and 0.1 mM EDTA.
4. RPMI complete culture medium (1 L) (Gibco-BRL, Grand Island, NY) supplemented with 50 mL of heat-inactivated Fetal Calf Serum (Sigma Chemical Co., St. Louis, MI), 1.19 g Hepes, 2.0 g of Sodium Bicarbonate, 3.5 µL of beta-mercaptoethanol, 50 mg Gentamicin Reagent Solution (Gibco-BRL, Grand Island, NY), and heat-inactivated Fetal Calf Serum (Sigma Chemical Co., St. Louis, MI) (see Notes 1 and 2).
5. Centrifuge allowing spinning 50 mL culture tubes at $500 \times g$.

Table 1
Effect of hydrogen peroxide-induced oxidative stress on secretory capacity of various cytokines from human peripheral blood lymphocytes

Analyte	IFN-gamma		IL-2		IL-8		TNF-alpha		IL-4		IL-5	
Cells/mL	10^4		5×10^4		10^4		5×10^3		5×10^5		10^6	
Treatment	H_2O_2	Control	H_2O_2	Control	H_2O_2	Control	H_2O_2	Control	H_2O_2	Control	H_2O_2	Control
Number of spots ± st. dev.	23.8 ± 1.6	73.2 ± 12.4	207.7 ± 7.5	329.2 ± 41.8	66.3 ± 9.5	131.3 ± 12.3	41.8 ± 21.7	66.2 ± 13.1	15.8 ± 4	17.3 ± 2.4	9.2 ± 2.2	7 ± 1.1
Cells/mL	10^5		2×10^5		5×10^4		5×10^4		2×10^6		5×10^6	
Treatment	H_2O_2	Control	H_2O_2	Control	H_2O_2	Control	H_2O_2	Control	H_2O_2	Control	H_2O_2	Control
Number of spots ± st. dev.	195.7 ± 13	282.7 ± 24.6	236.6 ± 111.6	693.3 ± 24.2	234.2 ± 23	347 ± 33.8	233.3 ± 12.6	401.8 ± 22.3	58.7 ± 9.5	52.8 ± 4.3	29.7 ± 3.3	28.7 ± 4.5

Secretory capacity was analyzed in ELISpot assays by counting the number of spots formed by secreted proteins on the bottom of the 96-well plate

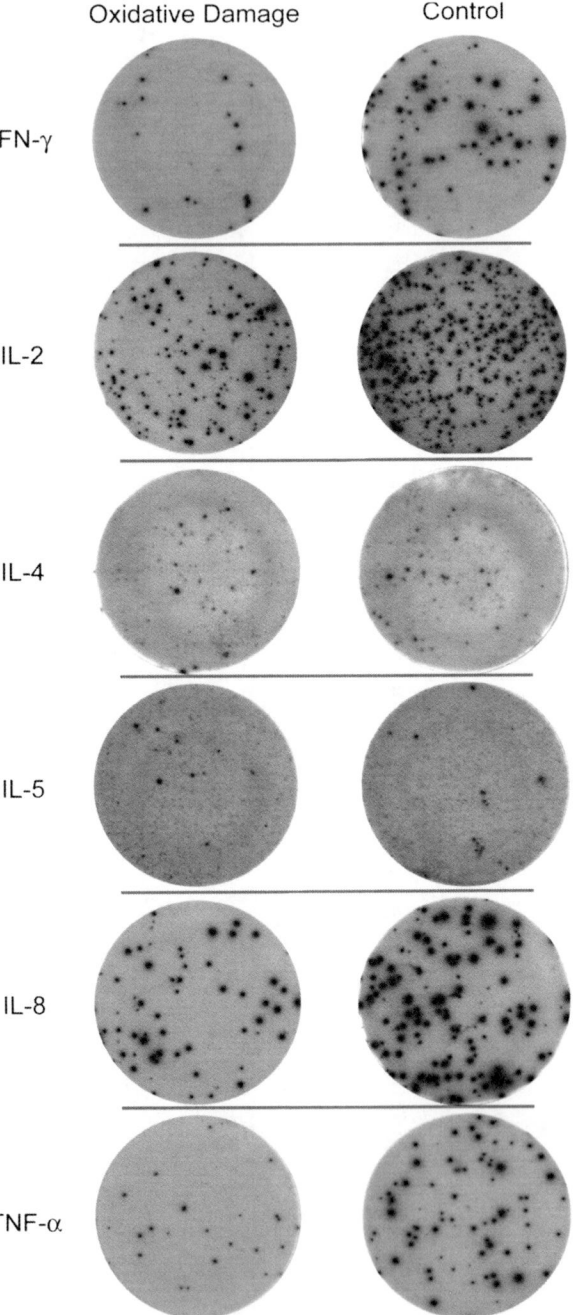

Fig. 1. Typical ELISpot images of secreted cytokines from peripheral blood lymphocytes. Note prominent inhibitory effect of oxidative stress on secretion of all cytokines except IL-4 and IL-5.

6. Hemacytometer (Thermo Fisher Scientific, Rochester, NY) to count lymphocytes under the microscope.

7. Trypan Blue Dye (Gibco BRL, Grand Island, NY).

8. Upright microscope equipped with bright-field illumination and phase contrast condenser.

2.2. Induction of Oxidative Stress

1. Reagent to induce oxidative stress: first prepare 0.5 mM solution of H_2O_2 by adding 0.5 µL of 30% H_2O_2 to 8.8 mL of Hank's Balanced Salt Solution (HBSS; Gibco BRL, Grand Island, NY). Then, prepare 5 µM solution of H_2O_2 by adding 100 µL of 0.5 mM H_2O_2 solution to 9.9 mL of HBSS. Store at 4°C. Calculate the working volume of Oxidative Stress reagent needed for the stimulation of lymphocytes in culture flasks.

2.3. ELISpot Assays

1. Commercially available ready-to-use ELISpot assay kits (R&D Systems, Inc.) to study the secretion of human IFN-γ (Cat # EL285), IL-2 (Cat # EL202), IL-4 (Cat # EL204), IL-5 (Cat # EL205), IL-8 (Cat # EL208), and TNF-α (Cat # EL210). Each kit includes a dry 96-well PVDF membrane-backed plate precoated with capture antibody, a concentrated solution of detection antibody, a concentrated solution of streptavidin-conjugated alkaline phosphatase, BCIP/NBT chromogenic substrate, wash and dilution buffers.

2. Mitogens to stimulate the release of cytokines from cultured PBMCs: Calcium Ionomycin (CaI, #C-7522, Sigma Chemical Co., St. Louis, MI), Phorbol 12-myristate 13-Acetate (PMA, #P-8139, Sigma Chemical Co., St. Louis, MI), Phaseous Vulgaris Red Kidney Bean Phytohemaglutinin (PHA, #L-3897, Sigma Chemical Co., St. Louis, MI).

3. Hand-held Nunc-Immuno™ 12 plate washer (Thermo Fisher Scientific, Rochester, NY).

4. Membrane removal device (MVS Pacific, Minneapolis, MN).

5. ELISpot plate reader QHub (MVS Pacific, Minneapolis, MN).

3. Methods

3.1. Isolation of Human Peripheral Blood Lymphocytes

1. Collect Blood samples from healthy donors in standard citrate-phosphate-dextrose unit bags (Leukopack, Memorial Blood Centers of Minnesota) and separate PBMCs using density centrifugation ($500 \times g$ for 30 min) of 25 mL of blood layered on 20 mL of 1.077 g/mL Ficoll-Paque Plus at 25°C (see Note 3).

2. Discard the upper plasma layer after centrifugation and transfer PBMCs into two sterile 50 mL tubes.

3. PBMCs were then resuspended in 45 mL of sterile PBS and centrifuged for 5 min at $500 \times g$.

4. Discard supernatant and resuspend the pellet in 10 mL of Red Blood Cells lysing solution and incubate for 5 min at room temperature.

5. After lysing, add sterile PBS to reach 50 mL graduation mark on the tube to resuspend PBMCs.

6. Centrifuge tubes for 5 min at $500 \times g$.

7. Discard supernatants and supplement 30–40 mL of RPMI complete medium with heat-inactivated Fetal Calf Serum, Hepes, Sodium Bicarbonate, beta-mercaptoethanol and Gentamicin Reagent Solution, and add to the tubes with PBMCs.

8. Mix cells 1:2 with Trypan Blue dye and pipette 10 μL of that mixture into each side of a hemacytometer under a coverslip (see Note 4). Count cells under the microscope using 20× lens and phase contrast condenser.

3.2. Induction of Oxidative Stress

1. Add H_2O_2 directly to PBMCs in culture media and incubate in 37°C/CO_2 humidified incubator for 30 min.

2. Transfer PBMCs from the incubator into a sterile hood, discard culture media with H_2O_2 and rinse cells three times with sterile culture media.

3. Prepare several serial dilutions of PBMCs and mix them with corresponding mitogens (see Note 5).

3.3. ELISpot Assays

1. Plate PBMCs (100 μL/well; 6 wells per group) into the ELISpot plates at following cell concentrations (see Note 6):
   (a) IFN-gamma assay: 10^4 and 10^5 cells/mL;
   (b) IL-2 assay: 5×10^4 and 2×10^5 cells/mL;
   (c) IL-8 assay: 10^4 and 5×10^4 cells/mL;
   (d) TNF-alpha assay: 5×10^3 and 5×10^4 cells/mL;
   (e) IL-4 assay: 5×10^5 and 2×10^6 cells/mL;
   (f) IL-5 assay: 10^6 and 5×10^6 cells/mL.

2. Stimulate PBMCs with mitogens added directly to cells in ELISpot plates and incubate in a CO_2 incubator at 37°C (see Notes 7 and 8). Use the following stimulation and incubation interval:
   (a) IFN-gamma, IL-2, IL-5 and IL-8: stimulate with a mixture of 0.5 μg/mL of Calcium Ionomycin and 50 ng/mL of PMA for 18 h;
   (b) IL-4: stimulate with 3 μg/mL of PHA for 18 h;
   (c) TNF-alpha: no mitogens added, incubate for 18 h.

3. After finishing the incubation, remove PBMCs from the plates by rinsing wells four times with wash buffer (see Notes 9 and 10).

4. Make working solutions of detection antibodies by mixing concentrated detection antibodies 1:120 with dilution buffer.

5. Add 100 μL of detection antibody working solution into each well and incubate ELISpot plates overnight at 4°C.

6. Wash plates three times with the wash buffer.

7. Prepare working solution of Streptavidin–Alkaline Phosphatase by mixing the concentrated stock solution 1:120 with corresponding dilution buffer.

8. Add 100 μL of Streptavidin–Alkaline Phosphatase working solution into each well and incubate for 2 h at room temperature.

9. Wash plates three times with wash buffer.

10. Add 100 μL of ready-to-use BCIP/NBT substrate into each well and incubate for 30–60 min at room temperature in a place protected from direct light.

11. Wash plates three times with distilled water and let them dry completely (see Note 11).

12. Quantify spots using automated ELISpot reader.

4. Notes

1. Sterilize RPMI complete culture medium and reagents that will be used to separate out the white blood cells through 0.2 μm sterile filter to allow their long-term storage.

2. When using fetal calf serum, it is important to heat inactivate the serum at 56°C for 30 min. After the heat inactivation, the serum should be filtered.

3. When layering ficoll, make sure that the blood does not mix with the ficoll to gain the best separation and the highest yield of PBMCs.

4. Overfilling hemacytometer with cell solution may result in inaccurate cell quantification. While counting cells on a hemacytometer, first find the middle square which contains 25 smaller squares and count cells in five of them. Calculate the average and multiply by 25 (total number of squares in that area), and then multiply by 2 (cell dilution factor), and then multiply by 10,000 to determine the number of cells in 1 mL of original cell suspension. The resulting number should be used for calculating serial dilutions of PBMCs.

5. Making serial dilutions of cells prevents overdevelopment of ELISpot plate and allows to develop a quantifiable number of spots that can be counted either manually or using automated ELISpot plate readers.

6. For better well-to-well reproducibility cells need to be mixed thoroughly before adding them into the wells. This may require shaking the tube with cells after filling every four wells in ELISpot plate.

7. Plates can be wrapped in aluminum foil to provide even heat distribution across the bottom of the ELISpot plates during their incubation. This helps to improve well-to-well spot consistency across the plate (described by Kalyuzhny and Stark [19] and in kit's insert). Aluminum foil also helps to reduce background staining. This is a very simple procedure which can be done as follows: before plating cells, ELISpot plate is placed onto 13 × 16 cm piece of aluminum foil (e.g., Reynolds Wrap Quality Aluminum Foil, Consumer Products Division of Reynolds Metal, Richmond, VA); after that cells are added into the wells and plate is covered with the lid, and edges of the foil are shaped loosely around the edges of the plate to wrap it. After finishing the incubation of the cells, the foil can be removed and either discarded or saved and used on the next ELISpot plate.

8. Shelves in the CO_2 incubator must be leveled to avoid moving cells toward one side of the well: this may produce under and overdeveloped parts of the well and hinder quantification of spots. It is also important to avoid disturbing cultured cells (e.g., by slamming the door of the incubator) during the incubation which may cause the development of weakly stained fuzzy spots.

9. Make sure that the height of prongs in the hand held plate washer is properly adjusted so prongs do not touch the membranes on the bottom of the ELISpot plate: PVDF membranes are fragile and are easily punctured by protruding prongs.

10. Between washes it is important to tap out the excess liquid in the well onto a paper towel to prevent diluting the sequential reagents added into the plate.

11. Plates must be dried completely before analysis because wet membranes appear dark and obscure the detection and quantification of spots.

Acknowledgments

Christopher Hartnett for assistance with the isolation of peripheral blood lymphocytes.

References

1. Flescher, E., Tripoli, H., Salnikow, K., and Burns, F. J. (1998) Oxidative stress suppresses transcription factor activities in stimulated lymphocytes. *Clin. Exp. Immunol.* **112**, 242–247.
2. Flescher, E., Bowlin, T. L., Ballester, A., Houk, R., and Talal, N. (1989) Increased polyamines may downregulate interleukin 2 production in rheumatoid arthritis. *J. Clin. Invest.* **83**, 1356–1362.
3. Czerkinsky, C. C., Nilsson, L. A., Nygren, H., Ouchterlony, O., and Tarkowski, A. (1983) A solid-phase enzyme-linked immunospot (ELISPOT) assay for enumeration of specific antibody-secreting cells. *J. Immunol. Methods* **65**, 109–121.
4. Sedgwick, J. D. and Holt, P. G. (1983) A solid-phase immunoenzymatic technique for the enumeration of specific antibody-secreting cells. *J. Immunol. Methods* **57**, 301–309.
5. Kalyuzhny, A. E. (2005) Chemistry and biology of the ELISPOT assay. *Methods Mol. Biol.* **302**, 15–31.
6. Tanguay, S. and Killion, J. J. (1994) Direct comparison of ELISPOT and ELISA-based assays for detection of individual cytokine-secreting cells. *Lymphokine Cytokine Res.* **13**, 259–263.
7. Pass, H. A., Schwarz, S. L., Wunderlich, J. R., and Rosenberg, S. A. (1998) Immunization of patients with melanoma peptide vaccines: immunologic assessment using the ELISPOT assay. *Cancer J. Sci. Am.* **4**, 316–323.
8. Asai, T., Storkus, W. J., and Whiteside, T. L. (2000) Evaluation of the modified ELISPOT assay for gamma interferon production in cancer patients receiving antitumor vaccines. *Clin. Diagn. Lab. Immunol.* **7**, 145–154.
9. Kamath, A. T., Groat, N. L., Bean, A. G., and Britton, W. J. (2000) Protective effect of DNA immunization against mycobacterial infection is associated with the early emergence of interferon-gamma (IFN-gamma)-secreting lymphocytes. *Clin. Exp. Immunol.* **120**, 476–482.
10. Schmittel, A., Keilholz, U., Thiel, E., and Scheibenbogen, C. (2000) Quantification of tumor-specific T lymphocytes with the ELISPOT assay. *J. Immunother.* **23**, 289–295.
11. Keane, N. M., Price, P., Stone, S. F., John, M., Murray, R. J., and French, M. A. (2000) Assessment of immune function by lymphoproliferation underestimates lymphocyte functional capacity in HIV patients treated with highly active antiretroviral therapy. *AIDS Res. Hum. Retroviruses* **16**, 1991–1996.
12. Chapman, A. L., Munkanta, M., Wilkinson, K. A., Pathan, A. A., Ewer, K., Ayles, H., Reece, W. H., Mwinga, A., Godfrey-Faussett, P., and Lalvani, A. (2002) Rapid detection of active and latent tuberculosis infection in HIV-positive individuals by enumeration of Mycobacterium tuberculosis-specific T cells. *AIDS* **16**, 2285–2293.
13. Eriksson, K., Nordstrom, I., Horal, P., Jeansson, S., Svennerholm, B., Vahlne, A., Holmgren, J., and Czerkinsky, C. (1992) Amplified ELISPOT assay for the detection of HIV-specific antibody-secreting cells in subhuman primates. *J. Immunol. Methods* **153**, 107–113.
14. Jakobson, E., Masjedi, K., Ahlborg, N., Lundeberg, L., Karlberg, A. T., and Scheynius, A. (2002) Cytokine production in nickel-sensitized individuals analysed with enzyme-linked immunospot assay: possible implication for diagnosis. *Br. J. Dermatol.* **147**, 442–449.
15. Pelfrey, C. M., Cotleur, A. C., Lee, J. C., and Rudick, R. A. (2002) Sex differences in cytokine responses to myelin peptides in multiple sclerosis. *J. Neuroimmunol.* **130**, 211–223.
16. Bienvenu, J., Monneret, G., Fabien, N., and Revillard, J. P. (2000) The clinical usefulness of the measurement of cytokines. *Clin. Chem. Lab. Med.* **38**, 267–285.
17. Okamoto, Y., Gotoh, Y., Tokui, H., Mizuno, A., Kobayashi, Y., and Nishida, M. (2000) Characterization of the cytokine network at a single cell level in mice with collagen-induced arthritis using a dual color ELISPOT assay. *J. Interferon. Cytokine Res.* **20**, 55–61.
18. Bailey, T., Stark, S., Grant, A., Hartnett, C., Tsang, M., and Kalyuzhny, A. (2002) A multidonor ELISPOT study of IL-1beta, IL-2, IL-4, IL-6, IL-13, IFN-gamma and TNF-alpha release by cryopreserved human peripheral blood mononuclear cells. *J. Immunol. Methods* **270**, 171–182.
19. Kalyuzhny, A. and Stark, S. (2001) A simple method to reduce the background and improve well-to-well reproducibility of staining in ELISPOT assays. *J. Immunol. Methods* **257**, 93–97.

Chapter 16

Cytokinesis-Block Micronucleus Cytome Assay in Lymphocytes

Philip Thomas and Michael Fenech

Abstract

The cytokinesis-block micronucleus cytome (CBMN cyt) assay is a new and comprehensive technique for measuring DNA damage, cytostasis, and cytotoxicity in different tissue types, including lymphocytes. DNA damage events are scored specifically in once-divided binucleated cells. These events include; (a) micronuclei (MNi), a biomarker of chromosome breakage and/or whole chromosome loss; (b) nucleoplasmic bridges (NPBs), a biomarker of DNA misrepair and/or telomere end-fusions; and (c) nuclear buds (NBUDs), a biomarker of elimination of amplified DNA and/or DNA repair complexes. Cytostatic effects are measured via the proportion of mono-, bi-, and multinucleated cells and cytotoxicity via necrotic and/or apoptotic cell ratios. The assay has been applied to the biomonitoring of *in vivo* exposure to genotoxins, *in vitro* genotoxicity testing and in diverse research fields, such as nutrigenomics and pharmacogenomics. It has also been shown to be important as a predictor of normal tissue and tumor radiation sensitivity and cancer risk. This protocol also describes the current established methods for culturing lymphocytes, slide preparation, cellular and nuclear staining, scoring criteria, data recording, and analyses.

Key words: Micronucleus, Lymphocyte, Cytome, Nucleoplasmic bridge, Nuclear bud, DNA damage, Cytotoxicity

1. Introduction

Micronuclei (MNi) originate from chromosome fragments or whole chromosomes that lag behind at anaphase during nuclear division (1–3). The cytokinesis-block micronucleus assay is the preferred method for measuring MNi in cultured human and/or mammalian cells because scoring is specifically restricted to once-divided binucleate cells, which are the cells that express MNi (3, 4). In this assay, once-divided cells are recognized by their binucleate appearance after blocking cytokinesis with cytochalasin-B, an inhibitor of microfilament ring assembly required for the completion of cytokinesis (3–5). Restricting microscopic analysis of

MNi in binucleate cells prevents confounding effects caused by suboptimal or altered cell division kinetics. This is a major variable in micronucleus assay protocols that do not distinguish between nondividing cells that cannot express MNi and dividing cells that can (6, 7). As a result of its reliability and reproducibility, the cytokinesis-block micronucleus assay has become one of the standard cytogenetic tests for genetic toxicology testing in human and mammalian cells.

Over the past 17 years, the cytokinesis-block micronucleus assay has evolved into a comprehensive method for measuring chromosome breakage, DNA misrepair, chromosome loss, nondisjunction, necrosis, apoptosis, and cytostasis (3, 8–14) This method is now also used to measure nucleoplasmic bridges (NPBs), a biomarker of dicentric chromosomes resulting from telomere end-fusions and/or DNA misrepair, and to measure nuclear buds (NBUDs), a biomarker of gene amplification (15–18). The "cytome" concept implies that every cell in the system studied is scored cytologically for its viability status (necrosis, apoptosis), its mitotic status (mono-, bi-, and multinucleated) and its chromosomal damage or instability status. For these reasons, it is now appropriate to refer to this technique as the cytokinesis block micronucleus cytome (CBMN Cyt) assay (Fig. 1).

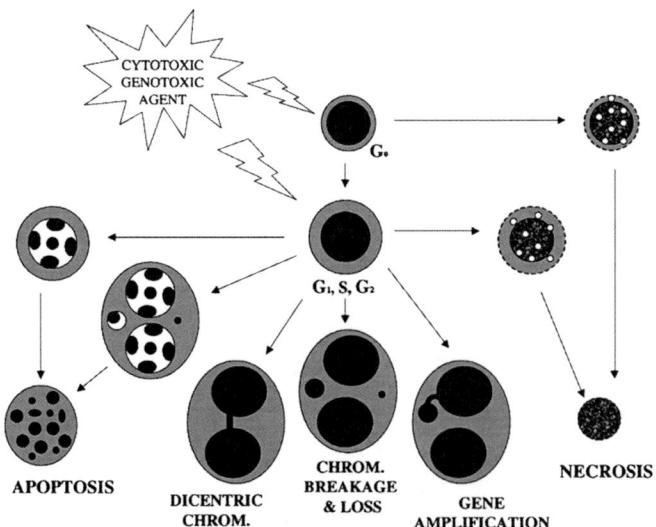

Fig. 1. The various possible fates of cultured cytokinesis-blocked cells following exposure to cytotoxic/genotoxic agents. Using these biomarkers within the CBMN assay, it is possible to measure the frequency of chromosome breakage (MNi), chromosome loss (MNi), chromosome rearrangement, e.g., dicentric chromosomes (NPB), gene amplification (nuclear buds), necrosis, and apoptosis. In addition, cytostatic effects are readily estimated from the ratio of MONO, BN, and MULT cells. Chromosome loss can be distinguished from chromosome breakage using pancentromeric probes or kinetochore antibodies. In addition, nondisjunction (malsegregation of chromosomes) can also be measured in BN cells using chromosome-specific centromeric probes.

2. Materials

2.1. Reagents

1. Ficoll–Paque, sterile. *Caution*: This product is very stable if bottle remains unopened; however, it can deteriorate if exposed to air for long periods of time. For best results, always date the bottles, use 100 mL bottles and remove from the bottle slowly using a needle and syringe without breaking the seal. Store at 4°C. Use at room temperature (20–22°C).
2. Hank's balanced salt solution (HBSS), sterile, with calcium and magnesium, without phenol red. Store at 4°C and use at room temperature.
3. RPMI 1640 medium, without l-glutamine, sterile liquid. Store at 4°C. Use at 37°C when preparing cultures.
4. Fetal bovine serum (FBS), heat-inactivated, sterile. Store frozen at –20°C. Thaw in a 37°C water bath before adding to the culture medium. Once thawed, FBS will remain stable at 4°C for 3–4 weeks. *Caution*: Avoid repeated refreezing and thawing of the FBS.
5. l-Glutamine solution, 200 mM sterile solution, cell-culture tested. Store frozen at –20°C in 1.1 mL aliquots. Thaw at room temperature before adding to the culture medium.
6. Sodium pyruvate solution, sodium pyruvate 100 mM, sterile solution. Store frozen at –20°C in 1.1 mL aliquots. Thaw at room temperature before adding to the culture medium.
7. Cytochalasin B (Cyt-B). *Caution*: This material is toxic and a possible teratogen. It must always be purchased in sealed vials. The preparation of this reagent must be carried out in a cytoguard cabinet and the following personal protection must be used: Tyvek gown, P2 dust mask, double nitrile gloves, and safety glasses.
8. Dimethyl sulphoxide (DMSO).
9. Phytohemagglutinin (PHA).
10. Isoton II used as a cellular diluent (Coulter Electronics, #8546719).
11. Zapoglobin (Coulter Electronics, #9366013).
12. DePex (or DPX) mounting medium.
13. Diff-Quik staining set. Consists of fixative, solution 1 (orange) and solution 2 (blue) stains can be reused and can last several months depending on usage. Diff-Quik is a modified Wright–Giemsa stain.
14. Methanol for fixing cells. This fixative can be used in place of the Diff-Quik fixative once it runs out.

15. 0.85% (w/v) isotonic saline, sterile, for the preparation of PHA solution: add 4.25 g sodium chloride to 500 mL Milli-Q water, autoclave at 121°C for 30 min, and then store at 4°C.
16. Sterile vacutainer blood tubes with lithium heparin as anticoagulant (e.g., Greiner vacuette tubes 9 mL).
17. Biological safety cabinet (e.g., Gelman Sciences Class II BH Series).
18. Cell counter (e.g., Coulter Electronics model ZB1). A hemocytometer can be used instead if an electronic cell counter is not available.
19. Cytocentrifuge (e.g., Shandon Cytocentrifuge from Thermo Electron Corporation, http://www.thermo.com).
20. Microscope slides, frosted end, 76×26 mm (3×1″), 1 mm thick, wiped with alcohol and allowed to dry before use.
21. Filter cards – Shandon 3×1″, thick, white, boxes of 200.
22. Cytocentrifuge cups – supplied with instrument from Thermo Electron Corporation. Must be clean and completely dry before assembly.
23. Coverslips, no. 1, 22×50 mm.
24. Tissues or Whatman no. 1 filter paper, 24 cm diameter.
25. Microscope with excellent optics for bright-field and fluorescence examination of stained slides at ×1,000 magnification (e.g., Leica DMLB, Nikon Eclipse 600).
26. Zapoglobin.
27. 70 mL polystyrene gamma sterile containers.
28. 10 mL graduated sterile pipettes.
29. Preferred tube for Ficoll separation – dependent on the blood volume: 50 mL polystyrene, sterile, conical, 15 mL polypropylene, sterile, conical, 10 mL polystyrene, yellow capped, sterile, V-bottom.
30. Preferred tube for culturing cells is dependent on the final culture volume. Use 6 mL polystyrene, vented cap, gamma-sterilized, round bottom culture tube, for 750 µL isolated lymphocyte culture.
31. Single-use 0.22 mm filter units, syringe-driven.
32. 0.22 mm hydrophobic filter.
33. 1% (w/v) hypochlorite solution for disinfecting cytocentrifuge cups. Change weekly.

2.2. Reagent Setup

2.2.1. Culture Medium

Prepare by adding 90 mL of RPMI 1640 medium with 10 mL of 10% FBS, 1 mL of 200 mM l-glutamine, and 1 mL of 100 mM sodium pyruvate solution. Prepare the culture medium in sterile tissue culture grade glass or plastic bottles. Smaller volumes of culture medium may be prepared depending on the number of cultures required. Culture medium can be stored for 1 week at 4°C before use. It is not usually necessary to sterile filter the prepared culture medium if rigorous aseptic technique is followed; however, to be absolutely certain, medium may be sterile-filtered using a 0.22 mm filter.

2.2.2. Cytochalasin B

Dissolve 5 mg solid in 8.33 mL DMSO to give a Cyt-B solution concentration of 600 µg/mL as follows:

1. Remove the Cyt-B vial from −20°C and allow to reach room temperature. Do not remove the seal.
2. Sterilize the top of the rubber seal with ethanol and allow the ethanol to evaporate.
3. Pipette 8.3 mL of DMSO into a 50 mL sterile Falcon tube.
4. Vent the vial seal with a 25 G needle and a 0.22 µm hydrophobic filter to break the vacuum.
5. Using a 5 mL sterile syringe and another needle, inject 4 mL of the 8.3 mL DMSO through the seal.
6. Mix contents gently. Cyt-B should dissolve readily in DMSO. Leave syringe and needle in place.
7. Remove the 4 mL from the vial into the syringe and eject into a labeled sterile TV10 tube.
8. Aspirate the remaining 4.3 mL of DMSO into the syringe and inject into the vial as before.
9. Remove this final volume from the vial and eject into the TV10 tube.
10. Mix and dispense 100 µL aliquots into sterile 5 mL polystyrene yellow capped tubes labeled Cyt B with the date.
11. Store at −20°C for up to 12 months. The vials of powder are guaranteed by Sigma for 2 years if stored at −20°C.

2.2.3. Phytohemagglutinin

The preparation of this reagent has been modified slightly from the instructions in the product insert to suit this protocol. 45 mg solid is dissolved in 20 mL sterile isotonic saline to make a solution of 2.25 mg/mL as follows:

1. Remove the PHA vial from the refrigerator and allow to reach room temperature. Do not remove the seal.
2. Sterilize the top of the rubber seal with ethanol and allow the ethanol to evaporate.
3. Pipette 20 mL of sterile isotonic saline into a 50 mL sterile Falcon tube.
4. Vent the PHA vial seal with a 25 G needle and a 0.22 µm hydrophobic filter to break the vacuum.

5. Using a 10 mL sterile syringe and another needle, inject 5 mL of the 20 mL isotonic saline into the vial and dissolve the solid.
6. Remove the 5 mL solution from the vial into the syringe and eject into another sterile 50 mL Falcon tube that is appropriately labeled.
7. Add another 5 mL of isotonic saline to the vial, using the syringe and mix gently.
8. Remove this volume from the vial and eject into the labeled Falcon tube.
9. Add the remaining 10 mL of isotonic saline to the labeled Falcon tube.
10. Dispense 500 µL aliquots into sterile, labeled 1.2 mL vials (cryovial type).
11. Store at 4°C for up to 6 months. The 500 µL aliquots are single-use only. Discard any remaining solution in vial. Discard any vial with signs of bacterial contamination. Slight turbidity may occur but this does not affect the activity of PHA.

3. Methods

3.1. Isolation and Counting of Lymphocytes (Timing Approximately 3 h)

1. Collect fresh blood by venipuncture into vacutainer blood tubes with lithium heparin anticoagulant. Keep blood tubes at room temperature (see Note 1). The volume taken is dependent on the number of cells needed. Usually, one can expect to collect up to 1×10^6 lymphocytes per 1 mL blood. The isolated cells are mainly lymphocytes, but may contain some monocytes. It is not practical or essential to determine the proportion of monocytes as they represent a minor fraction of the isolated cell population.

   Caution: All procedures until slide preparation must be carried out aseptically in a class II Biological Safety Cabinet.

2. Dilute whole blood 1:1 with room temperature HBSS and gently invert to mix.
3. Gently overlay diluted blood onto Ficoll–Paque using a ratio of 1:3 (e.g., 2 mL Ficoll–Paque: 6 mL diluted blood) being very careful not to disturb the interface.
4. Weigh and balance the centrifuge buckets before spinning the tubes at $400 \times g$ for 30 min at room temperature.
5. Remove the lymphocyte layer located at the interface of Ficoll–Paque and diluted plasma into a fresh tube, using a sterile, plugged, Pasteur pipette, taking care not to remove too much Ficoll–Paque.

6. Dilute the lymphocyte suspension with 3× its volume of HBSS at room temperature, mix gently, and then centrifuge at $180 \times g$ for 10 min.

7. Discard the supernatant and resuspend the cell pellet (slightly beige color) in 2× the volume removed of HBSS, using a Pasteur pipette, and then centrifuge at $100 \times g$ for 10 min at room temperature.

8. Discard the supernatant and gently resuspend the cells in 1 mL culture medium at room temperature, using a Pasteur pipette.

9. Count the cells using the Coulter Counter.

10. Set instrument settings for counting human lymphocytes (e.g., Coulter Counter Model ZB1; threshold: 8, attenuation: 1, aperture: 1/4, manometer: 0.5 mL).

11. Dilute 15 µL of cell suspension into 15 mL Isoton II and add five drops of zapoglobin to lyse any residual red blood cells.

12. Perform the count in triplicate to determine the mean. If the cell count is >15,000, count 7.5 µL cells instead, and then multiply by 2. Multiply the value by 2,000 (dilution factor of 1,000 and counting volume of 0.5 mL) to give the number of cells per mL.

13. To set the cells up at 1×10^6/mL in 750 µL culture medium, divide 750 (µL) by the number of cells per mL and multiply by 1. This gives the volume of cells to use in "µL." Make this volume up to 750 µL with medium.

    Example: mean cell count = 12,586

    $12,586 \times 2,000 = 25.2 \times 10^6$ cells per mL

    $750 \text{ µL} / 25.2 (\times 10^6) \times 1 (\times 10^6) = 29.8$ µL

    29.8 µL cell suspension + 720.2 µL medium = cell suspension at 1×10^6/mL in 750 µL.

    Pause point: Cells may be left in medium at this point at room temperature in the dark for 1–2 h with the tubes sealed.

3.2. Culture of Lymphocytes, Addition of Cytochalasin-B, and Harvesting of cells (Timing 72 h)

1. Using the above calculation (step 13), resuspend the cells at 1×10^6/mL into the culture medium. Add the calculated volume of medium into labeled 6 mL round-bottomed culture tubes, and then add the cells into the medium to make up a final volume of 750 µL. Set up duplicate cultures per subject and/or treatment studied.

2. Stimulate mitotic division of lymphocytes by adding PHA. Add 10 µL of the PHA solution to the 750 µL culture to give a final concentration of 30 µg/mL. Discard the remaining PHA after use (see Note 2).

3. Incubate the cells at 37°C with lids loose in a humidified atmosphere containing 5% CO_2 for exactly 44 h.

Critical step: Cyt-B must be added exactly 44 h after PHA stimulation (see Note 3).

4. Thaw out stock vial containing 100 µL solution of Cyt-B in DMSO at 600 µg/mL. *Caution*: Carry out all manipulations of the Cyt-B solutions in a fume hood with the base glass window down to chest level and use the following personal protection: Tyvek gown, double nitrile gloves and safety glasses. These precautions are necessary for protection against biohazardous and cytotoxic aerosols.

5. Aseptically add 900 µL of culture medium equilibrated to room temperature to the vial to obtain a 1,000 µL solution of 60 µg/mL Cyt-B.

6. Remove 56 µL of medium from the top of the 750 µL culture and replace with 56 µL of the 60 µg/mL Cyt-B solution to give a final Cyt-B concentration of 4.5 µg/mL.

7. Return cultures to the incubator and incubate the cultures for a further 28 h. Discard the remaining Cyt-B solution.

8. At 24–28 h after the addition of Cyt-B, harvest cells for slide preparation and scoring according to the procedure given below. The chosen harvest time should maximize the proportion of binucleate cells and minimize the frequency of mono- and multinucleated (three or more nuclei) cells.

3.3. Harvesting of Cells Using Cytocentrifugation (Timing Approximately 30 min)

1. Follow the manufacturer's instructions for assembly of the slides, filter-cards and cytocentrifuge cups within the cytocentrifuge rotor. Cytocentrifugation is performed at 18–20°C.

2. Prepare a concentration of cells that is sufficient to produce a monolayer of cells on each spot. It may be necessary to spin down a culture gently and take off a proportion of the supernatant to obtain an optimal concentration of cells on the slide following cytocentrifugation. In the case of lymphocyte cultures in round-bottomed tubes, it is only necessary to take off the supernatant (approximately 300 µL from a 1 mL culture), as the cells tend to settle in the incubator.

Caution: Loading and cytocentrifugation of cell culture sample (steps 2–7) must be carried out in a fume hood or preferably in an approved cytoguard cabinet. Appropriate safety protection including gloves must be worn.

3. Resuspend the cells well using a Pasteur or Gilson pipette to disaggregate the cells. It is best to suspend all the cultures to be harvested first, and then return to each one to resuspend just before loading into the sample cup.

4. Working quickly, add 100–200 µL of cell suspension to the well of each sample cup. The typical volume for cytokinesis blocked lymphocyte cultures is 120 µL. The required volume

may need slight adjustment depending on the concentration of cells in the culture and the optimal cell density for slide scoring, which is determined by trial and error.

5. Replace rotor lid until locked and spin down in cytocentrifuge at $52 \times g$ for 5 min.
6. *Press start*: The cytocentrifuge will run the program, stop and slow automatically.
7. Upon completion of spinning return the rotor to the fume hood and follow the manufacturer's procedure for opening each slide holder.

3.4. Drying, Fixing and Staining of Cells, and Slide Preparation (Timing Approximately 30 min)

1. Dismantle each slide holder, taking care not to smear the spots. Discard the filter cards into the biohazard bin. Place the sample cups in 1% hypochlorite solution to soak for at least 10 min before rinsing carefully six times in pure water and drying. Do not use direct heat to dry the cups.

   Caution: Appropriate protection, including nitrile gloves, must be worn while staining samples.

2. Place the slides horizontally on a slide tray and allow the cells to air-dry for exactly 10 min at room temperature. *Critical*: Do not air-dry slides for longer than 10 min, otherwise cellular and nuclear morphology is altered making it difficult to score the slides.

3. Place slides vertically in a dry staining rack and place in methanol or Diff-Quik fixative for 10–15 min. The methanol can be reused.

4. Transfer the rack without delay to Diff-Quik solution 1 (orange) and stain for 6 s while moving the rack back and forth. Increase staining time if required.

5. Transfer the rack without delay to Diff-Quik solution 2 (blue) and stain for another 6 s while moving the rack back and forth. Again, staining time can be increased if required.

   Critical step: Staining time is determined by trial and error depending on how fresh the stains are. The aim is to obtain optimal contrast between nuclear and cytoplasmic staining so that various biomarkers in the CBMN Cyt assay are easily and unequivocally scored.

6. Keeping the slides in the rack, wash the slides very briefly and gently with tap water and rinse with pure water, making sure that the slides are not destained by residual water droplets on cells (see Note 5).

7. Immediately place the slides face down on paper tissues (or better, Whatman no. 1 filter paper) to blot away any residual moisture. Do not place any pressure or rub on the cell spots.

8. Place the slides on a slide tray and allow to dry for about 10–15 min.

9. Examine the cells at ×100 and ×400 magnification to assess the efficiency of staining and the density of the cells, remembering that for the CBMN Cyt assay, it is necessary to have at least 1,000 binucleate cells to score. Restaining can be carried out at this stage if the staining is too light by repeating steps 2–8. If the cell density is too heavy or light, concentrate or dilute the cells as necessary and repeat the spinning and staining steps.

10. If the slides are satisfactory, seal the tubes or flasks containing the cultures and discard into a biohazard bin.

11. Leave the slides to dry completely for at least 30 min before putting coverslips on. *Pause point*: Slides can be left overnight at room temperature to dry.

3.5. Coverslipping and Storage

1. Place the slides to be coverslipped on tissue paper and set out one coverslip alongside each. *Caution*: Carry out coverslipping in a fume hood to avoid inhalation of organic solvent in DePex and leave slides until completely dry.

2. Put two large drops of DePex (use a plastic dropper) on each of the coverslips in the approximate area where the spots correspond.

   Caution: Wear nitrile gloves when applying DePex medium.

3. Invert the slide over the coverslip and allow the DePex to spread by capillary action. Slide the coverslip gently to and fro to expel any excess DePex and air bubbles. Ensure that the spots do not have air bubbles over them.

4. Wipe excess DePex from the edges of the slide and ensure that the medium or glass does not cover any of the frosted label area, as a coding label will not stick on these.

5. Place the slides on a tray and leave overnight in the fume hood to dry.

6. Store slides in slide boxes at room temperature and code with a sticky label over the frosted area before scoring.

3.6. Scoring Criteria

3.6.1. Criteria for Scoring Viable Mono-, Bi-, and Multinucleated Cells (Fig. 2a–c)

Frequency of viable mono-, bi-, and multinucleated cells is measured to determine cytostatic effects and the rate of mitotic division, which can be calculated using the nuclear division index. These cell types have the following characteristics:

1. Mono-, bi-, and multinucleated cells are viable cells with an intact cytoplasm and normal nucleus morphology containing one, two, and three or more nuclei, respectively.

2. They may or may not contain one or more MNi or NBUDs, and in the case of bi- and multinucleated cells they may or may

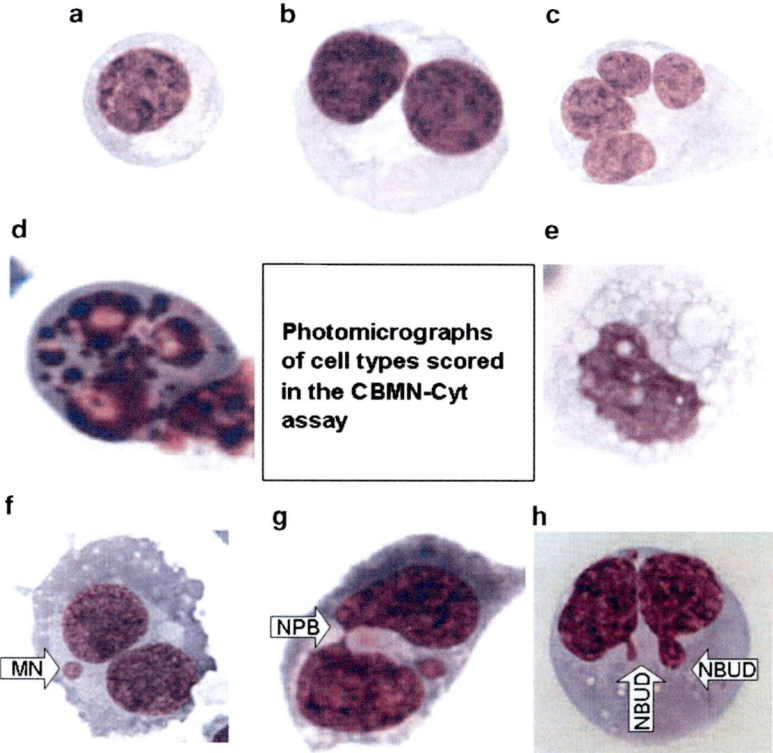

Fig. 2. Photomicrographs of the cells scored in the CBMN "cytome" assay. (a) Mononucleated cell; (b) binucleated cell; (c) multinucleated cell; (d) apoptotic cell; (e) necrotic cell; (f) BN cell containing MNi; (g) BN containing a NPB (and a MN); and (h) BN cell containing NBUDs. The ratios of mononucleated, BN, multinucleated, necrotic and apoptotic cells are used to determine mitotic division rate or NDI (a measure of cytostasis) and cell death (cytotoxicity). The frequency of BN cells with MNi, NPBs, or NBUDs provides a measure of genome damage and/or chromosomal instability. For a wider selection of photomicrographs of different types of cells and biomarkers scored in the CBMN Cyt assay, refer to Fenech et al. (19).

not contain one or more NPBs. Necrotic and apoptotic cells should not be included among the viable cells scored. On rare occasions, multinucleated cells with more than four nuclei are observed if cell-cycle time is much shorter than normal or the cytokinesis blocking time is too long.

3.6.2. Criteria for Scoring Apoptotic Cells (Fig. 2d)

Apoptotic lymphocytes are cells undergoing programmed cell death. They have the following characteristics:

1. Early apoptotic cells can be identified by the presence of chromatin condensation within the nucleus and intact cytoplasmic and nuclear membranes.
2. Late apoptotic cells exhibit nuclear fragmentation into smaller nuclear bodies within an intact cytoplasm/cytoplasmic membrane.

3. Staining intensity of the nucleus, nuclear fragments, and cytoplasm in both kinds of apoptotic cell is usually greater than that of viable cells.

3.6.3. Criteria for Scoring Necrotic Cells (Fig. 2e)

Necrosis is an alternative form of cell death that is thought to be caused by damage to cellular membranes, organelles, and/or critical metabolic pathways required for cell survival, such as energy metabolism. Necrotic lymphocytes have the following characteristics:

1. Early necrotic cells can be identified by their pale cytoplasm, the presence of numerous vacuoles (mainly in the cytoplasm and sometimes in the nucleus), damaged cytoplasmic membrane and a fairly intact nucleus.
2. Late necrotic cells exhibit the loss of cytoplasm and damaged/irregular nuclear membrane with only a partially intact nuclear structure and often with nuclear material leaking from the nuclear boundary.
3. Staining intensity of the nucleus and cytoplasm in both types of necrotic cell is usually less than that observed in viable cells.

3.6.4. Criteria for Selecting Binucleate Cells Suitable for Scoring MNi, NPBs, and NBUDs

The cytokinesis-blocked binucleate cells that may be scored for MNi, NPB, and NBUD frequency should have the following characteristics:

1. The cells should be binucleated.
2. The two nuclei in a binucleated cell should have intact nuclear membranes and be situated within the same cytoplasmic boundary.
3. The two nuclei in a binucleated cell should be approximately equal in size, staining pattern, and staining intensity.
4. The two nuclei within a binucleated cell may be attached by a nucleoplasmic bridge, which is no wider than 1/4th of the nuclear diameter.
5. The two main nuclei in a binucleated cell may touch but ideally should not overlap each other. A cell with two overlapping nuclei can be scored only if the nuclear boundaries of each nucleus are distinguishable.
6. The cytoplasmic boundary or membrane of a binucleated cell should be intact and clearly distinguishable from the cytoplasmic boundary of adjacent cells.

The cell types that should not be scored for the frequency of MNi, NPBs, and NBUDs frequency include mono- and multinucleated (with three or more nuclei) cells and cells that are necrotic or apoptotic.

3.6.5. Criteria for Scoring Micronuclei (Fig. 2f)

MNi are morphologically identical but smaller than nuclei. They also have the following characteristics:

1. The diameter of MNi in human lymphocytes usually varies between 1/16th and 1/3rd of the mean diameter of the main nuclei, which corresponds to 1/256th and 1/9th of the area of one of the main nuclei in a binucleate cell, respectively.
2. MNi are non-repaetile, and they can therefore be readily distinguished from artifacts, such as staining particles.
3. MNi are not linked or connected to the main nuclei.
4. MNi may touch but not overlap the main nuclei and the micronuclear boundary should be distinguishable from the nuclear boundary.
5. MNi usually have the same staining intensity as the main nuclei, but occasionally staining may be more intense.

3.6.6. Criteria for Scoring Nucleoplasmic Bridges (Fig. 2g)

A NPB is a continuous DNA-containing structure linking the nuclei in a binucleated cell. NPBs originate from dicentric chromosomes (resulting from misrepaired DNA breaks or telomere end fusions) in which the centromeres are pulled to opposite poles during anaphase. They have the following characteristics:

1. The width of an NPB may vary considerably, but usually does not exceed 1/4th of the diameter of the nuclei within the cell.
2. NPBs should also have the same staining characteristics as the main nuclei.
3. On rare occasions, more than one NPB may be observed within one binucleated cell.
4. A binucleated cell with a NPB may contain one or more MNi.
5. Binucleate cells with one or more NPBs and no MNi may also be observed.

It may be more difficult to score NPBs in binucleated cells with touching nuclei, and it is therefore reasonable to specify whether NPBs were scored in all binucleate cells regardless of the proximity of nuclei within a binucleate cell or whether they were scored separately in those binucleated cells in which nuclei were clearly separated and those binucleated cells with touching nuclei. There is not enough evidence yet to recommend scoring NPB only in binucleate cells in which nuclei do not touch.

3.6.7. Criteria for Scoring Nuclear Buds (Fig. 2h)

A NBUD represents the mechanism by which a nucleus eliminates amplified DNA and DNA repair complexes. NBUDs have the following characteristics:

1. NBUDs are similar to MNi in appearance with the exception that they are connected with the nucleus via a bridge that can be slightly narrower than the diameter of the bud or by a much thinner bridge depending on the stage of the extrusion process.

2. NBUDs usually have the same staining intensity as MNi.

3. Occasionally, NBUDs may appear to be located within a vacuole adjacent to the nucleus.

If it is difficult to determine whether the observed nuclear anomaly is a micronucleus touching the nucleus or a nuclear bud, it is acceptable to classify it as the latter.

A small protrusion of nuclear material from the nucleus without an obvious constriction between the nucleus and the protruding nuclear material should not be classified as a nuclear bud. This type of event is called a nuclear bleb. The significance of this type of event is unknown.

For a more comprehensive photographic gallery of the various cell types (see Note 6).

3.6.8. Slide Scoring for Cytostatic and Cytotoxic Outcomes

Slides should be coded before scoring by a person not involved in the experiment so that the person who scores the slides is not aware of the treatment conditions, individual or groups to which the cells on the slides belong. Slides are best examined at ×1000 magnifications using a good-quality bright-field or fluorescence microscope (e.g., Leica DMLB, Nikon Eclipse E 600). A score should be obtained for slides from each duplicate culture ideally from two different scorers using identical microscopes. The number of cells scored should be determined depending on the level of change (effect size) in the micronucleus index that the experiment is intended to detect and the expected standard deviation of the estimate. The optimal way to score slides in the CBMN cyt assay is to first determine the frequency of mono-, bi-, and multinucleated viable cells as well as the necrotic and apoptotic cells in a minimum of 500 cells; then the frequency of DNA damage biomarkers (MNi, NPBs and NBUDs) are scored in a minimum of 1,000 binucleated cells. For each slide, the following information should therefore be obtained:

1. The number of viable mono-, bi-, and multinucleated (with three or more nuclei) cells per 500 cells scored (from this information the nuclear division index can be derived; see Subheading 3.6.10.

2. The number of apoptotic cells per 500 cells.

3. The number of necrotic cells per 500 cells.

4. The number of MNi in at least 1,000 binucleate cells.

5. The frequency of binucleated cells containing MNi in at least 1,000 binucleate cells.

6. The frequency of binucleated cells containing NPBs in at least 1,000 binucleate cells.

7. The frequency of binucleated cells containing NBUDs in at least 1,000 binucleate cells.

The distribution of binucleated cells with zero, one or more MNi may be useful to record if abnormal distribution of MNi among cells may be expected owing to partial body radiation in the case of ionizing radiation exposure or if it is expected that the agent being tested might induce multiple MNi in affected cells (e.g., spindle poisons such as colcemid). Because most binucleate cells with more than one NPB or more than one NBUD are rare, it is usually not necessary to record the distribution of NPBs and NBUDs among binucleate cells.

The frequency of MNi and NBUDs in nondivided mononuclear cells can also be scored if the level of preexisting DNA damage needs to be measured. However, this approach cannot be used instead of the CBMN Cyt assay because (a) it does not account for DNA damage accumulated in the bulk of lymphocytes *in vivo* while circulating in the quiescent phase, (b) it cannot be used to measure NPBs, and (c) does not give the same results as scoring MN and NBUDS in binucleated cells (19, 20).

3.6.9. Number of Binucleate Cells that Should Be Scored

One of the most common questions is the number of binucleated cells to be scored in the CBMN Cyt assay. The accepted protocol is to score a minimum of 1,000 binucleate cells per treatment or time point although reports vary between 500 and 2,000 binucleate cells. An alternative approach is to keep on scoring binucleate cells until a fixed number of MNi are observed (e.g., 45 MNi). The latter has the advantage that more binucleate cells are scored when fewer MNi are induced, thus maintaining similar statistical power across different treatments. The main disadvantage is that more than 2,000 cells may have to be scored in cultures with low micronucleus frequency. In our experience, scoring 1,000 binucleate cells from each of the duplicate cultures (total 2,000 binucleate cells) always yields robust results.

3.6.10. Calculation of Nuclear Division Index

The nuclear division index (NDI) provides a measure of the proliferative status of the viable cell fraction. It is therefore an indicator of cytostatic effects and, in the case of lymphocytes, it is also a measure of mitogenic response, which is useful as a biomarker of immune function (21).

NDI is calculated according to the method of Eastmond and Tucker (22). Score 500 viable cells to determine the frequency of cells with 1, 2, 3, or 4 nuclei, and calculate the NDI using the formula:

$$NDI = (M1 + 2M2 + 3M3 + 4M4)/N,$$

where M1–M4 represent the number of cells with 1–4 nuclei, and N is the total number of viable cells scored (excluding necrotic and apoptotic cells). The NDI is a useful parameter for comparing the mitogenic response of lymphocytes and cytostatic effects of agents examined in the assay.

The lowest NDI value possible is 1.0, which occurs if all of the viable cells have failed to divide during the cytokinesis-block period and are therefore all mononucleated. If all viable cells completed one nuclear division and are therefore all binucleated, the NDI value is 2.0. An NDI value can only be greater than 2.0 if a substantial proportion of viable cells have completed more than one nuclear division during the cytokinesis-block phase and therefore contain more than two nuclei. For example, if 50% of viable cells are binucleated, 10% trinucleated, and 10% quadrinucleated, the NDI value is 2.2.

4. Notes

1. The published evidence available suggests that the storage of blood between 5 and 22°C for up to 24 h has no significant impact on baseline or radiation-induced micronucleus frequency (23). However, these observations need to be further verified and replicated. It is possible to perform the CBMN Cyt assay using cryopreserved lymphocytes, but there are conflicting reports on whether cryopreservation alters the frequency rate of MNi in binucleate cells (24, 25), which means that it is essential to optimize and verify that the freezing and thawing protocol used does not induce DNA damage.

2. The optimal PHA concentration for maximizing the proportion of binucleate cells should be verified for each batch of PHA using the binucleate frequency ratio and NDI to assess mitogenesis. The success of mitogen stimulation can be determined visually 24 h following PHA stimulation. If lymphocytes have been successfully stimulated, cultures will appear clumpy and grainy. Unstimulated cultures will appear cloudy and silty.

3. The critical aspect regarding the time of addition of Cyt-B is to ensure that it is added before the first mitotic cells start to appear so that all the observed binucleate cells that are captured are in fact once divided cells only. This is important because MNi tend to get lost in subsequent divisions and the micronucleus frequency in a second division cell is likely to be less than that in a first division cell after a genotoxic insult (9). Cyt-B may take up to 6 h before it starts to exert its cytokinesis-blocking action (unpublished observations), which means it should be added at least 6 h before cells start to enter M phase of the cell cycle. The optimal time to add Cyt-B with lymphocytes is usually 44 h after PHA stimulation; however, earlier addition of Cyt-B is acceptable if the culture conditions used cause an earlier than expected mitotic wave to occur.

To capture all once divided cells as binucleate cells, it is also essential to verify that the Cyt-B concentration is optimal to maximize the ratio of cytokinesis-blocked cells by doing a dose response across the concentration range of 2–10 μg/mL of Cyt-B using a 24 h cytokinesis-blocking time. The dose response should yield a plateau in response within the optimal concentration range for cytokinesis-blocking. Choose a concentration that is at least a dose point past the inflection point of the dose response and on the plateau. Experience has shown that concentrations of 4.5 and 6 μg/mL are optimal for isolated lymphocyte and whole blood lymphocyte cultures, respectively. These concentrations are usually also optimal for mammalian cell lines, but this should be checked for each cell line.

4. Hypotonic treatment for slide preparation is not recommendable because it may destroy necrotic cells and apoptotic cells, making them unavailable for assay. Inclusion of necrosis and apoptosis is important for the accurate description of mechanism of action and the measurement of cellular sensitivity to a chemical or radiation. Isolated lymphocyte culture assay or culture of cell lines does not require hypotonic treatment of cells for slide preparation, thus making it possible to preserve the morphology of both necrotic and apoptotic cells.

5. In our experience, most of the problems in the CBMN Cyt assay arise during slide preparation and staining. This is because the quality of the score depends on the quality of the slide. Main points to note: (1) avoid cell clumps by gently resuspending cells before harvest and transfer to slides; (2) maintain a moderate cell density so that it is relatively easy to identify cytoplasmic boundaries; and (3) stain only one slide initially to ensure that staining is optimal before staining the whole batch.

6. For a more comprehensive photographic gallery of the various cell types and biomarkers scored in the CBMN Cyt assay, refer to Fenech et al. (26).

References

1. Heddle, J. A. (1973) A rapid in vivo test for chromosomal damage. *Mutat Res* **18**, 187–90.
2. Schmid, W. (1975) The micronucleus test. *Mutat Res* **31**, 9–15.
3. Fenech, M. (2000) The in vitro micronucleus technique. *Mutat Res* **455**, 81–95.
4. Fenech, M., and Morley, A. A. (1986) Cytokinesis-block micronucleus method in human lymphocytes: effect of in vivo ageing and low dose X-irradiation. *Mutat Res* **161**, 193–8.
5. Carter, S. B. (1967) Effects of cytochalasins on mammalian cells. *Nature* **213**, 261–4.
6. Fenech, M. (1997) The advantages and disadvantages of the cytokinesis-block micronucleus method. *Mutat Res* **392**, 11–8.
7. Fenech, M. (2000) A mathematical model of the in vitro micronucleus assay predicts false

negative results if micronuclei are not specifically scored in binucleated cells or in cells that have completed one nuclear division. *Mutagenesis* **15**, 329–36.

8. Fenech, M., and Morley, A. (1985) Solutions to the kinetic problem in the micronucleus assay. *Cytobios* **43**, 233–46.
9. Fenech, M., and Morley, A. A. (1985) Measurement of micronuclei in lymphocytes. *Mutat Res* **147**, 29–36.
10. Degrassi, F., and Tanzarella, C. (1988) Immunofluorescent staining of kinetochores in micronuclei: a new assay for the detection of aneuploidy. *Mutat Res* **203**, 339–45.
11. Thomson, E. J., and Perry, P. E. (1988) The identification of micronucleated chromosomes: a possible assay for aneuploidy. *Mutagenesis* **3**, 415–8.
12. Farooqi, Z., Darroudi, F., and Natarajan, A. T. (1993) The use of fluorescence in situ hybridization for the detection of aneugens in cytokinesis-blocked mouse splenocytes. *Mutagenesis* **8**, 329–34.
13. Schuler, M., Rupa, D. S., and Eastmond, D. A. (1997) A critical evaluation of centromeric labeling to distinguish micronuclei induced by chromosomal loss and breakage in vitro. *Mutat Res* **392**, 81–95.
14. Fenech, M. (2002) Chromosomal biomarkers of genomic instability relevant to cancer. *Drug Discov Today* **7**, 1128–37.
15. Hoffelder, D. R., Luo, L., Burke, N. A., Watkins, S. C., Gollin, S. M., and Saunders, W. S. (2004) Resolution of anaphase bridges in cancer cells. *Chromosoma* **112**, 389–97.
16. Thomas, P., Umegaki, K., and Fenech, M. (2003) Nucleoplasmic bridges are a sensitive measure of chromosome rearrangement in the cytokinesis-block micronucleus assay. *Mutagenesis* **18**, 187–94.
17. Serrano-Garcia, L., and Montero-Montoya, R. (2001) Micronuclei and chromatid buds are the result of related genotoxic events. *Environ Mol Mutagen* **38**, 38–45.
18. Shimizu, N., Shimura, T., and Tanaka, T. (2000) Selective elimination of acentric double minutes from cancer cells through the extrusion of micronuclei. *Mutat Res* **448**, 81–90.
19. Fenech, M., Perepetskaya, G., and Mikhalevich, L. (1997) A more comprehensive application of the micronucleus technique for biomonitoring of genetic damage rates in human populations – experiences from the Chernobyl catastrophe. *Environ Mol Mutagen* **30**, 112–8.
20. Kirsch-Volders, M., and Fenech, M. (2001) Inclusion of micronuclei in non-divided mononuclear lymphocytes and necrosis/apoptosis may provide a more comprehensive cytokinesis block micronucleus assay for biomonitoring purposes. *Mutagenesis* **16**, 51–8.
21. Fenech, M. (2006) Cytokinesis-block micronucleus assay evolves into a "cytome" assay of chromosomal instability, mitotic dysfunction and cell death. *Mutat Res* **600**, 58–66.
22. Eastmond, D. A., and Tucker, J. D. (1989) Identification of aneuploidy-inducing agents using cytokinesis-blocked human lymphocytes and an antikinetochore antibody. *Environ Mol Mutagen* **13**, 34–43.
23. Lee, T. K., O'Brien, K., Eaves, G. S., Christie, K. I., and Varga, L. (1999) Effect of blood storage on radiation-induced micronuclei in human lymphocytes. *Mutat Res* **444**, 201–6.
24. Burrill, W., Levine, E. L., Hindocha, P., Roberts, S. A., and Scott, D. (2000) The use of cryopreserved lymphocytes in assessing inter-individual radiosensitivity with the micronucleus assay. *Int J Radiat Biol* **76**, 375–82.
25. O'Donovan, M. R., Freemantle, M. R., Hull, G., Bell, D. A., Arlett, C. F., and Cole, J. (1995) Extended-term cultures of human T-lymphocytes: a practical alternative to primary human lymphocytes for use in genotoxicity testing. *Mutagenesis* **10**, 189–201.
26. Fenech, M., Chang, W. P., Kirsch-Volders, M., Holland, N., Bonassi, S., and Zeiger, E. (2003) HUMN project: detailed description of the scoring criteria for the cytokinesis-block micronucleus assay using isolated human lymphocyte cultures. *Mutat Res* **534**, 65–75.

Chapter 17

Buccal Micronucleus Cytome Assay

Philip Thomas and Michael Fenech

Abstract

The Buccal Micronucleus Cytome (BMCyt) assay is a new minimally invasive system for studying DNA damage, chromosomal instability, cell death, and the regenerative potential of buccal mucosal tissue. This method is increasingly being used in molecular epidemiologic studies investigating the impact of nutrition, life-style factors, genotoxin exposure, and genotype on DNA damage and cell death. Biomarkers of this assay have been associated with increased risk for accelerated aging, cancer, and neurodegenerative diseases. This protocol describes the current established methods for buccal cell collection, slide preparation, cellular and nuclear staining, and scoring criteria.

Key words: Micronucleus, Buccal, Cytome, DNA damage, Cytotoxicity

1. Introduction

The regenerative capacity of tissues and organs within the body is fundamental to healthy aging. In the buccal mucosa, this is dependent on the number and division rate of regenerative cells (basal cells), their genomic stability, and propensity for cell death. The buccal mucosa is an easily accessible tissue that can be sampled in a minimally invasive manner without causing undue stress to participants. This method is increasingly being used in molecular epidemiologic studies investigating the impact of nutrition, life-style factors, genotoxin exposure, and genotype on DNA damage and cell death. The buccal mucosa provides a unique opportunity to study the regenerative capacity of epithelial tissue of ectodermal origin in humans. The assay has been used successfully to study DNA damage by scoring micronuclei (MNi) and/or using fluorescent probes to detect aneuploidy, chromosome breaks, and telomere length (1–5).

The buccal mucosa is a stratified squamous epithelium consisting of four distinct layers (6, 7). The stratum corneum or keratinized layer lines the oral cavity, comprising cells that are constantly being lost as a result of everyday activities such as mastication. Below this layer lies the stratum granulosum or granular cell layer and the stratum spinosum or prickle cell layer containing populations of both differentiated, apoptotic and necrotic cells. Beneath these layers are the rete pegs or stratum germinativum, containing actively dividing basal cells, which produce cells that differentiate and maintain the profile, structure, and integrity of the buccal mucosa (Fig. 1a).

Figure 1b illustrates diagrammatically the various cell types, nuclear anomalies, and possible inter-relationships following correlation analysis between the various cell types observed and scored in the BMCyt assay (1).

The BMCyt assay is a minimally invasive means of investigating events that identify changes in potential biomarkers that are reflective of DNA damage (MNi and/or nuclear buds), cellular proliferation potential (basal and/or binucleated cells), and/or cell death parameters (condensed chromatin, karyorrhectic, pyknotic, and karyolytic cells). These changes show distinct differences between the cytome profile within normal aging relative to that for premature aging clinical outcomes such as Down syndrome and Alzheimer's disease, and highlights the potential diagnostic value of the cytome approach for determining genome instability events (1, 2). In light of the fact that over 90% of cancers are epithelial in origin and that buccal mucosa is the site for oral cancer, buccal cell utilization has great epidemiologic potential as a means for genotoxic and cancer risk assessment (8). The following protocol describes the current established methods for buccal cell collection, slide preparation, cellular and nuclear staining, and scoring criteria.

2. Materials

2.1. Buccal Cell Sampling and Collection

1. Small-headed toothbrushes (2-cm head length).
2. Yellow-topped, 30-mL polystyrene containers.
3. Graduated, 10-mL sterile pipettes.
4. Milli-Q water (Milli-Q water purification system, Adelab Scientific, SA).
5. Buccal cell buffer at pH 7.0: 0.01 M Tris-HCl, 0.1 M ethylenediaminetetraacetic acid tetra Na salt, 0.02 M NaCl.

2.2. Buccal Cell Harvesting and Slide Preparation

1. Swinex filter holders.
2. Nylon net filters, 100 μm.

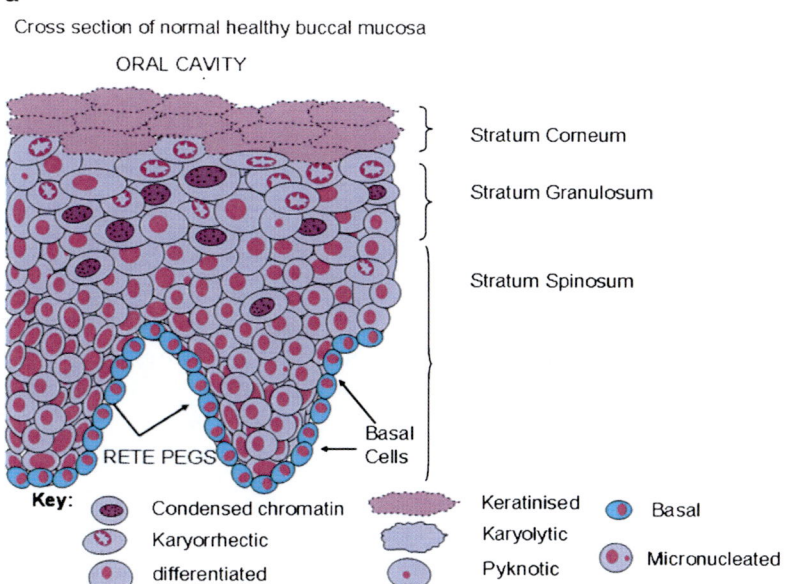

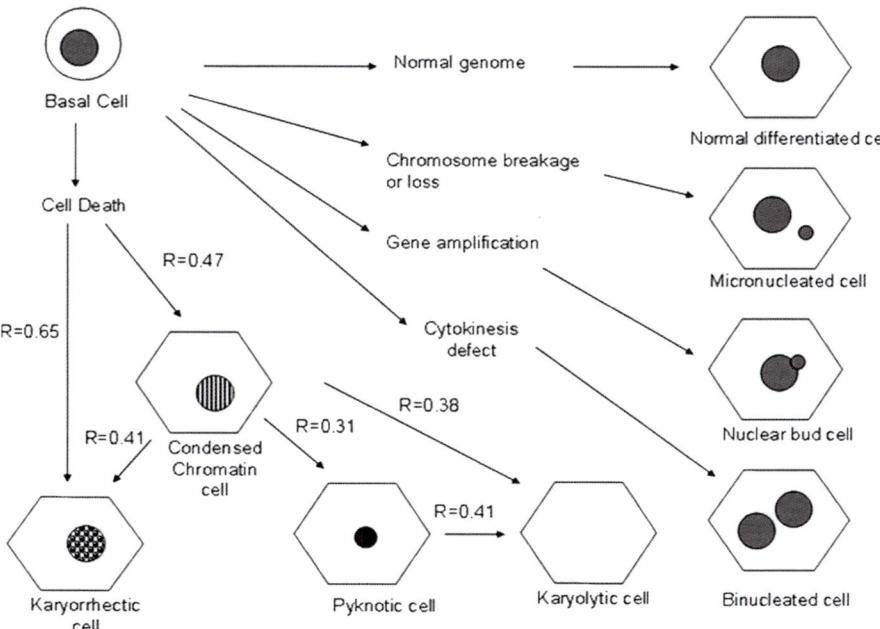

Fig. 1. (a) Diagrammatic representation of a cross-section of normal buccal mucosa of healthy individuals illustrating the different cell layers and possible spatial relationships of the various cell types. (b) Diagrammatic representation and possible inter-relationships between the various cell types observed in the buccal cytome assay based on the scheme proposed by Tolbert et al. (9). The R values refer to correlation factors for the frequency of the various cell types (1).

3. Cell counter (e.g., Coulter Electronics model ZB1). A hemocytometer can be used instead if an electronic cell counter is not available.
4. TV 10 polystyrene tubes.

5. Isoton II used as a cellular diluent (Coulter Electronics, #8546719).
6. Sterile plugged Pasteur pipettes 900 (22–23 cm).
7. Syringes 10 mL.
8. Needles 18G.
9. Counting vials 15 mL.
10. Cytocentrifuge (e.g., Shandon Cytocentrifuge from Thermo Electron Corporation, http://www.thermo.com).
11. Cytocentrifuge cups – supplied with instrument from Thermo Electron Corporation. Must be clean, rinsed 6 times in distilled or deionized water, and completely dried before assembly.
12. Filter cards – Shandon 3×1″, thick, white, boxes of 200 cards.
13. Microscope slides, frosted end, 76×26 mm (3×1″), 1-mm thick – wiped with alcohol and allowed to dry before use.
14. A Coplin jar unit holds 5 single 3×1" (75×25 mm) slides vertically or 10 slides placed back to back. Screw cap is made of white linerless polypropylene, which reduces solvent evaporation. It has a rectangular base, and is made of "800" soda-lime glass. Approximate inside dimensions: 70×30×30 mm (2.75×1.18×1.18″).
15. Coverslips, no. 1, 22×50 mm.
16. 0.01% (w/v) Hypochlorite solution for disinfecting cytocentrifuge cups. Change weekly.
17. Hand homogenizer (Wheaton Scientific, 0.1–0.15 mm gauge).
18. Dimethyl sulfoxide (DMSO).

2.3. Buccal Cell Staining

1. Ethanol/glacial acetic acid. Ethanol is mixed with glacial acetic acid in a 3:1 ratio. Fixative is to be made fresh each time and not stored.

   Caution: Acetic acid is corrosive, a respiratory irritant, and can cause serious burns. Fixative should be prepared in a fume hood or similar extraction cabinet, and the following personal protection used: Tyvek gown, double nitrile gloves, P2 dust mask, and safety glasses.

2. 50% Ethanol made up with Milli-Q deionized water (18.2 Ω resistivity).
3. 20% Ethanol made up with Milli-Q deionized water (18.2 Ω resistivity).
4. 5 M HCl.

   Caution: HCl is corrosive, a respiratory irritant, and can cause serious burns. The acid should be handled in a fume hood or similar extraction cabinet, and the following personal protection used: Tyvek gown, double nitrile gloves, P2 dust mask, and safety glasses. 5 M concentration should be made up fresh each time and not stored.

5. Schiff's reagent.

   Caution: Schiff's reagent is a skin, eye, and respiratory irritant, and should be handled in a fume hood or similar extraction cabinet, and the following personal protection should be used: Tyvek gown, double nitrile gloves, P2 dust mask, and safety glasses.

6. 0.2% (w/v) aqueous Light Green (Gurr's, cat. no. 06477).

7. DePex (or DPX) mounting medium.

2.4. Buccal Cell Scoring

1. Microscope with excellent optics for bright-field and fluorescence examination of stained slides at ×1,000 magnification (e.g., Leica DMLB, Nikon Eclipse 600).

2. Slide storage boxes.

2.5. Buccal Reagent Setup

1. Buccal cell buffer: To make 1 L of the buccal buffer: weigh 1.6 g of Tris-HCl, 1.2 g of ethylenediaminetetraacetic acid tetra sodium salt, and 37.2 g of NaCl and dissolve in 600 mL of Milli-Q water. Thoroughly dissolve the salts and make up the volume to 1,000 mL. Adjust pH to 7.0 and autoclave at 121°C for 30 min. The buffer will last for up to 3 months when stored at room temperature.

2. 5 M HCl: In order to determine the concentration of laboratory sourced HCl, the molarity is derived from the stated specific gravity. The specific gravity of BDH HCl used in our laboratory is 1.18 g/mL, which is the equivalent of 1,180 g/L. The molecular weight of HCl is 36.5. The molarity can therefore be determined by dividing the specific gravity by the molecular weight, resulting in an acidic molarity of 32.3 M (1,180/36.5). However, HCl assay specifications detailed on the accompanying product sheet indicate a 37% acidic solution. In order to determine the true molarity, we have to adjust for this percentage value. Our stock acid solution is found to have a molarity of 12 M by multiplying our initial value of 32.3 M by 0.37 (37%). In order to prepare a 200 mL working solution of 5 M HCl from our stock solution, the following calculation is performed: (200 mL (desired volume) ×5 M (desired concentration))/stock solution (12 M).

   This results in 83.3 mL of acid being required in our 200 mL working solution. *Caution*: Add 83.3 mL of acid slowly to 116.7 mL of water and not the reverse. This is important to avoid an exothermic reaction resulting in the generation of heat and potential splashing. This procedure should be performed in a well-ventilated fume hood with appropriate safety precautions.

3. Light Green cytoplasmic stain: 500 mL is prepared by dissolving 1 g of Light Green in 450 mL of Milli-Q water. When dissolved, make up to 500 mL and filter through Whatman No. 1 filter paper. Store in dark at room temperature where it should remain active for 3 years.

3. Methods

3.1. Buccal Cell Collection

1. Prior to buccal cell collection, the mouth is rinsed twice thoroughly with 100 mL of water each time to remove excess debris.
2. Each subject is provided with two 30 mL yellow-topped containers labeled LC (left cheek) and RC (right cheek), each containing 10 mL of buccal cell buffer.
3. Small-headed toothbrushes are rotated 10 times firmly against the inside of the cheek wall in a circular motion, starting from the middle and gradually increasing in circumference to produce an outward spiral effect. The reason for this motion is to enhance sampling over a greater area and to avoid continual erosion in a single population of cells. This is performed on each side of the cheek using a different brush (for sampling left and right areas of the mouth). It is important to keep the sampling method constant (see Note 1).
4. The head of the brush is then placed into the buffer container and rotated repeatedly such that the cells are dislodged and released into the buffer producing a cloudy suspension.

3.2. Buccal Cell Harvesting and Slide Preparation

1. Cells from both right and left cheeks are transferred into separate TV 10 centrifuge tubes and spun for 10 min at $400 \times g$.
2. The supernatant is aspirated off, leaving approximately 1 mL, and is replaced with 5 mL of buccal cell buffer. The cells are vortexed briefly.
3. The cells are re-spun at $400 \times g$ for 10 min. The supernatant is aspirated off and the cells are resuspended in 5 mL of buccal buffer.

   Critical: The best results are achieved after two washes in buccal cell buffer. This buffer helps to inactivate endogenous DNases present in the oral cavity and to remove bacteria and cell debris that could complicate scoring.
4. The supernatant is aspirated off and replaced with 5 mL of fresh buccal cell buffer.

5. The cell suspension is vortexed and then homogenized for 3 min in a handheld tissue homogenizer to increase the number of single cells in suspension.

6. Left and right cell populations are pooled in a 30 mL container before the cells are drawn up into a syringe using an 18G needle.

7. The cells are then passed into a TV 10 tube through a 100 μm nylon filter held in a Swinex holder. This removes large aggregates of unseparated cells that hinder slide preparation (see Note 2).

8. The cells are further spun at 400×g for 10 min and the supernatant removed. The cells are then resuspended in 1 mL of buccal cell buffer.

9. Count the cells using the Coulter Counter.

10. Set instrument settings for counting human buccal cells (e.g., Coulter Counter Model ZB1; threshold: 8, attenuation: 1, aperture: 1/4, manometer: 0.5 mL).

11. Dilute 300 μL of cell suspension into 15 mL of Isoton II.

12. Perform the count in duplicate to determine the mean cellular value.

13. The suspensions are prepared containing 80,000 cells/mL after calculated dilution with buccal cell buffer.

14. DMSO (50 μl/mL) is added to aid in cellular disaggregation and to obtain slide preparations with clearly separated cells.

15. Follow the manufacturer's instructions for assembly of the slides, filter-cards, and cytocentrifuge cups within the cytocentrifuge rotor.

16. Cytocentrifugation is performed at 18–20°C.

    Caution: Loading and cytocentrifugation of cell culture sample must be carried out in an approved cytoguard cabinet to avoid the possibility of infectious disease transfer from buccal cells. Appropriate safety protection including gloves must be worn.

17. Resuspend the cells well using a Pasteur or Gilson pipette to disaggregate the cells.

18. Add 120 μL of cell suspension to the well of each sample cup. The required volume may need slight adjustment depending on the concentration of cells in suspension and the optimal cell density for slide scoring, which is determined by trial and error.

19. Replace rotor lid until locked and spin down in cytocentrifuge at 52×g for 5 min.

20. Press Start. The cytocentrifuge will run the program, stop, and slow automatically.

21. Upon completion of spinning, return the rotor to the fume hood and follow the manufacturer's procedure for opening each slide holder.
22. The slides are air dried for exactly 10 min at room temperature (18–22°C).
23. The slides are fixed in ethanol: glacial acetic acid mix (3:1), followed by further air drying for 10 min.
24. Alternatively, if a cytocentrifuge is not available, first fix the cells in a suspension of 3:1 ethanol: acetic acid, and then drop 120–150 µL of cell suspension onto the slides, and allow to air dry for 10 min prior to staining.

3.3. Buccal Cell Staining and Microscopy

3.3.1. Feulgen Staining

1. Fixed slides (including a spare control slide) are treated for 1 min each in Coplin jars with 50 and 20% ethanol, and then washed with Milli-Q water for 2 min in Coplin jars.
2. The slides are placed in a Coplin jar of 5 M HCl for 30 min and then rinsed in running tap water for 3 min.

   Critical: Include a negative control with each batch to check for efficacy of 5 M HCl treatment by placing a sample slide in Milli-Q water for 30 min instead of in 5 M HCl.
3. The slides are drained (but not allowed to dry out) and placed in a room temperature Coplin jar of Schiff's reagent for 60 min in the dark at room temperature.
4. The slides are treated in running water for 5 min and rinsed well in Milli-Q.
5. The slides are stained in 0.2% Light Green for 20–30 s and rinsed well in Milli-Q water.
6. Immediately place the slides face-down on Whatman No. 1 filter paper to blot away any residual moisture. Do not place any pressure or rub on the cell spots.
7. Place the slides on a slide tray and allow to dry for about 10–15 min.
8. Examine the cells at ×100 and ×400 magnification to assess the efficiency of staining and the density of the cells. If the cell density is too heavy or light, concentrate or dilute the cells as necessary and repeat the spinning and staining steps.
9. Leave the slides to dry completely for at least 30 min before putting coverslips on.

   Pause point: The slides can be left overnight at room temperature to dry.
10. Place the slides to be coverslipped on tissue paper and set out one coverslip alongside each.

    Caution: Carry out coverslipping in a fume hood to avoid inhalation of organic solvent in DePex and leave the slides until completely dry.

11. Place two large drops of DePex (use a plastic dropper) on each of the coverslips in the approximate area where the spots correspond.

    Caution: Wear nitrile gloves when applying DePex medium.

12. Invert the slide over the coverslip and allow the DePex to spread. Slide the coverslip gently to and fro to expel any excess DePex and air bubbles. Ensure that the spots do not have air bubbles over them.

13. Wipe excess DePex from the edges of the slide and ensure that the medium or glass does not cover any of the frosted label area, as a coding label will not stick on these.

14. Place the slides on a tray and leave overnight in the fume hood to dry.

15. Store the slides in slide boxes at room temperature and code with a sticky label over the frosted area before scoring.

16. When observed under transmitted light microscopy, nuclei and MNi have a magenta coloration, while cytoplasm appears green (Fig. 2). In negative controls (i.e., no 5 M HCl treatment), the nuclei are not stained with the magenta color.

17. Fluorescence can be achieved with this stain when the cells are viewed under fluorescence with a far red filter (see Note 3).

3.4. Criteria for Scoring Cell Types in the Buccal Micronucleus Cytome Assay

The criteria for the various distinct cell types and nuclear anomalies scored in the BMCyt assay are mainly based on the criteria described by Tolbert et al. (9). These criteria are intended for classifying buccal cells into categories that distinguish between "normal" cells and cells that are considered "abnormal" based on cytological features that are indicative of abnormal nuclear morphology. These abnormal nuclear morphologies are thought to be indicative of DNA damage or cell death.

The buccal cells are classified into the following cytome cell types, as shown in Fig. 2.

3.4.1. Normal Basal Cells (Fig. 2a)

These are the cells from the basal layer. The nuclear to cytoplasm ratio is larger than that in differentiated buccal cells derived from basal cells. Basal cells have a uniformly stained nucleus and they are smaller in size when compared to differentiated buccal cells. No DNA containing structures apart from the nucleus are observed in these cells. The cytoplasm is typically stained a darker shade of green with Light Green compared to differentiated cells.

3.4.2. Normal "Differentiated" Cells (Fig. 2b)

These cells have a uniformly stained nucleus, which is usually oval or round in shape. They are distinguished from basal cells by their larger size and by a smaller nuclear to cytoplasmic ratio. No other DNA containing structures apart from the nucleus are observed in these cells. These cells are considered to be terminally differentiated relative to basal cells. No mitotic cells are observed in this population.

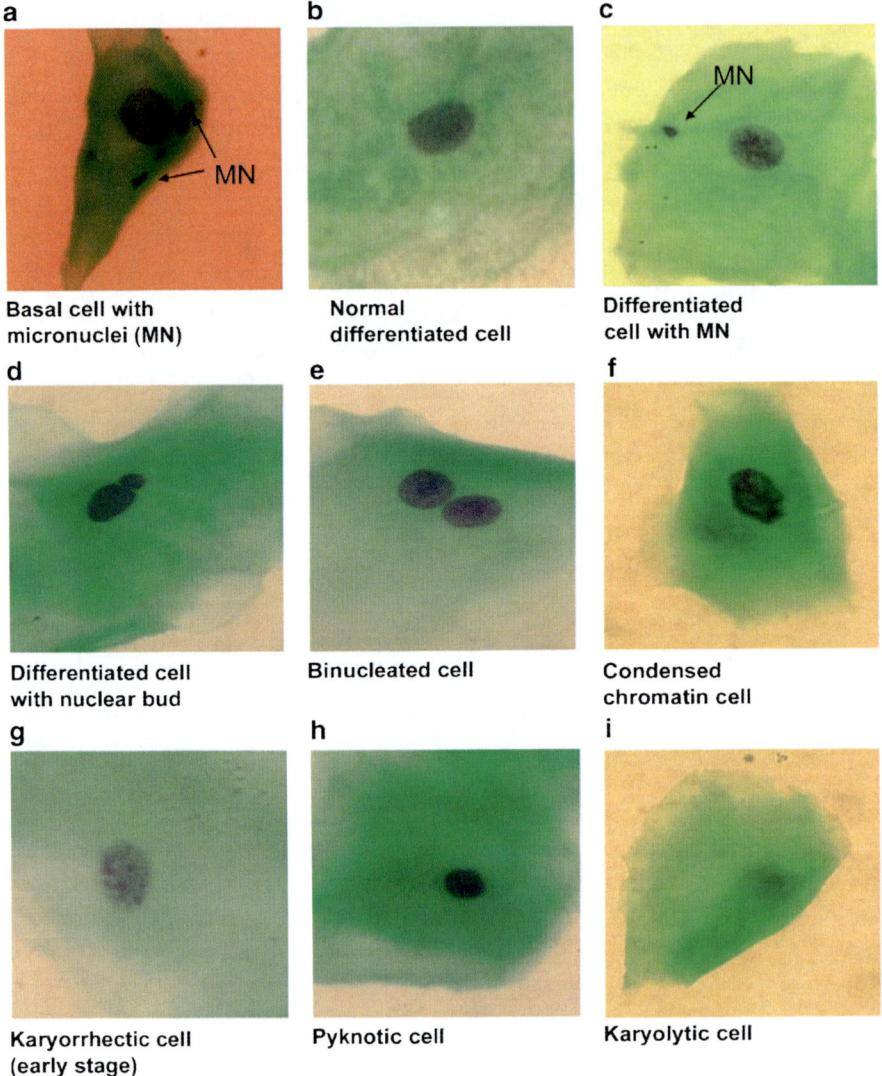

Fig. 2. Photomicrographs showing distinct buccal cell types as scored in the buccal micronucleus cytome assay.

3.4.3. Cells with Micronuclei (Fig 2a, c)

These cells are characterized by the presence of both a main nucleus and one or more smaller nuclei called MNi. The MNi are usually round or oval in shape and their diameter may range between 1/3 and 1/16 the diameter of the main nucleus. Cells with MNi usually contain only one micronucleus. It is possible but rare to find cells with more than 6 MNi. The nuclei in micronucleated cells may have the morphology of nuclei in normal cells or that of dying cells (i.e., condensed chromatin cells), although the latter are not scored to measure micronucleus frequency. The MNi must be located within the cytoplasm of the cells. The presence of MNi is indicative of chromosome loss or fragmentation occurring during previous nuclear division (10). MNi are scored only in basal and differentiated cells with uniformly stained nuclei. Cells with pyknotic, condensed chromatin, or karyorrhectic nuclei are not scored for MNi.

3.4.4. Cells with Nuclear Buds (Fig. 2d)

These cells have nuclei with an apparent sharp constriction at one end of the nucleus suggestive of a budding process, i.e., elimination of nuclear material by budding. In the original manuscript by Tolbert et al. (9), they were referred to as "broken egg" cells. The nuclear bud and the nucleus are usually in very close proximity and are apparently attached to each other. The nuclear bud has the same morphology and staining properties as the nucleus; however, its diameter may range from a half to quarter of that of the main nucleus. The mechanism leading to this morphology is not known but it may be due to elimination of amplified DNA and/or DNA repair complexes (11–13).

3.4.5. Binucleated Cells (Fig. 2e)

These cells have two nuclei instead of one. The nuclei are usually very close to each other and may be touching. The nuclei usually have the same morphology as that observed in normal cells. The significance of these cells is unknown but they may be indicative of failed cytokinesis following the last nuclear division. It has recently been shown that non-disjunction occurs with a higher frequency in binucleated cells that fail to complete cytokinesis rather than in cells that have completed cytokinesis (14). This recently identified mechanism is thought to be a cytokinesis checkpoint for aneuploid binucleated cells (14). The binucleate cell ratio may, therefore, prove to be an important biomarker for identifying individuals with cytokinesis defects, which could lead to higher than normal rates of aneuploidy (1).

3.4.6. Condensed Chromatin Cells (Fig. 2f)

These cells have nuclei with regions of condensed or aggregated chromatin exhibiting a roughly striated nuclear pattern in which the aggregated chromatin is intensely stained. In these cells, it is apparent that chromatin aggregates in some regions of the nucleus while being lost in other areas. When chromatin aggregation is extensive, the nucleus may appear to be fragmenting. These cells may be undergoing early stages of apoptosis, although this has not been conclusively proven. These cells may appear to contain MNi but should not be scored for MNi in the assay.

3.4.7. Karyorrhectic Cells (Fig. 2g)

These cells have nuclei that are characterized by the more extensive appearance of nuclear chromatin aggregation relative to condensed chromatin cells. They have a speckled nuclear pattern with apparent dissolution of nuclear material, which is indicative of fragmentation and eventual disintegration of the nucleus. These cells may be undergoing a late stage of apoptosis, but this has not been conclusively proven. These cells should not be scored for MNi in the assay. According to Tolbert et al. (9), fragmentation of the nucleus may rarely occur prior to advanced nuclear dissolution. These cells are also classified as karyorrhectic.

3.4.8. Pyknotic Cells (Fig. 2h)

These cells are characterized by a small shrunken nucleus, with a high density of nuclear material that is uniformly but intensely stained. The nuclear diameter is usually one to two-thirds of a nucleus in normal differentiated cells. The precise biological significance of pyknotic cells is unknown but it is thought that these cells may be undergoing a form of cell death; however, the precise mechanism remains unknown. They may represent an alternative stage of nuclear disintegration that is distinct from the condensed chromatin and karyorrhectic mechanisms.

3.4.9. Karyolytic Cells (Fig. 2i)

In these cells, the nucleus is completely depleted of DNA and is apparent as a ghost-like image that has no Feulgen staining. These cells thus appear to have no nucleus. It is probable that they represent a very late stage in the cell death process, but this has not been conclusively proven.

3.5. Scoring Method

Slides should be coded before scoring by an individual not involved in the experiment, so that the slide scorer is not aware of the treatment conditions and individual or groups to which the cells on the slides belong. Slides are best examined at x1000 magnification using a good-quality bright-field or fluorescence microscope (e.g., Leica DMLB and Nikon Eclipse E 600). The optimal way to score slides in the BMCyt assay is to first determine the frequency of all the various cell types in a minimum of 1,000 cells. Then the frequency of DNA damage biomarkers (MNi and nuclear buds) is scored in a minimum of 2,000 differentiated cells and 200 basal cells separately. For each slide, the following information should therefore be obtained:

1. The number of viable basal and differentiated cells per 1,000 cells scored.
2. The number of pyknotic, condensed chromatin, karyorrhectic, and karyolytic cells per 1,000 cells.
3. The number of binucleates per 1,000 cells scored.
4. The number of MNi and NBUDs in at least 2,000 differentiated cells and in 200 basal cells.
5. The frequency of basal cells and/or differentiated cells containing MNi.
6. The frequency of basal and/or differentiated cells containing NBUDs.

Preferably cells are scored using both bright-field and fluorescence microscopy. Cells containing MNi on bright field are confirmed as being positive for the biomarker by examining the cells also under fluorescence. The incidence of false positives can be minimized as DNA material such as nuclei and MNi fluoresce brightly red when viewed under fluorescence with a far red filter (emission wavelength range 580–620 nm).

4. Notes

1. It is important to note that the sampling method should be kept constant because the distribution of cell types may change depending on the sampling method. Repeated vigorous sampling may lead to collection of cells from the less differentiated basal layer (2).

2. We have performed experiments to investigate whether the filtering process in the preparation of the single cell suspension has adverse effects on cell population ratios. The larger cell aggregates that were filtered out reflected the same cellular population ratios as the eventual single cell suspension used in slide preparation for analysis. This is important in order to rule out selection against a particular cell type, which would have significant effects on both data analysis and interpretation.

3. One of the unique features of the staining technique used in this assay is that DNA material fluoresces when viewed under fluorescence with a far red filter (emission wavelength range 580–620 nm). This is important because cells containing MNi on bright field can be confirmed as being positive by examining the cells under fluorescence, and the nuclear texture which is important in classifying condensed chromatin and karyorrhectic cells may be easier to discern. This minimizes the incidence of false positives, thereby giving a more accurate assessment of DNA damage events.

   Previous studies have shown that false-positive results in micronucleus frequency as a result of using Romanowsky type stains can lead to inaccurate data interpretation. In a study investigating micronucleus frequency in relation to staining techniques in buccal mucosa of smokers against non-smokers, a four- to fivefold increase in MNi frequency in smokers was found using Leishman's stain. However, when a specific DNA fluorescent dye was used, there were no significant differences between these groups. Leishman's stain has been shown to increase the number of false positives as they positively stain keratin bodies that are often mistaken for MNi and is, therefore, not appropriate for this type of analysis (15).

References

1. Thomas, P., Harvey, S., Gruner, T., and Fenech, M. (2007) The buccal cytome and micronucleus frequency is substantially altered in Down's syndrome and normal ageing compared to young healthy controls. *Mutat Res* **638**, 37–47.
2. Thomas, P., Hecker, J., Faunt, J., and Fenech, M. (2007) Buccal micronucleus cytome biomarkers may be associated with Alzheimer's disease. *Mutagenesis* **22**, 371–379.
3. Thomas, P., and Fenech, M. (2007) Chromosome 17 and 21 aneuploidy in buccal cells is increased with ageing and in Alzheimer's disease. *Mutagenesis* **23**, 57–65.
4. Surralles, J., Autio, K., Nylund, L., Jarventaus, H., Norppa, H., Veidebaum, T., Sorsa, M., and Peltonen, K. (1997) Molecular cytogenetic analysis of buccal cells and lymphocytes from benzene-exposed workers. *Carcinogenesis* **18**, 817–823.
5. Titenko-Holland, N., Jacob, R. A., Shang, N., Balaraman, A., and Smith, M. T. (1998) Micronuclei in lymphocytes and exfoliated buccal cells of postmenopausal women with dietary changes in folate. *Mutat Res* **417**, 101–114.
6. Masters, B. R., Gonnord, G., and Corcuff, P. (1997) Three-dimensional microscopic biopsy of in vivo human skin: a new technique based on a flexible confocal microscope. *J Microsc* **185**, 329–338.
7. Veiro, J. A., and Cummins, P. G. (1994) Imaging of skin epidermis from various origins using confocal laser scanning microscopy. *Dermatology* **189**, 16–22.
8. Cairns, J. (1975) Mutation selection and the natural history of cancer. *Nature* **255**, 197–200.
9. Tolbert, P. E., Shy, C. M., and Allen, J. W. (1992) Micronuclei and other nuclear anomalies in buccal smears: methods development. *Mutat Res* **271**, 69–77.
10. Fenech, M., and Morley, A. A. (1986) Cytokinesis-block micronucleus method in human lymphocytes: effect of in vivo ageing and low dose X-irradiation. *Mutat Res* **161**, 193–198.
11. Fenech, M., and Crott, J. W. (2002) Micronuclei, nucleoplasmic bridges and nuclear buds induced in folic acid deficient human lymphocytes-evidence for breakage-fusion-bridge cycles in the cytokinesis-block micronucleus assay. *Mutat Res* **504**, 131–136.
12. Shimizu, N., Itoh, N., Utiyama, H., and Wahl, G. M. (1998) Selective entrapment of extrachromosomally amplified DNA by nuclear budding and micronucleation during S phase. *J Cell Biol* **140**, 1307–1320.
13. Shimizu, N., Kamezaki, F., and Shigematsu, S. (2005) Tracking of microinjected DNA in live cells reveals the intracellular behavior and elimination of extrachromosomal genetic material. *Nucl Acids Res* **33**, 6296–6307.
14. Shi, Q., and King, R. W. (2005) Chromosome nondisjunction yields tetraploid rather than aneuploid cells in human cell lines. *Nature* **437**, 1038–1042.
15. Nersesyan, A., Kundi, M., Atefie, K., Schulte-Hermann, R., and Knasmuller, S. (2006) Effect of staining procedures on the results of micronucleus assays with exfoliated oral mucosa cells. *Cancer Epidemiol Biomarkers Prev* **15**, 1835–1840.

Chapter 18

γ-H2AX Detection in Peripheral Blood Lymphocytes, Splenocytes, Bone Marrow, Xenografts, and Skin

Christophe E. Redon, Asako J. Nakamura, Olivier Sordet, Jennifer S. Dickey, Ksenia Gouliaeva, Brian Tabb, Scott Lawrence, Robert J. Kinders, William M. Bonner, and Olga A. Sedelnikova

Abstract

Measurement of DNA double-strand break (DSB) levels in cells is useful in many research areas, including those related to DNA damage and repair, tumorigenesis, anti-cancer drug development, apoptosis, radiobiology, environmental effects, and aging, as well as in the clinic. DSBs can be detected in the nuclei of cultured cells and tissues with an antibody to H2AX phosphorylated on serine residue 139 (γ-H2AX). DSB levels can be obtained either by measuring overall γ-H2AX protein levels in a cell population or by counting γ-H2AX foci in individual nuclei. Total levels can be obtained in extracts of cell populations by immunoblot analysis, and in cell populations by flow cytometry. Furthermore, with flow cytometry, the cell cycle distribution of a population can be obtained in addition to DSB levels, which is an advantage when studying anti-cancer drugs targeting replicating tumor cells. These described methods are used in genotoxicity assays of compounds of interest or in analyzing DSB repair after exposure to drugs or radiation. Immunocyto/immunohistochemical analysis can detect γ-H2AX foci in individual cells and is very sensitive (a single DSB can be visualized), permitting the use of extremely small samples. Measurements of γ-H2AX focal numbers can reveal subtle changes found in the radiation-induced tissue bystander response, low dose radiation exposure, and in cells with mutations in genomic stability maintenance pathways. In addition, marking DNA DSBs in a nucleus with γ-H2AX is a powerful tool to identify novel DNA repair proteins by their abilities to co-localize with γ-H2AX foci at the DSB site. This chapter presents techniques for γ-H2AX detection in a variety of human and mouse samples.

Key words: γ-H2AX, DNA damage, Immunofluorescence, Immunoblotting, Flow cytometry, Lymphocytes, Splenocytes, Bone marrow, Xenografts, Skin

1. Introduction

DNA double-strand breaks (DSBs) may be caused by a variety of factors, including exposure to chemical and environmental stresses, errors in cellular metabolism, and they may be formed as essential intermediates in programmed biological processes (1). Immediately after the formation of a DSB, histone H2AX, a protein component of chromatin, is phosphorylated on serine 139 (2). The phosphorylated form of H2AX is named γ-H2AX. During the 30 min following DSB formation, several hundred to over a thousand γ-H2AX molecules are formed along the chromatin adjacent to the DSB site to form a γ-H2AX focus (3). Virtually every DSB is represented by a γ-H2AX focus (4, 5). The γ-H2AX foci serve as a target for the recruitment of many DNA damage repair factors (6). After the DSB is repaired, the γ-H2AX molecules are dephosphorylated and the γ-H2AX foci disappear (7).

Compared to γ-H2AX-based methods, other measures of DNA DSB levels, which rely on analyzing DNA fragmentation, either by pulsed-field gel electrophoresis, comet assay, or DNA elution, are less sensitive. With ionizing radiation, high doses are required to produce amounts of DSBs detectable by DNA fragmentation assays. These doses are lethal for mammals and many other organisms (8–11). The numbers of γ-H2AX foci correlate directly with the amount of radiation received (5, 12).

Several antibodies directed against γ-H2AX are commercially available (20). Antibodies against γ-H2AX are used in two types of assays. The first type involves measurements of total γ-H2AX protein levels in a cell population (immunoblotting and flow cytometry). Immunoblotting measures the amount of γ-H2AX in protein extracts after homogenization of the cells or tissue (13, 14), while flow cytometry measures the total amount of γ-H2AX per cell in a population (10, 15, 16). Flow cytometry analysis of γ-H2AX levels also yields information on the cell cycle distribution of a population, which is useful when studying compounds that preferentially act on S-phase cells (17). The second type of assay involves immunocytochemical and immunohistochemical visualization of γ-H2AX foci in cultured cells and tissues, respectively. Despite the fact that these techniques are labor intensive and costly, they permit detection of individual DSBs. This is crucial when measuring the effects of low-dose radiation exposure, identifying weakly genotoxic compounds, and studying aging and genomic instability (reviewed in (1)). Both immunoblotting and flow cytometry techniques are less sensitive than microscopy. For example, techniques using microscopy can detect a single DSB, while immunoblotting and flow cytometry typically have detection limits of 2–20 DSBs per cell, depending on the cell type. Moreover, because γ-H2AX marks DSB sites, the use of microscopy permits the study of DNA repair in single cells, identifying repair factors that co-localize with γ-H2AX. Thus, the detection of γ-H2AX is a powerful and straightforward tool for the analysis of DNA DSB damage *in vivo*. The following protocols permit the detection of γ-H2AX in human and mouse tissues, attached and suspension cell cultures, and peripheral blood mononuclear cells (PBMCs).

2. Materials

2.1. Isolation of Suspension Cells

2.1.1. Isolation and Activation of Peripheral Blood Lymphocytes

1. For human blood collection, 4- or 6-mL anticoagulant sodium heparin-coated tubes (BD Biosciences, San Jose, CA).
2. For mouse blood collection, 0.5-mL anticoagulant EDTA-coated capillary tubes (Fisher Scientific, Billerica, MA).
3. 15-mL centrifuge tubes (Corning, Life Sciences, Lowell, MA).
4. Phosphate-buffered saline (PBS), pH 7.4.
5. Ficoll-Paque (GE Healthcare, Piscataway, NJ).
6. Suspension cell medium (SCM): RPMI medium (Invitrogen, Carlsbad, CA) supplemented with 10% fetal bovine serum (FBS; Atlanta Biologicals, Lorensville, GA) and 100 U/mL penicillin, 100 μg/mL streptomycin (Invitrogen).
7. Phytohemaglutinin A (PHA; Sigma-Aldrich, St. Louis, MO) stock solution at 1 mg/mL in water is used for lymphocyte activation. Store at –20°C.

2.1.2. Isolation and Activation of Splenocytes

1. 100-mm tissue culture dishes (BD Biosciences).
2. Tweezers.
3. 5-mL syringes (BD Biosciences).
4. 25-gauge needles (BD Biosciences).
5. SCM.
6. Activation cocktail which is at 2×: SCM containing 10 μg/mL concanavalin A (Con-A), 40 μg/mL lipopolysaccharide (LPS), and 300 U/mL interleukin-2 (IL-2) from mouse, all from Sigma-Aldrich. Stock solutions in water: Con-A 10 mg/mL, LPS 5 mg/mL, and IL-2 100,000 U/mL are stored at –20°C.

2.1.3. Isolation of Bone Marrow Cells

1. 1.6-mL graduated microcentrifuge tubes (CLP, San Diego, CA).
2. 3-mL syringes (BD Biosciences).
3. 25-gauge needles.
4. SCM.

2.2. Immunocytochemistry in Suspension Cells

1. 15-mL conical centrifuge tubes (Corning, Life Sciences).
2. Microscopic superfrost plus slides (Erie Scientific, Portsmouth, NH).
3. Coplin glass jars (Electron Microscopy Sciences, Hatfield, PA).
4. Gold-seal coverslips 18 × 18 mm (Electron Microscopy Sciences).

5. Liquid-repellent slide marker pen (Pap Pen) (Daido Sangyo, Tokyo, Japan).
6. 20% paraformaldehyde solution (Electron Microscopy Sciences) stored at room temperature (RT).
7. PBS, pH 7.4.
8. Centrifuge (model GS-6KR) with GH 3.8 swinging bucket rotor (Beckman Coulter, Fullerton, CA) (see Note 1).
9. Cytocentrifuge (model Shandon Cytospin 3, Fisher Scientific) and cytocentrifuge supplies.
10. 70% ethanol stored at −20°C must be cold at the time of application.
11. PBS-TT: PBS containing 0.5% Tween-20 and 0.1% Triton X-100. PBS-TT may contain 5% bovine serum albumin (BSA, Sigma-Aldrich) for blocking, or 1% BSA for diluting primary and secondary antibodies. PBS-TT is stored at RT for up to 1 month.
12. Primary antibodies: mouse monoclonal anti-γ-H2AX (Cat.# ab18311, Abcam, Cambridge, MA) for human cells (stored at −20°C), or rabbit polyclonal anti-γ-H2AX (Cat# NB100-384, Novus Biologicals, Littleton, CO) for mouse cells (stored at 4°C). Dilute before use with PBS-TT containing 1% BSA.
13. Secondary antibodies: goat anti-mouse Alexa Fluor 488-conjugated IgG (Cat# A11029) for mouse monoclonal anti-γ-H2AX, or Alexa Fluor 488-conjugated goat anti-rabbit IgG (Cat.# A11034) for rabbit polyclonal anti-γ-H2AX (both from Invitrogen). Secondary antibodies are stored at 4°C and diluted before use with PBS-TT containing 1% BSA.
14. 50 mg/mL RNase A (Sigma-Aldrich). Aliquoted stock solution is stored at −20°C and diluted with PBS to 0.5 mg/mL before use.
15. Vectashield mounting medium containing propidium iodide (PI) (Vector Laboratories, Burlingame, CA).
16. Laser scanning confocal microscope Nikon PCM 2000 (Nikon, Augusta, GA), or inverted microscope Olympus IX70 (Olympus America, Center Valley, PA).

2.3. Flow Cytometry in Peripheral Blood Lymphocytes

1. SCM containing RPMI medium (Invitrogen) supplemented with 10% FBS (Atlanta Biologicals) and 100 U/mL penicillin, 100 μg/mL streptomycin (Invitrogen).
2. PBS, pH 7.4, pre-chilled on ice.
3. 70% ethanol is stored at −20°C to be cold at the time of application.
4. 20% paraformaldehyde (Electron Microscopy Science). Prepare a 4% paraformaldehyde solution in PBS fresh for each experiment.

5. Triton X-100. Prepare a 0.25% Triton X-100 solution in ice-cold PBS fresh for each experiment and keep in the refrigerator until the time of application.
6. 15-mL conical centrifuge tubes (Corning, Life Sciences).
7. Centrifuge (model GS-6KR) with GH 3.8 swinging bucket rotor (Beckman Coulter) (see Note 2).
8. BSA (Sigma-Aldrich). Prepare a 1% solution in PBS fresh for each experiment.
9. Primary mouse monoclonal anti-γ-H2AX antibody (Cat# ab18311, Abcam) for human cells. Store at –20°C and dilute in PBS/1% BSA before use.
10. Secondary goat anti-mouse Alexa Fluor 488-conjugated IgG (Cat# A11029, Invitrogen). Store at 4°C and dilute in PBS/1% BSA before use.
11. RNase A (Sigma-Aldrich) stock solution (10 mg/mL in water). Aliquot and store at –20°C.
12. PI (Sigma-Aldrich) stock solution (1 mg/mL in water). Store at 4°C.
13. 5-mL BD Falcon™ round-bottom tubes (BD Biosciences).
14. Flow cytometer FACScan (BD Biosciences).

2.4. Western Blotting in Peripheral Blood Lymphocytes

1. PBS, pH 7.4, containing 10 mM NaF (Sigma-Aldrich). 1 M NaF stock solution in water is stored at 4°C.
2. 2× SDS protein gel loading solution (Quality Biologicals, Gaithersburg, MD).
3. Beta-mercaptoethanol (β-ME) (Sigma-Aldrich).
4. 1.6-mL graduated microcentrifuge tubes (CLP).
5. Novex 4–20% Tris–glycine polyacrylamide gels (Invitrogen).
6. Tris–glycine–SDS running buffer (10×) (Bio-Rad Laboratories, Hercules, CA).
7. Kaleidoscope prestained standards (Bio-Rad Laboratories)
8. Novex Xcell II mini cell system (Invitrogen).
9. Tris–glycine transfer buffer (25×) (Invitrogen).
10. PVDF membrane (0.2-μm pore size) (Invitrogen). The 0.45-μm pore size is also compatible for γ-H2AX detection.
11. TBS-T: Tris-buffered saline (TBS), pH 7.4 (Mediatech Inc, Manassas, VA) containing 0.05% Tween-20 (Invitrogen).
12. Non-fat milk powder (Bio-Rad Laboratories).
13. Methanol (Fisher Scientific).
14. 50-mL conical tubes (BD Biosciences).
15. PageBlue™ protein staining solution (Fermentas International Inc., Ontario, Canada).

16. Primary antibodies: mouse monoclonal anti-γ-H2AX (Cat# ab18311, Abcam) and rabbit polyclonal anti-H2AX (Cat# ab11175, Abcam). Dilute before use with blocking buffer (see Subheading 3).
17. Secondary antibodies: anti-mouse horseradish peroxidase (HRP)-conjugated IgG or anti-rabbit HPR-conjugated IgG (Cat# NA931V or NA934V, respectively, GE Healthcare).
18. Enhanced chemiluminescent (ECL) reagents (GE Healthcare).
19. High sensitivity chemiluminescent films (GE Healthcare).
20. Restoring Western blot stripping buffer (Pierce biotechnology, Rockford, IL).

2.5. Immunohistochemistry in Xenografts and Skin

1. Screw-cap cryovials (Fisher Scientific, Pittsburgh, PA).
2. Biotinylated monoclonal anti-γ-H2AX antibody JBW301 (Cat# 16-193, Upstate Biotechnology, Lake Placid, NY). Now the antibody can be obtained from Millipore (Billerica, MA).
3. Streptavidin–Alexa Fluor-488 conjugate (Cat# S32354 or S11223, Invitrogen).
4. Prolong gold with DAPI (Invitrogen).
5. Absolute ethanol (AAPER, Toronto, Ontario, Canada).
6. 10% neutral buffered formalin (NBF; Richard Allen Scientific, Kalamazoo, MI).
7. Xylene, histology grade (Fisher Scientific).
8. Paraffin Tissue-Tek VIP (Electron Microscopy Sciences).
9. Temno 2N2711X tru-cut 18-gauge biopsy needles (Allegiance Healthcare, McGaw Park, IL).
10. Vision BondMax autostainer (Leica Microsystems, Bannockburn, IL) and ancillary reagents: Bond dewax, Bond wash solution, Bond ER1 solution.
11. Leica RM2255 automated microtome (Nussloch, Germany).
12. Accu-Edge low-profile microtome blades (Sakura Finetek, Torrance, CA).
13. Water bath (Triangle Biomedical Science, Durham, NC).
14. Superfrost plus slides (Erie Scientific).
15. Incubator (model 10-140, Quality Lab, Chicago, IL).
16. Leica DM-5000 fluorescence microscope equipped with Phase Contrast Head, 20× HCX Plan Fluotar Phase 2 objective, Leica EL6000 external light source, Leica DM5000B fluorescent light source, and Leica filter systems A4 and L5.
17. Retiga 2000R CCD Camera and Image Pro or Q Capture Pro software (Q-Imaging, Tucson, AZ).

3. Methods

3.1. Immunocytochemistry Detection of γ-H2AX in Peripheral Blood Lymphocytes

This protocol describes the immunocytochemical detection of γ-H2AX in mammalian suspension cells, including post-mitotic or activated peripheral blood lymphocytes, splenocytes, and bone marrow (Fig. 1). It can be applied for *in vivo* studies monitoring patient response to radiotherapy or drug treatment (18), or animal model experiments, as well as for *ex vivo* experiments with mammalian material (see Note 3).

3.1.1. Cell Isolation

3.1.1.1. Isolation and Activation of Peripheral Blood Lymphocytes

1. Dilute blood samples collected by venipuncture or cardiac puncture into anticoagulant tubes with 1 vol. of PBS.
2. Layer 1 vol. of the blood–PBS mix on the top of 2 vol. of Ficoll-Paque in a centrifuge tube.
3. Centrifuge the solution at $700 \times g$ for 25 min at RT (the breaking function of the centrifuge must be deactivated for this step).
4. Collect the layer containing lymphocytes.
5. Wash the lymphocytes twice with 10 mL of SCM by centrifuging at $600 \times g$ for 5 min at RT.

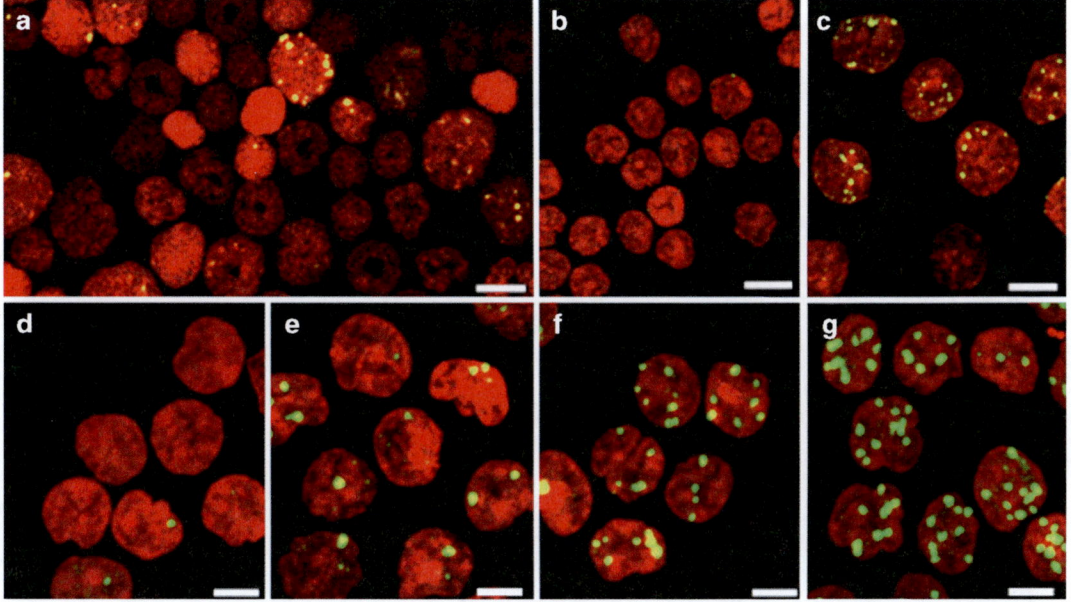

Fig. 1. Representative images of γ-H2AX staining in suspension cells isolated from mice and humans. (**a**) Unirradiated mouse bone marrow; (**b**, **c**) Splenocytes, non-activated (**b**) and at 2 days post-activation (**c**). Splenocyte activation leads to DSB formation resulting from replication stress (19); (**d–g**) Human blood lymphocytes, unirradiated (**d**) and 30 min after 0.1 Gy-irradiation (**e**), 0.4 Gy-irradiation (**f**) and 0.6 Gy-irradiation (**g**). *Green*, γ-H2AX; *red*, DNA stained with PI. Bar is 5 μm. Magnification, 1,000×.

6. Dilute the lymphocytes to 4×10^6 cells/mL in SCM; they can be processed immediately, or maintained at 37°C for up to 3 days.

7. Lymphocytes may be activated by adding 20 μg/mL of PHA to the 2× activation cocktail used for spleen cells.

3.1.1.2. Isolation and Activation of Spleen Cells

1. Sacrifice mouse, remove the spleen aseptically, and place it in 100-mm tissue culture dishes.
2. While holding the spleen with tweezers, and using a 5-mL syringe containing 3 mL of SCM, disperse spleen cells by pushing the SCM into one side of the spleen through a 25-gauge needle.
3. Repeat the procedure by turning the spleen 180° and applying SCM to disperse more cells.
4. Cells are stored on ice in SCM for up to 3–4 h, or if necessary until fixation.
5. Dilute splenocytes to 4×10^6 cells/mL in SCM. Contrary to peripheral lymphocytes, the splenocytes have to be used on the day of their isolation from the spleen.
6. For splenocyte activation, add 2 mL of spleen culture (4×10^6 cells/mL) to 2 mL of 2× activation cocktail.

3.1.1.3. Isolation of Bone Marrow Cells

1. Sacrifice mouse and remove femur and tibia by cutting through the bones at the ankle and near the pelvis.
2. Trim the muscles and fat from the bones, and then separate the two bones by cutting through the knee joint.
3. Using a 3-mL syringe containing 1 mL of SCM, flush the bone marrow cells into a microfuge tube by inserting a 25-gauge needle in the openings of the bone ends.
4. Pipette the cells gently to remove clumps; they can be stored on ice at 4°C for up to 3–4 h, or if necessary until fixation.

3.1.2. Cell Fixation and Preparation for Staining

1. Transfer a 2-mL aliquot of cells (4×10^6 cells/mL) to a 15-mL conical centrifuge tube.
2. Fix with 220 μL of 20% paraformaldehyde solution (2% final concentration).
3. Vortex briefly.
4. Incubate for 20 min at RT.
5. Add 10 mL of PBS and mix briefly.
6. Centrifuge for 5 min at $600 \times g$ at RT and discard the supernatant.
7. Repeat washing by centrifugation twice more with 10 and 5 mL of PBS.
8. Resuspend the pellet in PBS at a concentration of 2×10^6 cells/mL (see Note 4).

9. Spot cell samples onto slides by cytospining 200–300 μL of cell suspension for 4 min at $80 \times g$ at RT (see Note 5).

3.1.3. Immunocytochemistry

1. After cytospin, dry the specimens for 5 min at RT.
2. Re-hydrate with PBS for 15 min.
3. Place in a −20°C pre-chilled glass Coplin jar and incubate with −20°C pre-chilled 70% ethanol at room temperature for at least 20 min (see Note 6).
4. Wash the slides in PBS for 15 min.
5. Block in PBS-TT containing 5% BSA for 30 min at 20°C.
6. Wash in PBS once for 5 min.
7. Incubate the specimens with the primary antibody at a 500× dilution in PBS-TT containing 1% BSA for 2 h at RT. One of the following antibodies is used: mouse monoclonal anti-γ-H2AX or rabbit polyclonal anti-γ-H2AX (see Note 7).
8. Wash in PBS three times for 5 min each.
9. Incubate the specimens with secondary antibody at a 500× dilution in PBS-TT containing 1% BSA for 1 h at RT. The secondary antibody is one of the following, depending on the primary antibody: Alexa Fluor 488-conjugated goat anti-mouse IgG or Alexa Fluor 488-conjugated goat anti-rabbit IgG.
10. Wash in PBS three times for 5 min each.
11. Incubate the specimens with 0.5 mg/mL RNase A in PBS at 37°C for 20 min.
12. Wash in PBS twice for 5 min each.
13. Apply mounting medium with PI and coverslip. The edges of the coverslip should be sealed with nail polish (see Note 8).
14. Perform microscopy.

3.1.4. Imaging and Image Analysis

1. Count the γ-H2AX foci directly using a fluorescent microscope. We typically count foci in 50–100 nuclei.
2. A confocal microscope is used to capture whole nuclei. Each confocal image is taken with increments of 0.5 μm with z-sections condensed, so all detectable foci are visible in a single plane.
3. γ-H2AX foci are counted by eye in images opened in Photoshop or Paint Shop Pro software.
4. Alternatively, image-counting or intensity measuring software can be used, such as the Image Pro 6.2 Analyzer (Media Cybernetics, Bethesda, MD), IPLab (BD), or Image Quant (Molecular Dynamics).

3.2 Detection of γ-H2AX in Peripheral Blood Lymphocytes by Flow Cytometry

This protocol describes the detection of γ-H2AX by flow cytometry in *ex vivo* and *in vivo* samples such as human lymphocytes isolated from peripheral blood (Fig. 2).

1. Use 10×10^6 lymphocytes per sample (see Note 9) at 1×10^6 cells/mL. Lymphocytes are maintained in SCM containing 10% FBS for up to 3 days.
2. After treatment, transfer the cell culture to a 15-mL conical tube. All steps described below are performed in 15-mL conical tubes.
3. Wash the cells once with 1 mL of ice-cold PBS, i.e., cells are centrifuged at $500 \times g$ for 5 min at 4°C, resuspended in 1 mL of ice-cold PBS and centrifuged again right away at $500 \times g$ for 5 min. All the washes indicated below are performed the same way.
4. Fix the cells with 1 mL of 4% paraformaldehyde for 10 min at RT.
5. Wash the cells once with 1 mL of ice-cold PBS (see step 3).
6. Permeabilize the cells with 1 mL of pre-chilled (−20°C) 70% ethanol for 20 min at RT or overnight at 4°C (see Note 10).

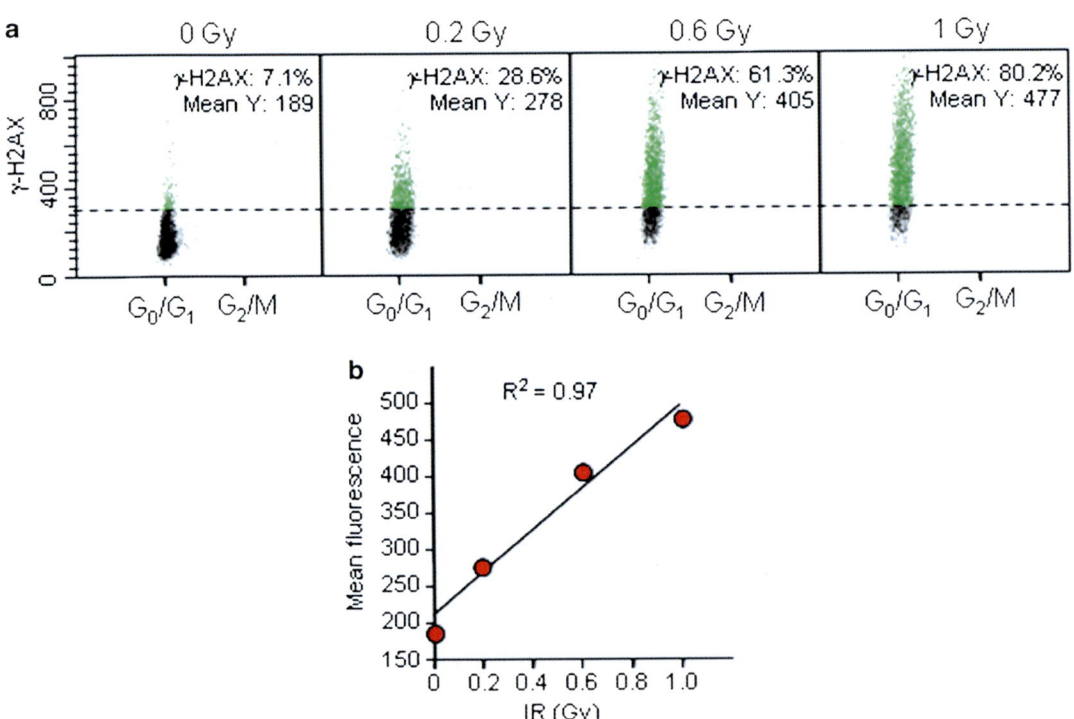

Fig. 2. Flow cytometry analysis of γ-H2AX and DNA content in irradiated post-mitotic human primary lymphocytes. (a) Cells were analyzed 30 min post-irradiation. γ-H2AX-positive cells are shown in *green*. Numbers are percentages of γ-H2AX positive cells and mean fluorescence intensity of γ-H2AX (mean Y). (b) Quantification of mean fluorescence intensity data shown in (a).

7. Wash the cells twice with 1 mL of ice-cold PBS.

8. Further permeabilize the cells with 1 mL of ice-cold 0.25% Triton X-100 for 5 min on ice.

9. Wash the cells once with 1 mL of ice-cold PBS.

10. Incubate the cells with 200 μL of mouse monoclonal anti-γ-H2AX antibody at 250× dilution in PBS/1% BSA for 1 h at RT. No agitation is needed.

11. Wash the cells once with 1 mL of ice-cold PBS.

12. Incubate the cells with 200 μL of goat anti-mouse Alexa Fluor-488 antibody at 250× dilution in PBS/1%BSA for 30 min at RT. No agitation is needed. At this step and the next ones, protect the cells from light with foil.

13. Wash the cells once with 1 mL of ice-cold PBS.

14. Resuspend the cells in 500 μL of a solution containing 50 μg/mL of PI and 0.5 mg/mL of RNase A. This solution is obtained by a 20× dilution of stock solutions, 1 mg/mL PI and 10 mg/mL RNase A, in PBS.

15. Transfer the cells to a 5-mL BD Falcon™ round-bottom tube and incubate for a few minutes at RT before analyzing by flow cytometry in FL2-A (PI) versus FL1-H (γ-H2AX). The samples can be stored for 2 h in the refrigerator before the analysis. For the data shown in Fig. 2, the following settings were used with the CellQuest software (Tampa, FL): FSC (voltage E00, Amp gain 2.14, linear mode), SSC (voltage 381, Amp gain 1.00, linear mode), FL1-H (voltage 790, Amp gain 1.00, linear mode, see Note 11), FL2-H (voltage 500, Amp gain 1.00, linear mode), FL2-A (Amp gain 1.00, linear mode), and FL2-W (Amp gain 2.79, linear mode). For acquisition, open three windows: FSC versus SSC (removes cellular debris during the analysis), FL2-A versus FL2-W (removes the cell doublets during the analysis), and FL2-A versus FL1-H.

3.3. Detection of γ-H2AX in Peripheral Blood Lymphocytes by Western Blotting

This protocol describes the detection of overall γ-H2AX levels relative to total H2AX in human peripheral blood lymphocytes. The procedure is simple and inexpensive and could be used in most laboratories. However, compared to microscopy, this Western blotting-based detection of γ-H2AX is less sensitive, which should be considered, especially when measuring effects of low-dose radiation.

3.3.1. Protein Extraction

1. Add 5×10^5 cells/mL of lymphocytes to 1.6-mL microcentrifuge tubes (see Note 12).

2. Centrifuge the cells at $2,000 \times g$ for 5 min at 4°C.

3. Wash the cells with PBS containing 10 mM of NaF.

4. Centrifuge the cells at 2,000×*g* for 5 min at 4°C and discard the supernatants.

5. Add 30 µL of 1× SDS protein gel loading solution containing 2.5% β-ME. Add β-ME to the loading solution right before use.

6. Boil the specimens for 10 min.

7. Chill the specimens on ice for 5 min (see Note 13).

8. Vortex briefly.

9. Centrifuge the specimens at 16,000×*g* for 5 min at 4°C.

3.3.2. SDS-Polyacrylamide Gel Electrophoresis

1. These instructions assume the use of the Novex Xcell II mini cell system. Prepare 1× Tris–glycine–SDS running buffer by diluting 100 mL of 10× Tris–glycine–SDS running buffer with 900 mL of water. Add the running buffer to the chamber.

2. Load 15 µL of supernatant samples and 15 µL of pre-stained molecular weight standard to a 4–20% Tris–glycine polyacrylamide gel.

3. Complete the assembly of the gel system and connect the power supply.

4. Perform electrophoresis at 150 V for 1.5 h, until blue dye line runs off the gel.

3.3.3. Western Blotting

1. These instructions assume the use of the Novex Xcell II blot module system. Prepare 1× transfer buffer by mixing 40 mL of 25× Tris–glycine transfer buffer, 760 mL of water, and 200 mL of methanol. The transfer buffer is made before the transfer step and pre-chilled at 4°C.

2. Carefully remove the PVDF membrane sandwiched between the two filter papers, soak with methanol for few seconds, and then wet the membrane in transfer buffer. Soak the filter papers in transfer buffer. Wash six sponges with water and remove all bubbles, then soak in transfer buffer.

3. Disconnect the gel system from the power supply and disassemble it. Remove and discard the top and bottom parts of the gel. Soak the gel in transfer buffer.

4. Put three sponges on the cassette and carefully lay the first wet filter paper on the sponge.

5. Lay the gel on the top of the paper.

6. Carefully lay the PVDF membrane on top of the gel.

7. Lay the second wet filter paper on the top of the membrane, insuring that no bubbles are trapped in the resulting sandwich.

8. Put three sponges on the filter paper.

9. Place the cassette into the transfer tank. Make sure the orientation is correct (the anode is closer to the membrane than the gel).
10. Complete the assembly of the transfer system and connect the power supply. Transfer the protein at 50 V for 1 h.
11. Prepare blocking buffer by diluting 2.5 g of non-fat milk with 50 mL of TBS-T.
12. Once the transfer is complete, take the cassette out of the tank and carefully disassemble. Remove the top three sponges and the filter paper.
13. Remove and incubate the PVDF membrane in 50 mL of blocking buffer at RT for 1 h.
14. Wash the gel with water for 5 min and incubate it in PageBlue™ protein staining solution overnight. After staining, wash the gel with water several times until protein bands are clear.
15. Prepare a 500× dilution of the anti-γ-H2AX primary mouse monoclonal antibody solution in blocking buffer in a 50-mL conical tube.
16. Carefully put the membrane in the tube and, rotating the tube, incubate at 4°C overnight (see Note 14). Make sure the protein side of the membrane is facing the inside of the tube.
17. Take the membrane out of the tube and wash three times for 5 min each with TBS-T.
18. Prepare a 15,000× dilution of the secondary antibody solution (anti-mouse HRP-conjugated IgG) in TBS-T. Incubate the membrane in the secondary antibody solution at RT for 1 h.
19. Discard the secondary antibody solution and wash the membrane six times for 5 min each with TBS-T.
20. These instructions assume the use of ECL Western blotting detection reagents from GE Healthcare. Prepare the detection reagent by mixing 1 mL of detection reagent 1 with 1 mL of detection reagent 2.
21. Drain the excess TBS-T from the washed membrane and place on a sheet of plastic wrap. Cover the membrane with the detection reagent and incubate at RT for 5 min.
22. Drain the excess of detection reagent and place the membrane in a clear plastic file. Wrap the membrane and gently smooth away any air bubbles.
23. Place the wrapped membrane in an X-ray film cassette.
24. Place a sheet of high sensitivity chemiluminescent detection film on top of the membrane. Exposure typically takes a few seconds to a few minutes.

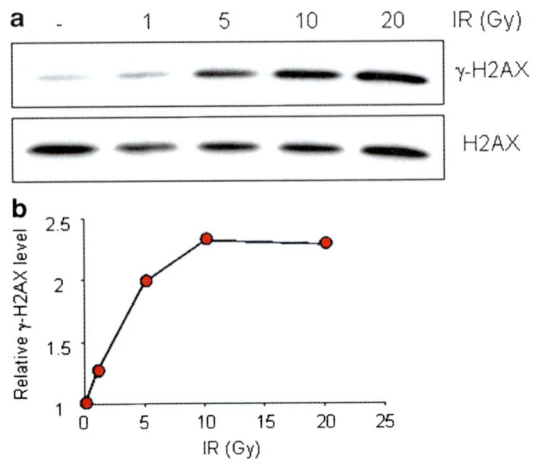

Fig. 3. Western blot analysis of γ-H2AX in irradiated post-mitotic human primary lymphocytes. (**a**) Dose-dependent increase of γ-H2AX level. (**b**) Quantification of relative induction of γ-H2AX shown in (**a**). γ-H2AX levels were normalized by total H2AX expression levels.

25. Once a satisfactory exposure of the γ-H2AX has been obtained, strip the signal off of the membrane and re-blot with the anti-H2AX antibody.
26. Wash the membrane for 5 min with TBS-T.
27. Incubate the membrane with stripping buffer at RT for 10 min on a shaker.
28. Wash the membrane twice for 5 min each with TBS-T.
29. Prepare a 4,000× dilution of the anti-H2AX primary rabbit polyclonal antibody solution (anti-rabbit HRP-conjugated IgG) in blocking buffer in a 50-mL conical tube. Repeat the incubation and development steps described above. An example of the results produced is shown in Fig. 3.

3.4. Immunohistochemical Detection of γ-H2AX in Xenografts and Skin

Fluorescent immunostaining of tissues prepared by touch printing, or frozen tissue sectioning was addressed in our earlier publication (14). Here, we present a method of fluorescent γ-H2AX staining in formalin-fixed, paraffin-embedded biopsies of skin and xenografts (Fig. 4), which retain better cellular and tissue morphology than frozen tissues. The presented assay overcomes the problems of high autofluorescence and generally poor detection of γ-H2AX in paraffin sections.

Biotinylated monoclonal anti-γ-H2AX antibody is used as the detector, and Alexa Fluor 488-conjugated streptavidin serves as the reporter. Specimens for this assay were collected according to standard operating procedures employed by SAIC-Frederick, for biopsy materials. This method was developed to detect drug-induced γ-H2AX changes in xenografts, and applied to evaluate the performance of topoisomerase 1 inhibitors in a nude mouse

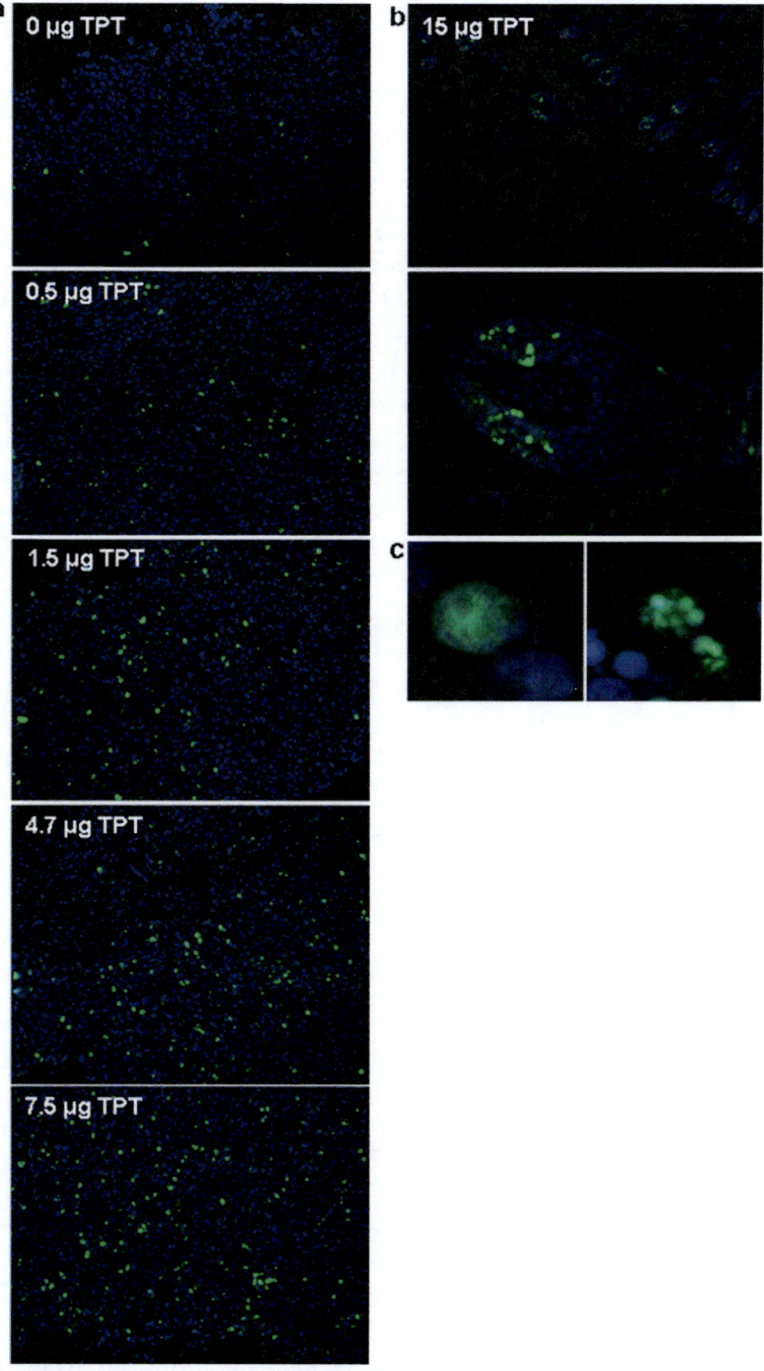

Fig. 4. Representative images of γ-H2AX staining in paraffin sections. (a) Dose–response to topotecan (TPT) in a mouse xenograft model (human melanoma) in nude mice. Mice were dosed one time with TPT. At approximately 4 h post-dosing, mice were anesthetized, a skin flap was cut, and the xenograft was biopsied. (b) Xenografts showing skin and hair follicle response at 15 mg/kg (mouse weight) TPT treatment, 4 h post-dosing. *Lower Panel* shows enhanced images. (c) Two patterns of immunofluorescence in paraffin sections, pan-staining (*left*) and focal pattern (*right*). Image analysis is based on the scoring of γ-H2AX-positive cells. *Green*, γ-H2AX; *blue*, DNA stained with DAPI. Magnification, 200×.

3.4.1. Specimen Collection

Our procedures are optimized for handling an 18-gauge needle biopsy, but have been successfully employed for pieces of resected tumors from human patients and mouse xenografts (see Note 15).

1. Prior to collection of the tumor piece, label a 2-mL screw-cap cryovial for each tumor piece to be collected and place the vials into a Dewar flask containing either dry ice/acetone or liquid nitrogen to pre-cool the vials.
2. Collect the specimen without a perfusion step. We have successfully obtained specimens from patients under local anesthesia and mice under general anesthesia (isoflurane gas). Needle biopsies may be directly inserted into the cryovial. Touching the tip of the free-tissue end of the biopsy will cause it to adhere to the vial wall, and the needle may then be withdrawn, leaving the frozen biopsy in the tube.
3. For xenografts, the excised tumor should be quartered (for xenografts staged in the 200 mg size range) by cutting with fine scissors, and then pieces are placed into the cryovials.
4. The vials should be placed into dry ice and held at $-80°C$ until processing. In our laboratory, specimens are frozen within 5 min of collection. Speed is critical to specimen integrity.
5. Skin biopsies from mice are collected using the same procedures, except that small surgical scissors are employed to snip a small piece of skin for analysis. On a nude mouse, the preferred biopsy area is the snout in the vicinity of the vibrissae.

3.4.2. Specimen Fixation

1. Frozen specimens should be thawed only long enough to dry the outside of the vials with a paper towel.
2. Fix specimens by immersion in 10% NBF, in a sealed vial, for at least 16 h and up to 96 h at RT.
   (a) Each specimen is fixed in a separate vial and must be completely immersed in the NBF. Vials are labeled to correspond to the original specimen vial.
   (b) Alternatively, the 10% NBF may be added to the original specimen vial if they are to be processed immediately.

3.4.3. Paraffin Embedding

1. Remove the biopsy from the vial and place it between two pads in a cassette to be processed (a screened cassette may also be used) (see Note 16).
2. Place the cassette into 70% ethanol to begin the paraffin embedding sequence.

3. Paraffin embedding at RT is as follows:
   (a) 70% ethanol – 1 × 30 min
   (b) 80% ethanol – 2 × 30 min
   (c) 95% ethanol – 2 × 30 min
   (d) 100% ethanol – 3 × 30 min
   (e) 100% xylene – 2 × 30 min
   (f) Paraffin – 3 × 45 min at 60°C
   (g) Paraffin – 1 × 30 min at 60°C.
4. Fill an embedding mold with paraffin and place the biopsy in the mold.
5. Place an embedding cassette on top of the mold and move the specimen to a cooling plate to harden.

3.4.4. Microtomy

1. Select the specimen block for analysis; also select a control tissue block and a calibrator tissue block.
2. Cut 5-μm sections using a Leica RM2255 automated microtome and Accu-Edge low-profile microtome blades.
3. After the sections are cut, float them on a water bath set at 46°C, and then mount on Superfrost plus slides.
4. Dry the sections overnight in an incubator set at 37°C.

3.4.5. Slide Preparation

1. Select sections for mounting starting at section number 5. Prepare the first and fifth section for γ-H2AX staining. Prepare the third section for H&E staining according to standard methods. Place the second and fourth sections on slides and store at 2–8°C until needed. Maximum storage time under these conditions is 30 days.
2. Set up calibrator specimens for day-to-day variability in specimen processing and staining. For these calibrators, treat mice with a single dose of topotecan (1.5 or 4.7 mg/kg) or vehicle (water) prepared such that injections are 0.1 mL per 10 g body weight (see Note 17).
3. Handle biopsies and test specimens identically. Embed biopsies from three different dosage levels in the same paraffin block for processing.

3.4.6. Immunohistochemistry

Steps are performed using the Vision Biosystems instrument. The steps may be duplicated manually for small sample sets, but temperature control is important. PBS with 0.05% Tween may be substituted for the Bond wash and diluents.

1. De-paraffinization. Apply Bond dewax solution at 72°C, for 30 s, followed by two Bond dewax rinses (also at 72°C).
   (a) Rinse three times with absolute ethanol.

　　　　(b) Wash twice in Bond wash solution.

　　　　(c) Incubate for 5 min at RT with fresh Bond wash solution.

　　2. Antigen rescue. This step is performed using the Vision Biosystems instrument.

　　　　(a) Rinse the slide twice with Bond ER1 solution.

　　　　(b) Add fresh citrate buffer (pH 6) and heat to 100°C, holding that temperature for 10 min.

　　　　(c) Cool the slide to RT in the citrate solution for 12 min.

　　　　(d) Rinse the slide three times with Bond wash solution, then add fresh Bondwash, and incubate for 3 min at RT.

　　3. Stain with primary monoclonal biotinylated anti-γ-H2AX antibody JBW301 (working concentration is 10 μg/mL in Bond diluent) (see Note 18).

　　　　(a) Rinse the slide three times with Bond wash solution.

　　　　(b) Apply primary antibody twice with 30-min incubation at RT both times.

　　　　(c) Rinse the slide two more times with Bond wash, 5 min per change.

　　4. Develop with streptavidin–Alexa Fluor-488 conjugate (working concentration is 10 μg/mL in Bond diluent).

　　　　(a) Add the conjugate and incubate for 30 min at RT.

　　　　(b) Rinse the slide twice with Bond wash, 5 min per change.

　　　　(c) Rinse the slide with deionized water, 5 min per change.

　　5. Rinse in 0.05% Tween–PBS twice, 5 min per change.

　　6. Apply mounting medium with DAPI and coverslip the slide while it is still wet.

　　　　(a) A semi-permanent mount is created by sealing the edges of the coverslip with nail polish.

　　　　(b) Keep the slides in the dark at 2–8°C for a few hours before imaging. Long-term storage (30 days) at –80°C is possible, although we often observe a rise in background fluorescence over time (see Note 19).

3.4.7. Imaging

　　1. Perform this step in the phase-contrast mode on a Leica DM5000 microscope (see Note 20) that is equipped with the Retiga 2000R monochrome camera. Calibrate the camera using the auto white-balance function on a section of tissue lacking cells; for H&E stained slides, use Retiga Liquid Crystal RGB color filter.

　　2. Image and analyze at least three fields per needle biopsy section, and many more for larger tissue pieces.

3.4.8. Image Analysis

Quantitative analysis of the acquired images can be performed in several ways; however, there are limitations imposed by the nature of the tissue being analyzed, specifically the thickness of the specimen (see Note 21).

1. Observe the nuclear staining and view any evidence of cytoplasmic staining as artifactual background (generally characterized as a diffuse green stain). Compensate for this background by decreasing the exposure time of the camera.
2. Perform quantitation at 200×, where there is essentially one plane of focus, and use a high resolution camera to allow higher magnification of the image on the computer screen.

4. Notes

1. Similar equipment with the mentioned capabilities can be used throughout.
2. The use of a fixed rotor is not recommended because following centrifugation, paraformaldehyde-fixed cells tend to stick to the side of the tube, which may result in cell loss.
3. Human peripheral blood lymphocytes were obtained from healthy donors by venipuncture from the NIH blood bank in accordance with NIH regulations. Mouse peripheral blood lymphocytes were obtained by cardiac puncture; mouse splenocytes and bone marrow cells were properly collected at the time of mouse sacrifice, according to standard operating procedures employed by SAIC-Frederick.
4. Fixed cells can be stored in PBS overnight at 4°C before spotting on slides and immunostaining.
5. The same procedure is used to prepare different cell types for fixation, cytospining, and immunostaining.
6. Cytospin specimens on slides can be stored overnight (up to a week) in 70% ethanol by placing the jar at 4°C.
7. To hold a small volume of solution on the cytospin preparations, use a Pap Pen to draw a waterproof circle around the samples. This allows the use of as little as 100 µL of solution per slide.
8. Once the slides have been stained with the fluorescent probe, they should be kept out of sunlight and viewed in rooms with minimal overhead fluorescent lighting. The slides can be stored at −20°C for several months.
9. If fewer cells are used, use 1.5-mL Eppendorf tubes instead of 15-mL conical tubes to facilitate the visualization of the pellets resulting from the centrifugation steps.

10. The specimens can be stored in 70% ethanol by placing the tubes at 4°C for several days but less than a week.

11. If γ-H2AX fluorescence intensity is important, the FL1-H channel should be switched into logarithmic mode and the FL1-H voltage adjusted to set up the cells between 10^0 and 10^1.

12. This protocol can be adapted for many other cell types. However, lower cell concentrations should be used because too much extracted DNA might prevent proper electrophoresis. This problem, however, can be overcome by sonicating the samples.

13. Boiled samples can be stored at 4°C for few days. If stored, the samples should be boiled again before the SDS-polyacrylamide gel electrophoresis (SDS-PAGE) step.

14. The primary antibody reaction can be done at RT for 1 h. The primary antibody reaction can also be done using a hybridization bag (SealPak pouches, KAPAK, Minneapolis, MN) instead of a 50-mL conical tube.

15. The method of selecting sections for mounting on slides and staining will promote evaluation of non-overlapping fields of the tumor for γ-H2AX analysis, while the H&E section will overlap both fields, aiding in orientation and interpretation.

16. We place several biopsies into a single paraffin block to control staining variability within a treatment group. For example, all specimens from vehicle-treated xenograft specimens could be stained in a single block.

17. Topotecan is an efficient upregulator of γ-H2AX in many human tumor cell lines and xenografts, with optimal sampling time being 1–4 h after drug injection, with the maximum signal at 1 h post-injection; for skin biopsies, the optimal sampling time is somewhat later due to a lag associated with distribution, with maximum signal reached at 4 h post-dosing. We evaluate skin effects by counting the number of positive nuclei per hair follicle.

18. When dealing with critical reagents such as streptavidin–Alexa Fluor-488 and biotinylated JBW301, it is advantageous to order custom-made lots of material with lot-specific custom-made antibodies. This approach decreases assay variability due to lot-to-lot performance differences. These lots can be aliquoted and stored at −80°C for up to a year. Preferred storage for primary antibodies is in 50% glycerol and buffer according to the manufacturer's specification.

19. It is best to store the slides at 2–8°C overnight after addition of the Prolong Gold/DAPI mounting medium to allow permeation of all nuclei.

20. The use of phase to select fields for image analysis has two advantages: (a) fields that have multiple cells and are phase dense are selected; (b) selection of fields is now semi-random, i.e., not based on the expected signal, limiting experimental bias. Other imaging systems (camera, softwares, etc.) can be used but will need further optimization by the users.

21. A 5-μm thick section has many planes of focus at high magnification, making analysis either visually or by imaging software challenging. Tumor sections have multiple overlapping nuclei, also making quantitation a challenge. At early timepoints after drug exposure (1–2 h), it is possible to count foci in nuclei. However, many nuclei will have progressed to a dispersed γ-H2AX signal by 2 h post-dosing, and there will be only rare distinct foci at 4 h post-dosing. A more useful approach is to count the number of positive nuclei in a field. When analyzing skin biopsies, we have documented that scoring the three most positive hair follicles per tissue section gives representative and reproducible results across treatment groups and experiments performed with topoisomerase 1 inhibitors. Finally, it is possible to train a computerized imaging system to analyze images using nuclear area algorithms, normalizing the γ-H2AX green channel fluorescence to the co-localized DAPI blue channel fluorescence.

Acknowledgments

This work was funded by the Intramural Research Program of the National Cancer Institute, Center for Cancer Research, NIH. B.T., S.L., and R.K. were funded by NCI Contract N01-CO-12400. Human blood samples were obtained from paid healthy volunteers who gave written informed consent to participate in an IRB-approved study for the collection of blood samples for *in vitro* research use. The protocol is designed to protect subjects from research risks as defined in 45CFR46 and to abide by all internal NIH guidelines for human subjects research (protocol number 99-CC-0168). NCI-Frederick is accredited by AAALAC International and follows the Public Health Service Policy for the Care and Use of Laboratory Animals. Animal care was provided in accordance with the procedures outlined in the "Guide for Care and Use of Laboratory Animals" (National Research Council, 1996; National Academy Press; Washington, DC). All studies were conducted according to an approved animal care and use committee protocol.

References

1. Bonner, W. M., Redon, C. E., Dickey, J. S., Nakamura, A. J., Sedelnikova, O. A., Solier, S., and Pommier, Y. (2008) GammaH2AX and cancer, *Nat. Rev. Cancer* **8**, 957–967.
2. Rogakou, E. P., Pilch, D. R., Orr, A. H., Ivanova, V. S., and Bonner, W. M. (1998) DNA double-stranded breaks induce histone H2AX phosphorylation on serine 139, *J. Biol. Chem.* **273**, 5858–5868.
3. Rogakou, E. P., Boon, C., Redon, C., and Bonner, W. M. (1999) Megabase chromatin domains involved in DNA double-strand breaks in vivo, *J. Cell. Biol.* **146**, 905–916.
4. Sedelnikova, O. A., Rogakou, E. P., Panyutin, I. G., and Bonner, W. M. (2002) Quantitative detection of (125)IdU-induced DNA double-strand breaks with gamma-H2AX antibody, *Radiat. Res.* **158**, 486–492.
5. Rothkamm, K., and Lobrich, M. (2003) Evidence for a lack of DNA double-strand break repair in human cells exposed to very low X-ray doses, *Proc. Natl. Acad. Sci. U. S. A.* **100**, 5057–5062.
6. Paull, T. T., Rogakou, E. P., Yamazaki, V., Kirchgessner, C. U., Gellert, M., and Bonner, W. M. (2000) A critical role for histone H2AX in recruitment of repair factors to nuclear foci after DNA damage, *Curr. Biol.* **10**, 886–895.
7. Chowdhury, D., Keogh, M. C., Ishii, H., Peterson, C. L., Buratowski, S., and Lieberman, J. (2005) Gamma-H2AX dephosphorylation by protein phosphatase 2A facilitates DNA double-strand break repair, *Mol. Cell.* **20**, 801–809.
8. Kohn, K. W. (1991) Principles and practice of DNA filter elution, *Pharmacol. Ther.* **49**, 55–77.
9. Olive, P. L., and Banath, J. P. (2006) The comet assay: a method to measure DNA damage in individual cells, *Nat. Protoc.* **1**, 23–29.
10. Banath, J. P., Macphail, S. H., and Olive, P. L. (2004) Radiation sensitivity, H2AX phosphorylation, and kinetics of repair of DNA strand breaks in irradiated cervical cancer cell lines, *Cancer Res.* **64**, 7144–7149.
11. Kiltie, A. E., and Ryan, A. J. (1997) SYBR Green I staining of pulsed field agarose gels is a sensitive and inexpensive way of quantitating DNA double-strand breaks in mammalian cells, *Nucleic Acids Res.* **25**, 2945–2946.
12. Redon, C. E., Dickey, J. S., Bonner, W. M., and Sedelnikova, O. A. (2009) Gamma-H2AX as a biomarker of DNA damage induced by ionizing radiation in human peripheral blood lymphocytes and artificial skin, *Adv. Space Res.* **43**, 1171–1178.
13. Pilch, D. R., Redon, C., Sedelnikova, O. A., and Bonner, W. M. (2004) Two-dimensional gel ana.lysis of histones and other H2AX-related methods, *Methods Enzymol.* **375**, 76–88.
14. Nakamura, A., Sedelnikova, O. A., Redon, C., Pilch, D. R., Sinogeeva, N. I., Shroff, R., Lichten, M., and Bonner, W. M. (2006) Techniques for gamma-H2AX detection, *Methods Enzymol.* **409**, 236–250.
15. Ismail, I. H., Wadhra, T. I., and Hammarsten, O. (2007) An optimized method for detecting gamma-H2AX in blood cells reveals a significant interindividual variation in the gamma-H2AX response among humans, *Nucleic Acids Res.* **35**, e36.
16. Olive, P. L., and Banath, J. P. (2004) Phosphorylation of histone H2AX as a measure of radiosensitivity, *Int. J. Radiat. Oncol. Biol. Phys.* **58**, 331–335.
17. Olive, P. L., Banath, J. P., and Keyes, M. (2008) Residual gammaH2AX after irradiation of human lymphocytes and monocytes in vitro and its relation to late effects after prostate brachytherapy, *Radiother. Oncol.* **86**, 336–346.
18. Sedelnikova, O. A., and Bonner, W. M. (2006) GammaH2AX in cancer cells: a potential biomarker for cancer diagnostics, prediction and recurrence, *Cell Cycle* **5**, 2909–2913.
19. Tanaka, T., Kajstura, M., Halicka, H. D., Traganos, F., and Darzynkiewicz, Z. (2007) Constitutive histone H2AX phosphorylation and ATM activation are strongly amplified during mitogenic stimulation of lymphocytes, *Cell Prolif.* **40**, 1–13.
20. Dickey, J.S., Redon, C.E., Nakamura, A.J., Baird, B.J., Sedelnikova, O.A., and Bonner, W.M. (2009) H2AX: functional roles and potential applications, *Chromosoma* **118**, 683-692.

Chapter 19

Immunologic Detection of Benzo(a)pyrene–DNA Adducts

Regina M. Santella and Yu-Jing Zhang

Abstract

The binding of chemical carcinogens to DNA is well established as the initiating step in the process of carcinogenesis. While early studies in animals or cells in culture took advantage of radiolabeled model carcinogens such as benzo(a)pyrene, interest in measuring DNA damage levels in humans necessitated the development of alternative methods. Among these, immunologic methods using polyclonal or monoclonal antibodies to carcinogen–DNA adducts have proven extremely useful in monitoring human exposure as well as being applicable to animal and cell culture studies. Here we describe the use of antibodies for immunohistochemical analysis of tissue sections, biopsies, or intact cells and for quantitation of carcinogen binding in DNA isolated from blood and tissues by enzyme-linked immunosorbent assays.

Key words: DNA adducts, Immunohistochemistry, Polycyclic aromatic hydrocarbons, Benzo(a)pyrene, Benzo(a)pyrene diol epoxide, DNA adducts, Immunofluorescence, Immunoperoxidase

1. Introduction

It is well established that carcinogen binding to DNA is the initiating step in the process of carcinogenesis. Numerous studies have investigated the formation of DNA adducts in tissues of animals treated *in vivo* as well as in cell culture studies. Early studies took advantage of radiolabeled carcinogens such as aflatoxin B_1 or benzo(a)pyrene (BP) to determine levels of DNA damage by isolation of DNA and quantitation of radioactivity. However, interest in measuring DNA damage in humans necessitated the development of alternative methods. Among these are immunoassays using polyclonal or monoclonal antibodies to particular carcinogen–DNA adducts. A limitation of this approach is that the adduct must be well-characterized and it must be possible to synthesize an immunizing antigen. For DNA adducts, two types of antigens have been used (1). DNA can be highly modified

(>1%) and then complexed with methylated bovine serum albumin before injection in an adjuvant. Lower modification levels in the DNA tend not to be antigenic. Alternatively, the monoadduct in the ribose form can be coupled to a carrier protein through the adjacent hydroxides on the sugar. Once a specific and sensitive antibody has been developed, it can be used both in competitive enzyme-linked immunosorbent assays (ELISA) on DNA isolated from blood or tissue samples (2) and for immunohistochemical studies of adducts in intact tissues, exfoliated buccal or bladder cells, or white blood cells (3–5).

While a number of antibodies have been developed against DNA adducts, here, we concentrate on those recognizing the primary DNA adduct formed in humans by exposure to BP. *In vivo* metabolism of BP by the cytochrome P450 and epoxide hydrolase enzymes results in the generation of a highly reactive diol epoxide ($7\beta,8\alpha$-dihydroxy-$9\alpha,10\alpha$-epoxy-7,8,9,10-tetrahydrobenzo(a) pyrene (BPDE-I)) that binds to DNA. Both polyclonal (6) and monoclonal (7) antibodies have been developed that recognize BPDE–DNA, but only the monoclonal antibodies are commercially available. While the antibodies have been validated against radioactive measurement of DNA damage levels, these studies were done with cells treated in culture with [^{3}H]BP (8). It was later learned that both the polyclonal and monoclonal antibodies recognize the DNA adducts of several other polycyclic aromatic hydrocarbon (PAH) diol epoxides in addition to that of BP, including those from chrysene and benzanthracene (9, 10). The affinity constants for these different structurally similar diol epoxide adducts differ from that for BPDE–DNA.

Humans are exposed to complex mixtures of PAH that differ in their levels of specific compounds depending upon the source of exposure. Many of these PAHs can be metabolized to diol epoxides and bind to DNA. Since they will all be recognized by the antibody to BPDE–DNA but with different affinities, absolute adduct levels cannot be determined. However, measurement of this class of adducts as a whole is still a useful marker of exposure.

Thus, a major limitation of the immunohistochemical method for detection of PAH–DNA adducts is that only a semiquantitative estimation of adduct levels can be obtained. Attempts have been made to obtain more quantitative data by treating cells in culture with various amounts of [^{3}H]BP and determining adduct levels by radioactivity (11). Then, aliquots of the radiolabeled cells can be stained at the same time the test samples are stained and a standard curve of relative staining intensity versus adduct level generated based on radioactivity. This standard curve can be used to estimate adduct levels in the test samples using the relative staining intensity, but only if they solely contain BP adducts.

Described here are immunofluorescence and immunoperoxidase methods for the detection of BPDE–DNA adducts in frozen

Immunologic Detection of Benzo(a)pyrene–DNA Adducts   273

or paraffin tissue sections or slides containing smeared cells. With appropriate microscopes for quantitation of staining, relative staining intensity can be determined and adduct levels between samples compared. When used on biological samples from humans with exposure to complex mixtures of PAH, the assay is a general, semiquantitative indicator of total PAH–DNA adducts but cannot provide absolute adduct levels. The major advantages of the immunohistochemical method for DNA adduct detection are the elimination of the need to isolate DNA and the applicability to very small amounts of sample.

2. Materials

2.1. Sample Preparation

As indicated, some of the following reagents are needed specifically for the preparation of smeared, frozen or paraffin samples.

1. Precoated slides (Superfrost/Plus Microscope Slides Precleaned, Fisher Scientific, Pittsburg, PA)
2. Methanol
3. Acetic Acid
4. Acetone
5. 100, 95, and 70% ethanol
6. Xylene (see Note 1)
7. ImmEdge Pen (Vector Laboratories Inc., Burlingame, CA)
8. Tissue-Tek® O.C.T. Compound (Electron Microscope Sciences, Hatfield, PA)

2.2. Immunoperoxidase Detection

1. Phosphate buffered saline (PBS): prepare 10× stock with 1.37 M NaCl, 27 mM KCl, 100 mM Na_2HPO_4, 18 mM KH_2PO_4, pH 7.4. Autoclave before storage. Prepare working solution by dilution of one part with nine parts of water.
2. Tris buffer: 10 mM Tris-HCl, 1 mM EDTA, pH 7.5.
3. RNase: 100 µg/mL in Tris buffer (R6513, Sigma-Aldrich, St. Louis, MO).
4. Proteinase K: 10 µg/mL in Tris buffer (P6556, Sigma-Aldrich, St. Louis, MO).
5. HCl: 4 N HCl.
6. Tris base: 50 mM Trizma®.
7. 1% Triton X-100 in PBS.
8. Blocking solution: ABC kit (Vectastain Elite Mouse IgG ABC Kit, PK-6102, Vector Laboratories Inc., Burlingame, CA).
9. Hydrogen peroxide: 0.3% in methanol, made fresh daily.

10. Primary antibody: 5D11 generated against BPDE-I-DNA (7). Available from Santa Cruz Biotechnology, Trevigen, Hycult Biotechnolgy bv, GenWay Biotech, Cell Sciences, and ABR-Affinity BioReagents. Prepare according to recommendations (see Note 2).
11. Secondary antibody: ABC kit (Vectastain Elite Mouse IgG ABC Kit, PK-6102, Vector Laboratories Inc., Burlingame, CA). Prepare according to the manufacturer's instructions.
12. ABC reagent: ABC complex solution from ABC kit (Vectastain Elite Mouse IgG ABC Kit, PK-6102, Vector Laboratories Inc., Burlingame, CA). Prepare according to manufacturer's instructions 30 min before use.
13. Diaminobenzidine (DAB) reagent: (Vectastain DAB kit, SK-4100 Vector Laboratories Inc., Burlingame, CA) (see Notes 3 and 4).
14. Harris hematoxylin.
15. Permount.

2.3. Immunofluorescence Detection

1. PBS: prepare 10× stock with 1.37 M NaCl, 27 mM KCl, 100 mM Na_2HPO_4, 18 mM KH_2PO_4, pH 7.4. Autoclave before storage. Prepare working solution with by dilution of one part with nine parts of water.
2. Tris buffer: 10 mM Trizma® (T1503, Sigma-Aldrich, St. Louis, MO), 1 mM EDTA, 0.4 M NaCl, pH 7.5.
3. RNase: 100 µg/mL in Tris buffer (R6513, Sigma-Aldrich, St. Louis, MO).
4. Proteinase K: 10 µg/mL in Tris buffer (P6556, Sigma-Aldrich, St. Louis, MO).
5. HCl: 4 N HCl.
6. Tris base: 50 mM Trizma® (T1503, Sigma-Aldrich, St. Louis, MO).
7. Blocking buffer: 1.5% goat serum (v/v) in 1× PBS.
8. Primary antibody: 5D11 generated against BPDE-I-DNA (7). Available from Santa Cruz Biotechnology, Hycult Biotechnology bv, Trevigen, GenWay Biotech, Cell Sciences, and ABR-Affinity BioReagents. Dilute according to recommendations (see Note 2).
9. Goat anti-mouse IgG conjugated to fluorescein isothiocyanate (FITC) (ICN Pharmaceuticals, Inc. Aurora, OH) (see Notes 2 and 5).
10. DAPI: 4′,6-diamidino-2-phenylindole (DAPI) 1 µg/mL in PBS (Polysciences, Warrington, PA) (see Note 5).
11. ProLong Antifade Kit (P7481, Invitrogen, Carlsbad, CA).

3. Methods

3.1. Sample Preparation

Frozen or paraffin sections can be stained as well as smeared cells such as buccal or white blood cells. Frozen sections and smeared cells are air dried and fixed in cold methanol and acetic acid (3:1) solution for 5 min (4°C) followed by cold acetone (4°C) for an additional 5 min, and then air dried before staining.

Paraffin-embedded sections must be deparaffinized before staining. Do not allow the slides to dry at any time during this procedure.

1. Melt paraffin in 54–58°C incubator for 40 min–1 h.
2. Transfer the slides into Coplin jar with xylene and incubate for 3×5 min at room temperature (see Note 6).
3. Transfer the slides into Coplin jar with 100% ethanol for 2×5 min at room temperature.
4. Transfer the slides into Coplin jar with 95% ethanol for 2×5 min at room temperature.
5. Transfer the slides into Coplin jar with 70% ethanol for 2×5 min.
6. Wash the slides in water for 5 min twice.

Use ImmEdge Pen to outline sample on the slide.

3.2. Immunoperoxidase Detection

1. Wash the slides with 1× PBS in Coplin jar on a shaker for 5 min, twice.
2. Place the slides in a covered slide holder, add sufficient RNase solution to cover cells or sections, and incubate for 1 h at 37°C in a covered humidified chamber.
3. Wash the slides with 1× PBS in Coplin jar on a shaker for 5 min, twice.
4. Place the slides in a covered humidified chamber, add proteinase K solution to cover cells or sections, and incubate for 10 min at room temperature.
5. Wash the slides with 1× PBS in Coplin jar on a shaker for 5 min, twice.
6. After treatment, check the slides under the microscope to ensure cells remain intact.
7. Denature DNA by treatment with 4N HCl for 10 min at room temperature.
8. Add 50 mM Tris base solution in Coplin jar to neutralize for 5 min at room temperature.

9. Wash the slides with 1× PBS in Coplin jar on a shaker for 5 min, three times.
10. Incubate the slides with 0.3% H_2O_2–Methanol for 30 min at room temperature to quench endogenous peroxidase activity.
11. Wash the slides with 1× PBS in Coplin jar on a shaker for 5 min, three times.
12. Add blocking solution from ABC kit prepared as described by the manufacturer and incubate for 45 min at 37°C in a covered humidified chamber.
13. Remove the blocking solution, add primary antibody 5D11 diluted in the blocking solution to cover cells or sections, and incubate overnight at 4°C in a covered humidified chamber.
14. The following day, wash the slides with 1× PBS in Coplin jar on a shaker for 5 min, three times.
15. Add biotinylated secondary antiserum (ABC kit) and incubate for 30 min at room temperature in a covered humidified chamber.
16. Add ABC reagent (ABC kit) and incubate for 30 min at room temperature in a covered humidified chamber.
17. Wash the slides with 1× PBS in Coplin jar on a shaker for 5 min, three times.
18. Rinse the slides with 1% Triton X-100 for 30 s at room temperature.
19. Wash the slides with 1×PBS in Coplin jar for 5 min, three times.
20. Add freshly prepared DAB solution (DAB kit) and incubate for 2–6 min at room temperature.
21. Wash the slides in tap water in Coplin jar for 5–10 min, three times.
22. If desired, counterstain with Harris-hematoxylin solution for 1 min.
23. Dehydrate the slides with a series of ethanol solutions: 70% ethanol for 5 min, 95% ethanol for 5 min, two times, and 100% ethanol for 5 min, twice.
24. Clean with xylene for 5 min, three times.
25. Mount the slides with Permount. Be sure to clean the slides with xylene to remove excess Permount.

3.3. immunofluorescence Detection

1. Wash slides with 1× PBS in Coplin jar on a shaker for 5 min, twice.
2. Place the slides in a covered humidified chamber, add sufficient RNase solution to cover cells or sections, and incubate for 1 h at 37°C.

3. Wash the slides with 1× PBS in Coplin jar on a shaker for 5 min, twice.
4. Place the slides in a covered holder, add proteinase K, and incubate for 10 min at room temperature.
5. Wash the slides with 1× PBS in Coplin jar for 5 min, twice.
6. After treatment, check the slides under the microscope to ensure cells remain intact.
7. Denature DNA by treatment with 4N HCl for 10 min at room temperature.
8. Add 50 mM Tris base solution in Coplin jar to neutralize for 5 min at room temperature.
9. Wash the slides with 1× PBS in Coplin jar for 5 min, twice.
10. Add blocking solution (10% normal goat serum in PBS) and incubate for 45 min at 37°C in a covered humidified chamber.
11. Remove the blocking solution, add primary antibody 5D11 diluted in the blocking solution to cover cells or sections, and incubate overnight at 4°C in a covered humidified chamber.
12. The following day, wash the slides with 1× PBS in Coplin jar for 5 min, twice.
13. Add FITC-labeled secondary antiserum (1:150 dilution in blocking buffer) and incubate for 45 min at 37°C in the dark to prevent fading (see Note 5).
14. Wash the slides with 1× PBS in Coplin jar for 5 min, twice.
15. Counterstain with DAPI, for 10 min at room temperature in the dark (see Note 5).
16. Wash the slides with 1× PBS in Coplin jar for 5 min, twice.
17. Add cover glass with antifade solution.

4. Notes

1. Xylene is a hazardous volatile chemical and must be used in a fume hood.
2. It may be necessary to titrate the primary or secondary antibody to determine which dilution provides the strongest signal with least background. This can be done using positive and negative controls, such as cells treated in culture with or without BP. However, cells must have a functioning cytochrome P450 system to activate BP to the reactive compound, BPDE. Alternatively, BPDE can be used to treat cells directly. It can be purchased from NCI's Chemical Carcinogen Repository (Chemical Carcinogen Reference Standard Repository, Kansas City, MO). BPDE is a highly reactive chemical carcinogen and

must be handled with care. It hydrolyses rapidly in aqueous solutions to the non-hazardous BPDE-tetrol.

3. DAB is a suspected carcinogen. Care should be taken in handling and disposing.

4. In the presence of nickel ions, the precipitate formed by DAB is purplish blue rather than brown. Use of nickel may enhance sensitivity of staining. DAB Substrate Kit (Vector Catalog No. SK-4100) contains nickel chloride.

5. FITC and DAPI are light sensitive and should be kept in the dark. Diluted solutions should not be stored for reuse.

6. A series of Coplin jars should be set up for each step (e.g., three jars with xylene) and the slides serially transferred from one jar to the next. Solutions should be changed every 300–500 slides. But if not used extensively, the solutions should not be kept for more than 1 month.

References

1. Santella, R.M. (1999) Immunologic methods for detection of carcinogen-DNA damage in humans. *Cancer Epidemiol. Biomarkers Prev.* **8**, 733–739.

2. Gammon, M.D., Santella, R.M., Neugut, A.I., Eng, S.M., Teitelbaum, S.L., Paykin, A. et al. (2002) Environmental toxins and breast cancer on Long Island. I. Polycyclic aromatic hydrocarbon DNA adducts. *Cancer Epidemiol. Biomarkers Prev.* **11**, 677–685.

3. Hsu, T.M., Zhang, Y.J., Santella, R.M. (1997) Immunoperoxidase quantitation of 4-aminobiphenyl- and polycyclic aromatic hydrocarbon-DNA adducts in exfoliated oral and urothelial cells of smokers and nonsmokers. *Cancer Epidemiol. Biomarkers Prev.* **6**, 193–199.

4. Motykiewicz, G., Malusecka, E., Grzybowska, E., Chorazy, M., Zhang, Y.J., Perera, F.P. et al. (1995) Immunohistochemical quantitation of polycyclic aromatic hydrocarbon-DNA adducts in human lymphocytes. *Cancer Res.* **55**, 1417–1422.

5. Chen, S.Y., Wang, L.Y., Lunn, R., Tsai, W.Y., Lee, P.H., Lee, C.S. et al. (2002) Polycyclic aromatic hydrocarbon-DNA adducts in liver tissues of hepatocellular carcinoma patients and controls. *Int. J. Cancer* **99**, 14–21.

6. Poirier, M.C., Santella, R.M., Weinstein, I.B., Grunberger, D., Yuspa, S.H. (1080) Quantitation of benzo[a]pyrene-deoxyguanosine adducts by radioimmunoassay. *Cancer Res.* **40**, 412–416.

7. Santella, R.M., Lin, C.D., Cleveland, W.L., Weinstein, I.B.. (1984) Monoclonal antibodies to DNA modified by a benzo[a]pyrene diol epoxide. *Carcinogenesis* **5**, 373–377.

8. Santella, R.M., Weston, A., Perera, F.P., Trivers, G.T., Harris, C.C., Young, T.L. et al. (1988) Interlaboratory comparison of antisera and immunoassays for benzo(a)pyrene-diol-epoxide-I-modified DNA. *Carcinogenesis* **9**, 1265–1269.

9. Santella, R.M., Gasparro, F.P., Hsieh, L.L. (1987) Quantitation of carcinogen-DNA adducts with monoclonal antibodies. *Prog. Exp. Tumor Res.* **31**, 63–75.

10. Weston, A., Trivers, G., Vahakangas, K., Newman, M., Rowe, M. (1987) Detection of carcinogen-DNA adducts in human cells and antibodies to these adducts in human sera. *Prog. Exp. Tumor Res.* **31**, 76–85.

11. Zhang, Y.J., Hsu, T.M., Santella, R.M. (1995) Immunoperoxidase detection of polycyclic aromatic hydrocarbon-DNA adducts in oral mucosa cells of smokers and nonsmokers. *Cancer Epidemiol. Biomarkers Prev.* **4**, 133–138.

Chapter 20

Non-invasive Assessment of Oxidatively Damaged DNA: Liquid Chromatography-Tandem Mass Spectrometry Analysis of Urinary 8-Oxo-7,8-Dihydro-2′-Deoxyguanosine

Vilas Mistry, Friederike Teichert, Jatinderpal K. Sandhu, Rajinder Singh, Mark D. Evans, Peter B. Farmer, and Marcus S. Cooke

Abstract

The ability to non-invasively assess DNA oxidation and its repair, has significant utility in large-scale, population-based studies. Such studies could include the assessments of: the efficacy of antioxidant intervention strategies, pathological roles of DNA oxidation in various disease states and population or interindividual differences in antioxidant defence and DNA repair. The most popular method, to non-invasively assess oxidative insult to the genome is by the analysis of urine for 8-oxo-7,8-dihydro-2′-deoxyguanosine (8-oxodG), using chromatographic techniques or immunoassay procedures. The provenance of extracellular 8-oxodG remains a subject for debate. However, previous studies have shown that factors, such as diet and cell death, do not appear to contribute to extracellular 8-oxodG, leaving processes, such as the repair of DNA and/or the 2′-deoxyribonucleotide pool, as the sole source of endogenous 8-oxodG. The method in this chapter describes a non-invasive approach for assessing oxidative stress, via the efficient extraction of urinary 8-oxodG using a validated solid-phase extraction procedure. Subsequent analysis by liquid chromatography-tandem mass spectrometry provides the advantages of sensitivity, internal standardisation, and robust peak identification, and is widely considered to be the "gold standard".

Key words: Oxidative stress, DNA damage, DNA repair, Chromatography, Solid phase extraction, Urine, Liquid chromatography, Mass spectrometry

1. Introduction

Methods for high throughput, non-invasive assessment of DNA oxidation have potentially wide applications in a basic or applied biological science context, but would also facilitate transfer of such analyses to a clinical setting. The most widely measured product of DNA oxidation is 8-oxo-7,8-dihydro-2′-deoxyguanosine (8-oxodG) [1]. Although 8-oxodG has been examined in various extracellular matrices, e.g., serum, saliva and faecal matter, urine remains by far the most widely examined matrix (2), providing a non-invasive biomarker of oxidative stress, with a number of advantages over other biomarkers of oxidative stress. Being non-invasive, the assay represents less of an ethical issue than, for example, blood-based assessments of oxidative stress, in particular when sampling from vulnerable groups is required. Many of the issues of adventitious damage, associated with the study of cellular 8-oxodG (3), are circumvented by the analysis of urine. Urine is easily collected, transported, and stored, with 8-oxodG reported to be stable in urine, −20°C, for over 10 years (4). No special storage conditions are required, for example, no preservatives are necessary, which makes previously collected samples eminently suitable for analysis. Compared to blood, for example, large volumes of urine can been collected, allowing multiple analyses to be performed on the same sample, providing greater information.

There are many reports of the analysis of urinary 8-oxodG in various populations and pathologies. These analyses have been performed despite a lack of understanding of the precise source(s) of this lesion in extracellular matrices (5). This area remains under close scrutiny, in our laboratory and others, with DNA repair, cell turnover and diet considered the major potential sources of urinary 8-oxodG, however, emerging data are increasingly ruling out the significance of the latter two sources (6, 7). Concerning the repair origins for 8-oxodG, the most likely source would be the activity of the Nudix hydrolases activity (e.g. NUDT1 (MTH1) whose product, 8-oxodGMP, following dephosphorylation, yields 8-oxodG (8)). Thus, there is potentially more biological meaning to urinary 8-oxodG to be discovered, besides simply being a marker of nucleic acid oxidation (9). As 8-oxodG is a marker for oxidative stress, and can be detected in abundance (10), it can be used to monitor disease processes, such as ageing (9), cancer (11), and cardiovascular disease (12).

The two major analytical approaches that have been used to examine urinary 8-oxodG are chromatography and immunoassay. Each method has its own advantages and disadvantages. The more complex technology, costs, user training, and lower throughput of the former are outweighed by their more rigorous separation and compound identification capabilities. In contrast, immunoassay has higher sample throughput potential, requires less user training and capital outlay for equipment, but potentially suffers from a lack of specificity and requires the availability of sufficient amounts of appropriately characterised antibodies. The chromatographic analysis of urinary 8-oxodG has, to date, largely been undertaken using HPLC coupled with electrochemical detection, in most cases using column switching devices to provide effective sample clean-up (2). Gas-chromatography mass spectrometry (GC-MS) following HPLC pre-purification of urine has also received increasing attention, not least as it allows the examination of a wide range of damage products (13, 14). More recently, HPLC coupled to tandem mass spectrometry (LC-MS/MS) has gained popularity and appears to be the method of choice for the analysis of 8-oxodG, and in time, other lesions, in urine (15, 16). The procedure described here is based on the methodology reported by Teichert et al. (17) which involves the solid-phase clean-up of urine prior to LC-MS/MS analysis. The procedure, avoids the need for derivatisation of 8-oxodG as encountered with GC-MS and provides good separation of the urinary extract, showing sufficient sensitivity while retaining the advantage of mass selectivity for more rigorous sample identification.

2. Materials

2.1. Synthesis of Isotopically Labelled 8-oxodG {[$^{15}N_5$] 8-oxodG}

1. Water (see Note 1).
2. [$^{15}N_5$]-2′-deoxyguanosine (>98% ^{15}N; Spectra Stable Isotopes, Columbia, MD, USA).
3. Reaction buffer: 20 mM sodium phosphate buffer, pH 7.0, Chelex-treated. For Chelex treatment, add 0.05 g Chelex®-100 resin (Biotechnology grade, 00-20 mesh, sodium form, Bio-Rad, cat. no. 143-2832) per millilitre buffer, stir for 1 h, allow the resin to settle, and then decant carefully the buffer into a fresh container. Alternatively, pellet the resin by centrifugation and decant the buffer. Store buffer at 4–6°C until needed.
4. Copper sulphate, 20 mM, dissolved in water. This reagent can be stored for several weeks at 4°C.
5. Sodium ascorbate, 170 mM, dissolved in water. This reagent should be prepared immediately before use and not stored.
6. Hydrogen peroxide: 30% v/v solution, use straight from the bottle as supplied.
7. Catalase: dissolve 1 mg catalase (bovine liver; 10,700 units/mg solid; Sigma cat. no. C-40) per 1 mL of Chelex-treated, 20 mM phosphate buffer, pH 7.0. Prepare in sufficient quantities and use it on the same day.

2.2. Purification of [$^{15}N_5$] 8-oxodG

1. HPLC system capable of mobile phase flow rates of at least 5 mL/min and with a UV-visible diode array detection facility.
2. Centrifugal vacuum evaporator (speedvac) or freeze drier.
3. UV-visible spectrophotometer.
4. HPLC column: Columbus, 5 μM C_8 semi-preparative column, 250 × 10 mm (Phenomenex, Macclesfield, UK).

5. Mobile phase: 10% (v/v) methanol in ultrapure water or commercially obtained HPLC-grade water.

6. Standard solutions of 2′-deoxyguanosine (dG) and 8-oxodG (Sigma Chemical Co., Poole, UK, cat. no. D7145 and H5653, respectively), 50 μM prepared in 10% (v/v) methanol in water (mobile phase). Alternatively, more concentrated stocks can be prepared in water and diluted to 50 μM using the mobile phase.

2.3. Urine Collection, Creatinine Analysis, and Solid Phase Extraction

1. Vacuum manifold (see Note 2), pump, and trap.
2. Plate reader for 96-well plates, with filter for measurements at 490 nm.
3. [$^{15}N_5$]8-oxodG – synthesised in house, or obtained commercially (e.g. Cambridge Isotope Laboratories, [$^{15}N_5$]8-oxo-2′-deoxyguanosine, cat no. NLM-6715).
4. Deionised water.
5. HPLC grade water.
6. HPLC grade methanol.
7. HPLC grade acetonitrile.
8. Solid-phase extraction columns: Waters Oasis HLB, 1 cm^3, 30 mg (Waters Ltd., Elstree, UK).

2.4. LC-MS/MS Analysis of Urine Extracts

1. LC-Tandem mass spectrometry system (see Note 3).
2. Synergi Fusion-RP 80A C18 (4 um, 250 × 2.0 mm) attached to a Synergi Fusion-RP 80A C18 (4 mm, 4.0 × 2.0 mm) guard column and KrudKatcher disposable pre-column (0.5 mm) filter (Phenomenex, Macclesfield, Cheshire, UK).
3. Mobile phase: 0.1% aqueous acetic acid: methanol (85:15, v/v).
4. Standard 8-oxodG tuning solution: (10 pmol/μL) 8-oxodG dissolved in mobile phase.

3. Methods

The use of isotopically labelled 8-oxodG is critical to this procedure, to account for sample recovery and any instrumental differences in the ionisation of samples from one run to the next during the process of electrospray ionisation. We have synthesised our own isotopically labelled 8-oxodG, based on the method previously reported by Singh et al. (18), however, this compound can now be obtained commercially (see Subheading 2.3). We report the synthesis procedure for the benefit of those laboratories that, for whatever reason, may wish to make their own standard.

3.1. Synthesis of [$^{15}N_5$]8-oxodG

1. Prepare a 1 mg/mL solution of [$^{15}N_5$]dG in Chelex-treated, 20 mM sodium phosphate buffer, pH 7.0.
2. To the [$^{15}N_5$]dG solution add, sequentially and rapidly, vortexing between additions, copper solution, sodium ascorbate, and hydrogen peroxide (final concentrations, 1.2, 10, and 370 mM, respectively).
3. Incubate the reaction mixture at ambient temperature for 20 min (see Note 4), followed by the addition of 300 µL catalase solution.
4. The reaction mixture can be frozen at –20°C, or preferably –80°C, at this point if required.

3.2. Purification of [$^{15}N_5$]8-oxodG

1. Purification of [$^{15}N_5$]8-oxodG is performed by HPLC using sequential injections of the reaction mixture onto the semi-preparative HPLC column. The mobile phase flow rate is 5 mL/min. Verification of 8-oxodG (and dG) retention time is performed by preliminary injection of unlabelled standards (see Note 5). Eluting peaks are monitored by UV absorption at 245 and 254 nm, for 8-oxodG and dG, respectively.
2. Collect fractions at the appropriate retention time (see Note 6) and dry in a speedvac or freeze-drier. Dried material is then reconstituted by thorough vortexing in 1 mL water, and reconstituted aliquots are then pooled and dried once more.
3. Dried material is reconstituted by thorough vortexing in 0.5 or 1 mL water.
4. Final quantification is performed by determining the concentration of 8-oxodG by UV absorbance at 245 nm, $\varepsilon = 12,300$ M^{-1}cm^{-1} (see Note 7).
5. Aliquot concentrated material and store at –80°C. Aliquots of concentrated material can be diluted, further aliquotted and stored at –80°C as desired. A working solution of 2.0 pmol/µL is appropriate for routine urinary 8-oxodG analysis.

3.3. Urine Collection and Creatinine Analysis

1. Collect spot urine samples from subject groups. Samples are usually early morning, first void, midstream urine sample (see Note 8). Samples can be stored at –20 or –80°C until analysis, without significant degradation of the analyte (4).
2. Each urine should be analysed for creatinine concentration as a normalisation factor. This may be done using a commercially available colorimetric assay, based on the Jaffe alkaline picrate method, in a 96-well plate format (Metra® Creatinine, Quidel Corp., San Diego, CA, USA) (see Note 9).

3.4. Solid Phase Extraction

1. Centrifuge urine sample at 16,000 × g for 10 min in a benchtop microcentrifuge. A schematic overview of this procedure is presented in Fig. 1.

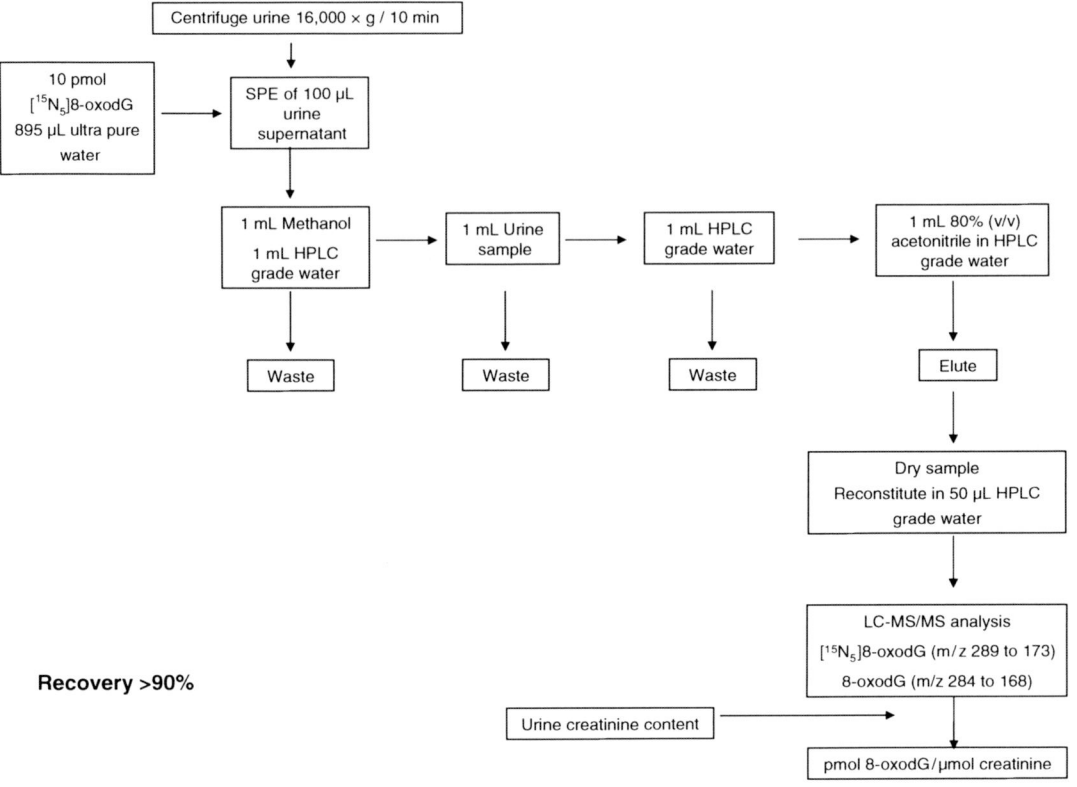

Fig. 1. Summary of solid-phase extraction procedure and LC-MS/MS analysis.

2. To 100 μL urine supernatant, in a 1.5 mL Eppendorf tube, add 5 μL of 2.0 pmol/μL [$^{15}N_5$]8-oxodG (i.e. 10 pmol).

3. Make-up urine to 1 mL with deionised water and vortex to mix.

4. Use one SPE column per sample and dispose after use (see Note 10).

5. Pre-condition column packing material as follows, with vacuum applied, run liquid to waste (see Note 11).

6. Mount HLB column on vacuum manifold.

7. Wet HLB column with 1 mL methanol. The flow rate should be approximately 1 mL/min for all steps.

8. Prime HLB column with 1 mL HPLC grade water.

9. Pass diluted urine through the column, run to waste and dry column under vacuum.

10. Wash unbound material from the column with 1 mL HPLC grade water, run to waste and dry column under vacuum.

11. Elute 8-oxodG from column with 1 mL 80% (v/v) acetonitrile in HPLC grade water, collecting eluent into an appropriately labelled tube.

12. Either freeze eluent at this stage at −20°C, or −80°C, or proceed immediately to drying of samples in a speedvac.

3.5. LC-MS/MS Analysis of Urine Extracts

1. The mass spectrometer is tuned using the standard 8-oxodG tuning solution (see Note 12).
2. Reconstitute dried urine extracts in 50 μL water with thorough vortexing.
3. Transfer reconstituted extracts into HPLC vials containing low volume inserts and inject 10 μL on to the analytical column.
4. The column is eluted isocratically with mobile phase at a flow rate of 120 μL/min.
5. Selected reaction monitoring (SRM) analysis is performed for the $[M+H]^+$ ion to oxidised base $[B+H_2]^+$ transitions of 8-oxodG (m/z 284 to 168) and the stable isotope internal standard $[^{15}N_5]$8-oxodG (m/z 289 to 173) (see Note 13).
6. Quantification of 8-oxodG in each urine sample is determined from the ratio of the peak area of 8-oxodG to that of the internal standard in the same sample:

$$Q_{anal.} = (A_{anal.}/A_{istd.}) \times Q_{istd.}$$

where $Q_{anal.}$ and $Q_{istd.}$ correspond to the amounts (pmol) of the analyte and internal standard respectively, and $A_{anal.}$ and $A_{istd.}$ correspond to the peak areas of the analyte and internal standard, respectively. The quantity of 8-oxodG is then corrected for creatinine content yielding final values of pmol 8-oxodG/μmol creatinine. An example chromatogram is shown in Fig. 2 (see Note 14).

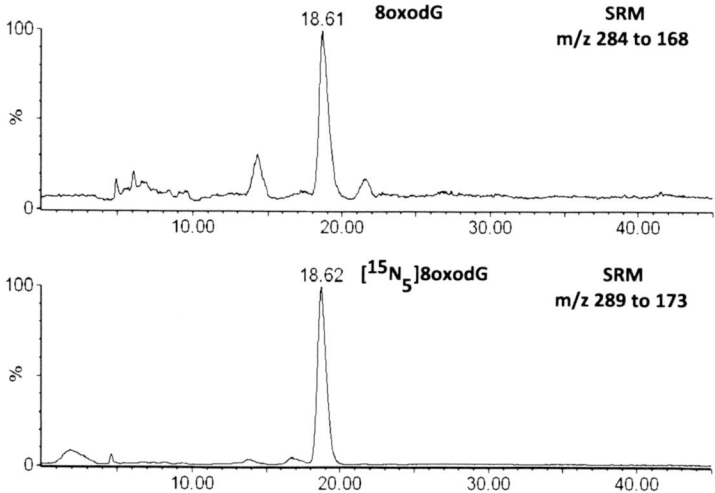

Fig. 2. Typical chromatograms, for the transitions m/z 289 to 173 and m/z 284 to 168 for the stable isotope internal standard $[^{15}N_5]$8-oxodG and 8-oxodG, respectively, obtained from solid-phase extract of healthy human urine, using the procedure outlined in this chapter.

4. Notes

1. Unless specified otherwise, throughout this procedure "water" refers to ultrapure water with resistance ≥18.2 MΩ and the content of organic matter <5 ppb.

2. A vacuum pump allows for better consistency and control of vacuum, compared to a water aspirator, but the latter can be used if a vacuum pump is not available.

3. The LC-MS/MS instrument used will depend on the individual laboratory performing this assay. The system used in method development, which we continue to use, consists of a Waters Alliance 2695 separations module connected to a Micromass Quattro Ultima (Waters Ltd., Manchester, UK) tandem quadrupole mass spectrometer with an electrospray interface. Specific details regarding the operation of each system are beyond the scope of this chapter.

4. Incubation at 37°C, compared to room temperature, has a negligible impact upon yield. Also, prolonging the incubation time does not appear to significantly modify the yield of 8-oxodG.

5. In addition to retention time, the identification of 8-oxodG is based upon on comparison of spectral properties [spectral overlay and absorbance ratios (250/260 nm; 280/260 nm) compared to the standard run under the HPLC purification conditions] to authentic, commercially obtained unlabelled standard. Further verification of stable isotope labelling, product identity, and impurity of unlabelled material is done by mass spectrometry. It is important to verify that labelled material is in vast excess, such that any unlabelled material, typically <2% of total material (depending on the exact purity of the starting material) does not contribute to the levels of endogenously measured 8-oxodG in each sample during internal standardisation. Collectively, this information should assure the user of purity and identity of the internal standard.

6. The majority of the material remaining at the end of the reaction is [$^{15}N_5$]dG. We collect this unmodified [$^{15}N_5$]dG for recycling in subsequent syntheses.

7. Additionally, we determine the concentration using mass spectrometry. This is done by injecting an appropriately diluted sample of [$^{15}N_5$]8-oxodG on to the LC-MS/MS under the conditions used for urinary 8-oxodG analysis (see Subheading 3.5). Such a sample also contains a known

amount of accurately quantified unlabelled 8-oxodG. Comparison of peak areas of labelled to unlabelled material allows the quantification of the labelled standard.

8. Spot urine samples can be used from other times of the day; however, this may result in higher intra-individual variability (19).

9. We generally use a hospital routine Chemical Pathology Service; however, the ELISA assay quoted in the protocol yields creatinine concentrations that compare very favourably with values obtained via an autoanalyser.

10. In our experience, processing more than three cartridges at any one time leads to poor control of flow rate, but this issue is probably dependent on the pump and manifold used.

11. The Oasis HLB cartridges are designed to function even if the column bed is allowed to dry out between steps. Leaving the column bed wet confers no advantage in terms of the final values obtained for urinary 8-oxodG.

12. In our laboratory, the 8-oxodG tuning solution is introduced into the mass spectrometer by continuous infusion at a flow rate of 10 μL/min with a Harvard model 22 syringe pump (Havard Apparatus Ltd., Edenbridge, UK).

13. LC-MS/MS instrument parameters for the system used in our laboratory – temperature of the electrospray source is maintained at 110°C and the desolvation temperature at 350°C. Nitrogen gas is used as the desolvation gas (650 L/h) and the cone gas set to zero. The capillary voltage is 3.20 kV, and the cone and RF1 lens voltages are 42 and 30 V, respectively. The collision gas is argon (indicated cell pressure 2.0×10^{-3} mbar) and the collision energy set at 12 eV. The dwell time is 200 ms and the resolution was two m/z units at peak base. Under these conditions, the limit of detection for the optimised analysis of urinary 8-oxodG by LC-MS/MS SRM on our system is 5 fmol on column (signal-to-noise ratio, S/N, = 4) for pure 8-oxodG standard.

14. Calculations should take any dilution or concentration factors into account, e.g. urinary 8-oxodG level (pmol/μmol creatinine) = $[(A_{anal.}/A_{istd.}) \times 2.0 \text{ pmol} \times 5]/$μmol creatinine.

In our experience, <5% of samples analysed are likely to have peaks that interfere with either the endogenous 8-oxodG or the internal standard peaks. In the vast majority of cases, where we have subjected these samples to re-extraction, the issue of interfering peaks has been resolved.

Acknowledgements

RS and PBF are supported by the U.K. Medical Research Council. Some development of this methodology in the laboratory of MDE and MSC was supported by using a University of Leicester miscellaneous income fund held by MSC. RS, MDE, PBF, and MSC are partners of ECNIS (Environmental Cancer Risk, Nutrition and Individual Susceptibility), a network of excellence operating within the European Union 6th Framework Program, Priority 5: "Food Quality and Safety" (Contract No. 513943). JKS is supported by (ECNIS, Contract No. FOOD-CT-2005-513943); FT is supported by Cancer Research UK (CRUK Programme Grant C325/A6691) and Experimental Cancer Medicine Centre Network (ECMC Grant C325/A7241).

The formation of a European-centred laboratory consortium, the European Standards Committee on Urinary (DNA) Lesion Analysis (ESCULA) is enabling an examination of methodology and issues surrounding the analysis of urinary 8-oxodG in the first instance. Further information about this consortium, including activities and membership can be obtained by contacting Dr. M.S. Cooke (msc5@le.ac.uk; http://escula.org).

References

1. Kasai, H. (1997) Analysis of a form of oxidative DNA damage, 8-hydroxy-2'-deoxyguanosine, as a marker of cellular oxidative stress during carcinogenesis. *Mutat. Res.* **387**, 147–163.
2. Bogdanov, M.B., Beal, M.F., McCabe, D.R., Griffin, R.M., and Matson, W.R. (1999) A carbon column-based liquid chromatography electrochemical approach to routine 8-hydroxy-2'-deoxyguanosine measurements in urine and other biologic matrices: a one-year evaluation of methods. *Free Radic. Biol. Med.* **27**, 647–666.
3. ESCODD, Gedik, C.M. and Collins, A. (2005). Establishing the background level of base oxidation in human lymphocyte DNA: results of an interlaboratory validation study. *FASEB J* **1**, 82–84.
4. Loft, S., Svoboda, P., Kasai, H., Tjonneland, A., Vogel, U., Moller, P., Overvad, K. and Raaschou-Nielsen, O. (2006). Prospective study of 8-oxo-7,8-dihydro-2'-deoxyguanosine excretion and the risk of lung cancer. *Carcinogenesis* **27**, 1245–1250.
5. Wu, L.L., Chiou, C.-C., Chang, P.-Y., and Wu, J.T. (2004) Urinary 8-OHdG: a marker of oxidative stress to DNA and a risk factor for cancer, atherosclerosis and diabetics. *Clin. Chim. Acta* **339**, 1–9.
6. Cooke, M. S., Evans, M. D., Dove, R., Rozalski, R., Gackowski, D., Siomek, A., Lunec, J., and Olinski, R. (2005) DNA repair is responsible for the presence of oxidatively damaged DNA lesions in urine. *Mutat. Res.* **574**, 58–66.
7. Siomek, A., Tujakowski, J., Gackowski, D., Rozalski, R., Foksinski, M., Dziaman, T., Roszkowski, K., and Olinski, R. (2006) Severe oxidatively damaged DNA after cisplatin treatment of cancer patients. *Int. J. Cancer* **119**, 2228–2230.
8. Cooke, M.S., Lunec, J., and Evans, M.D. (2002) Progress in the analysis of urinary oxidative DNA damage. *Free Radic. Biol. Med.* **33**, 1601–1614.
9. Haghdoost, S., Sjolander, L., Czene, S., and Harms-Ringdahl, M. (2006) The nucleotide pool is a significant target for oxidative stress. *Free Radic. Biol. Med.* **41**, 620–626.
10. Loft, S. and Poulsen, H.E. (1996). Cancer risk and oxidative DNA damage in man. *J. Mol. Med.* **74**, 297–312.

11. ESCULA, Marcus, S.C., Olinski, R. and Loft, S. (2008) Measurement and meaning of oxidatively modified DNA lesions in urine. *Cancer Epidemiol. Biomarkers Prev.* **1**, 3–14.
12. Haghdoost, S., Maruyama, Y., Pecoits-Filho, R., Heimburger, O., Seeberger, A., Anderstam, B., Suliman, M.E., Czene, S., Lindholm, B., Stenvinkel, P. and Harms-Ringdahl, M. (2006). Elevated serum 8-oxo-dG in hemodialysis patients: a marker of systemic inflammation? *Antioxid. Redox Signal.* **8**(11–12), 2169–2173.
13. Rozalski, R., Siomek, A., Gackowski, D., Foksinski, M., Gran, C., Klungland, A., and Olinski, R. (2004) Diet is not responsible for the presence of several oxidatively damaged DNA lesions in mouse urine. *Free Radic. Res.* **11**, 1201–1205.
14. Ravanat, J.L., Guicherd, P., Tuce, Z., and Cadet, J. (1999) Simultaneous determination of five oxidative DNA lesions in human urine. *Chem. Res. Toxicol.* **12**, 802–808.
15. Hu, C.W., Wang, C.J., Chang, L.W., and Chao, M.R. (2006) Clinical-scale high-throughput analysis of urinary 8-oxo-7,8-dihydro-2'-deoxyguanosine by isotope-dilution liquid chromatography-tandem mass spectrometry with on-line solid-phase extraction. *Clin. Chem.* **52**, 1381–1388.
16. Weimann, A., Belling, D., and Poulsen, H.E. (2001) Measurement of 8-oxo-2'-deoxyguanosine and 8-oxo-2'-deoxyadenosine in DNA and human urine by high performance liquid chromatography-electrospray tandem mass spectrometry. *Free Radic. Biol. Med.* **30**, 757–764.
17. Teichert, F., Verschoyle, R.D., Greaves, P., Thorpe, J.F., Mellon, J.K., Steward, W.P., Farmer, P.B., Gescher, A.J., and Singh, R. (2009). Determination of 8-oxo-2'-deoxyguanosine and creatinine in murine and human urine by liquid chromatography-tandem mass spectrometry: application to chemoprevention studies. *Rapid Commun. Mass Spectrom.* **23**, 258–266.
18. Singh, R., McEwan, M., Lamb, J.H., Santella, R.M., and Farmer, P.B. (2003) An improved liquid chromatography/tandem mass spectrometry method for the determination of 8-oxo-7,8-dihydro-2'-deoxyguanosine in DNA samples using immunoaffinity column purification. *Rapid Commun. Mass Spectrom.* **17**, 126–134.
19. Miwa, M., Matsumaru, H., Akimoto, Y., Naito, S. and Ochi, H. (2004). Quantitative determination of urinary 8-hydroxy-2'-deoxyguanosine level in healthy Japanese volunteers. *Biofactors* **22**, 249–253

Chapter 21

Assessing Sperm DNA Fragmentation with the Sperm Chromatin Dispersion Test

José Luis Fernández, Dioleyda Cajigal, Carmen López-Fernández, and Jaime Gosálvez

Abstract

The sperm cell has evolved to transmit a paternal haploid genome to the oocyte and form a new embryo. Therefore, it is essential that the integrity of this genome be evaluated as part of the standard semen analysis. The assessment of DNA fragmentation is consequently considered as an important parameter of sperm quality. The Sperm Chromatin Dispersion (SCD) test is a simple, fast, and reliable procedure to determine the frequency of sperm cells with fragmented DNA, and this may be confidently performed with the Halosperm® kit. Unfixed sperm cells are immersed in an agarose microgel on a slide, incubated in an acid unwinding solution that transforms DNA breaks into single-stranded DNA, and then in a lysing solution to remove protamines. After staining, the spermatozoa without fragmented DNA shows nucleoids with big halos of spreading of DNA loops, whereas those with fragmented DNA appear with a small or no halo. This may be visualized using fluorescence microscopy or with the standard bright-field microscope, without the requirement of more complex or expensive instrumentation. This procedure is very versatile, and being a diffusion-like assay with only a lysis protocol, may be usefully adapted for other species. Moreover, simultaneous determination of aneuploidies may be accomplished in the same sperm cell.

Key words: SCD, Sperm, DNA fragmentation, Fertility, Aneuploidy

1. Introduction

The standard analysis of human sperm quality in the clinic mainly comprises the determination of sperm quantity and concentration, morphological abnormalities of head, midpiece and tail, and motility (1). The essential purpose of the spermatozoa is the transmission of a haploid genome from the male to the oocyte, but current analysis seems mainly centred in the carrier, but not in the content. Logically, given its relevance for fertility, the assessment of the DNA integrity from the spermatozoa should be considered as an important new parameter to complement the conventional semen analysis (2).

Several studies reveal that infertile males have a higher frequency of sperm cells with DNA fragmentation than fertile controls. Moreover, those samples of lower quality according to the standard seminal analysis, also tend to have a higher proportion of spermatozoa with fragmented DNA (3–6). Nevertheless, some individuals with normal standard parameters may show elevated levels of sperm cells with DNA fragmentation. Furthermore, studies performed in samples subjected to *in vitro* fertilization (IVF) and intracytoplasmatic sperm injection into the oocyte (ICSI) demonstrate that the frequency of sperm cells with fragmented DNA in the sample may influence fertility, by negatively affecting the fertilization rate of the oocyte, the embryo quality, the blastocyst rate, the implantation rate and the pregnancy outcome (7–11). Moreover, the assessment of sperm DNA integrity is important as a complement to the study of sperm quality, providing relevant information in most andrological pathologies, as is the case of varicocele (12), genital infections (13), or cancer (14).

Several techniques have been applied to analyze DNA fragmentation in spermatozoa, mainly the sperm chromatin structure assay (SCSA), the terminal deoxynucleotidyl transferase-mediated nick end labeling (TUNEL) or the *in situ* nick translation (ISNT) assays, and the single-cell gel electrophoresis (SCGE) or comet assay (4, 5, 7, 10).These procedures cannot be performed routinely in the conventional semen analysis laboratory, since they are complex, difficult to implement, time consuming, or relatively expensive.

Recently, the Sperm Chromatin Dispersion (SCD) test has been developed and is available as a kit (Halosperm®); this is a very simple, rapid, and accurate procedure to determine sperm DNA fragmentation in any basic laboratory (15). The sperm cells are embedded in an agarose microgel on a slide, incubated in an acid solution that denatures the DNA exclusively in those spermatozoa with fragmented DNA, and then in a lysing solution that removes the protamines, so the DNA loops tightly packed in the nucleus are spread producing DNA halos emerging from a central core. After staining, the spermatozoa without fragmented DNA show nucleoids with big halos of spreading of DNA loops, whereas those with fragmented DNA appear with a small or no halo.

Fluorescence *In Situ* Hybridization (FISH) with labeled DNA probes may be performed on the sperm cells previously processed for the SCD test and immersed in the dried microgel (16). This allows to determine the presence of DNA fragmentation and aneuploidy simultaneously in each the sperm cell, and to study possible correlations between DNA fragmentation and chromosomal abnormalities.

2. Materials

2.1. SCD

2.1.1. Reagents and Technical Equipment

1. Bright-field or epifluorescence microscope with appropriate filters and objectives.
2. 4°C fridge.
3. 90–100°C and 37°C incubation bath(s).
4. Plastic gloves.
5. Lancet.
6. Glass slide covers (18 × 18 or 22 × 22 mm).
7. Micropipettes.
8. Trays for horizontal incubations.
9. Distilled water. Ethanol 100%.
10. Microwave oven and fume hood.
11. Halosperm kit® (Halotech DNA SL, Madrid, Spain; Conception Technologies, San Diego, USA). The kit contains coated slides, Eppendorf tubes with low-melting point agarose, a tube with 1 mL of acid denaturation solution (HCl 0.08 N) and a bottle with 125 mL of lysis solution.

2.1.2. Staining Solutions

2.1.2.1. Staining for Brightfield Microscopy

1. Wright's solution (Merck).
2. Phosphate buffer solution (PBS) pH 6.88 (Merck).
3. Mounting medium: Eukitt® (Panreac).

2.1.2.2. Staining for Fluorescence Microscopy

1. Fluorochromes for DNA staining.
2. Antifading: Vectashield (Burlingame, CA).

2.2. FISH

1. For assessing aneuploidy, a mix of directly labeled DNA probes is usually employed for human alphoid centromeric regions of X chromosome (DXZ1 Locus, SpectrumGreen labelled), Y chromosome (DYZ3 locus, SpectrumOrange labelled), and chromosome 18 (D18Z1 locus, SpectrumAqua labelled) (Vysis, Inc., Downer's Grove, IL). They are supplied with their specific hybridization buffer, and with information on preparation, incubation, washing, and detection.
2. Fixing solution: 10% formaldehyde in PBS pH 6.88.
3. Washing of fixing solution: PBS.
4. Alkaline denaturing solution: NaOH 0.05 N in distilled water.
5. DNA probe washing solutions: 50% formamide/2× SSC, pH 7 and 2× SSC, pH 7, both at 44°C.
6. Ethanol baths 70, 90, and 100%.

7. Counterstaining-Antifading solution: DAPI (1 µg/mL) in Vectashield. Store at 4°C, in the dark. DAPI is a potential carcinogen.

3. Methods

3.1. Standard SCD Procedure

3.1.1. Microgel Embedding of Cell Suspension (See Note 1)

1. While preparing the sperm sample, 10 mL of the lysing solution provided in the kit must be deposited in a tray with dimensions slightly higher than that of a conventional glass slide, and covered with aluminum foil, to be warmed at room temperature (22°C), but avoiding light exposure.
2. Dilute the semen sample in culture medium, sperm extender or PBS, to a concentration of 5–10 million per mL (see Note 2).
3. An Eppendorf tube with low-melting point agarose which is provided in the kit is put through a float. This float should be at the level of the top of the tube. Then, the float is left in a water bath at 90–100°C until the agarose dissolves, i.e., around 5 min. Alternatively, melt the agarose in a microwave oven.
4. Transfer the agarose containing Eppendorf tube, with the float, to a water bath at 37°C and leave for 5 min until the temperature is even (see Note 3).
5. Add 25 µL of the semen sample to the agarose Eppendorf tube and gently mix with the micropipette, avoiding the production of air bubbles.
6. Deposit the cell suspension from the agarose Eppendorf tube onto the coated side of a slide provided in the kit, and then cover it with a glass coverslip, avoiding trapping air bubbles. A drop of 14 or 20 µL for an 18 × 18 or 22 × 22 mm coverslip, respectively, is recommended.
7. Place the slide horizontally on a cold surface, for example, a metal or glass plate precooled at 4°C. Place the cold plate with the slide in the fridge at 4°C for 5 min, to allow the agarose to solidify.

3.1.2. Incubations in Solutions

1. Prepare the acid denaturant solution while the slide is in the fridge. For this purpose, mix 80 µL of the HCl solution from the kit with 10 mL of distilled water, and place in an incubation tray with dimensions slightly higher than that of the glass slide.
2. Remove the coverglass by sliding it gently. Immediately immerse the slide into the acid denaturation solution in a horizontal position, leave it to incubate for 7 min at room temperature (22°C) (see Note 4).
3. Wearing gloves, pick up the slide with the help of a lancet and immerse horizontally in the incubation tray containing 10 mL

of tempered lysis solution, leave it to incubate for 25 min (see Notes 5 and 6).

4. To wash the lysing solution, pick the slide up and immerse it horizontally for 3 min into a tray containing abundant distilled water.

5. Immerse the slide horizontally for 2 min into a tray with abundant 100% ethanol.

6. Afterward, leave the slide to dry horizontally, at room temperature, or in an oven at 37°C. After drying, the processed slides may be stored in archive boxes at room temperature in the dark for several months or immediately stained (see Note 7).

3.1.3. Staining

1. Prepare the fresh dye solution for bright-field microscopy by mixing the Wright solution with PBS (1:1).

2. Cover the dried microgel with a layer of the dye solution. Keep the slide horizontal for 10–15 min, blowing on it from time to time.

3. Decant the dye solution and briefly and smoothly wash the slide in tap water, then air dry. The coloring level must be checked under the microscope (see Notes 8 and 9).

4. Once the desired level of coloration is achieved and the slide is perfectly dried, it can be mounted in a permanent mounting medium like Eukitt® if desired.

3.1.4. Sperm Analysis and Classification

1. Examine the sample using a 100× immersion oil objective. The study of a minimum of 500 spermatozoa per sample is recommended, adopting the criteria of Fernández et al. (15) (Fig. 1a–f):

   (a) *Spermatozoa with big halo*: those whose halo width is similar to or higher than the minor diameter of the core (Fig. 1a, a′).

   (b) *Spermatozoa with medium-sized halo*: their halo size is between those with large and with small halo (Fig. 1b, b′).

   (c) *Spermatozoa with small halo*: the halo width is similar or smaller than 1/3 of the minor diameter of the core (Fig. 1c, c′).

   (d) *Spermatozoa without halo*: (Fig. 1d, d′).

   (e) *Spermatozoa without halo-degraded*: those that show no halo and present a core irregularly or weakly stained (Fig. 1e, e′).

   "Others": cell nuclei which do not correspond to spermatozoa. One of the morphological characteristics that distinguish them is the absence of tail. They are recorded but not included in the final result.

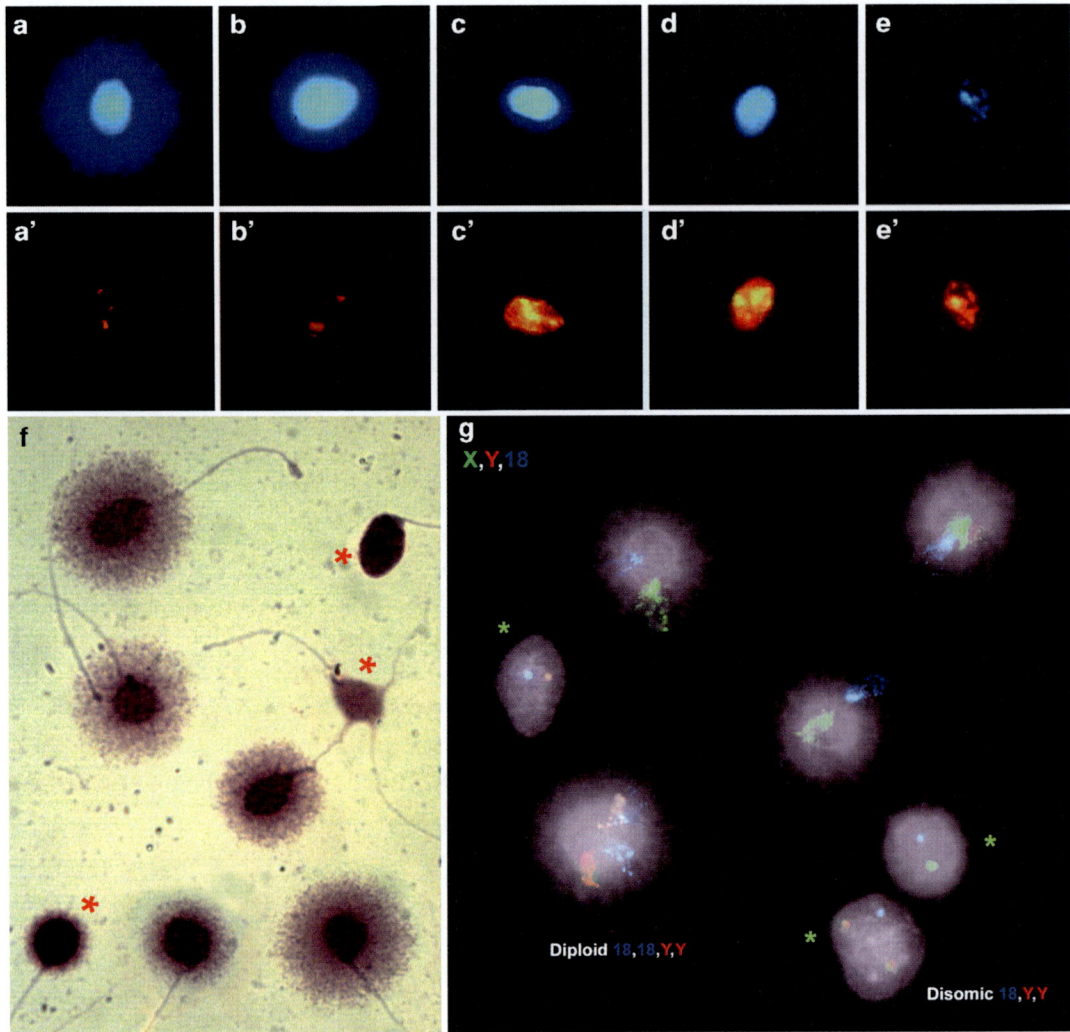

Fig. 1. Human sperm cells processed with the SCD test. (**a–e**) and (**a′–e′**): representative images from the different categories of halo sizes, observed after SCD test processing and subsequent DBD-FISH with a human whole genome probe, Cy3-labeled (*red*, **a′–e′**), to detect DNA breakage. DNA was DAPI counterstained (*blue*, **a–e**) to visualize the nucleoids and their halo of spreading of DNA loops. (**a, a′**): big halo; (**b, b′**): medium-size halo; (**c, c′**): small halo; (**d, d′**): without halo; (**e, e′**): without halo and degraded. Nucleoids showing small halo, without halo, and without halo-degraded (DAPI) contain fragmented DNA (DBD-FISH). (**f**) Wright staining of SCD-processed sperm cells allows an accurate visualization of the halo sizes under the conventional bright-field microscope. Those marked with asterisks have fragmented DNA. (**g**) Conventional FISH may be performed after SCD processing, allowing to evaluate DNA fragmentation (halo size) and aneuploidy, simultaneously in the same sperm cell. In the image, the three sperm showing asterisk contain fragmented DNA. FISH with a cocktail of DNA probes for centromeric alphoid sequences for 18 (*blue*), X (*green*), and Y (*red*) chromosomes, evidence a diploid sperm without fragmented DNA (big halo) and a sperm disomic for the Y chromosome and with fragmented DNA (without halo).

2. The results are expressed as percentage of each category, the percentage of spermatozoa with fragmented DNA being the sum of those with small halo, without halo and without halo-degraded (see Note 10).

3.2. Sequential SCD-FISH

The FISH procedure is performed on sperm cells previously processed for the SCD test and immobilized in the dried microgel (Fig. 1g).

1. After SCD processing do not stain the dried slide, but incubate it with the fixing solution for 12 min.
2. Wash the slide in excess of phosphate buffer for 1 min.
3. Deposit the double-stranded DNA probe mix for the alphoid centromeric regions of chromosomes X, Y, and 18 with its specific hybridization buffer in an Eppendorf tube, 20 µL total volume.
4. Denature the DNA probe mix at 70°C for 8 min in a water bath, and then immediately put on ice for 2–3 min.
5. Denature sperm DNA on the slide by incubation in alkaline denaturing solution for 15–25 s (see Note 11).
6. Immerse slide for dehydration in graded ethanol solutions (70–90–100%) for 5 min each, and then air-dry.
7. Pipette the denatured DNA probes on the microgel, cover with a glass cover slip and incubate overnight at 37°C in darkness, in a moist chamber lined with two sheets of wet filter paper.
8. Remove the glass coverslip by gently immersing the slides vertically in a Coplin jar containing isotonic saline solution at room temperature.
9. Wash the slides in Coplin jars containing 50% formamide/2× SSC pH 7 for 8 min, followed by 2× SSC pH 7 for 5 min, both at 44°C (see Note 12).
10. Counterstain DNA by pipetting 20 µL of Counterstaining-Antifading solution, and cover with a 24×60 mm glass coverslip. Avoid trapping air bubbles.
11. Examine the slides with an epifluorescence microscope, equipped with a triple-band pass filter and with monochrome filters for DAPI, SpectrumGreen, SpectrumOrange and SpectrumAqua for improved signal resolution.
12. Examine the sample using a 100× immersion oil objective. The study of 3,000–10,000 spermatozoa per sample is recommended, following the criteria of Muriel et al. (16) (see Note 13).

4. Notes

1. The purpose of microgel embedding is to provide an inert support to sperm cells, so they can be processed in a suspension-like environment, but on a slide. This way they can be

easily handled to be incubated in the solutions, avoiding centrifugations. Moreover, possible DNA fragments of relative high size that would be removed to the medium if cells were lysed in suspension, would be retained in the agarose matrix.

2. The sperm solvent should not be extremely dense to facilitate spreading. Both fresh samples and samples directly frozen in liquid nitrogen may be used. In practice, checking cell density by phase-contrast microscopy may be enough to adapt and obtain an appropriate cell concentration. Cell density within the agarose matrix should be not excessively high in order to avoid the overlapping of sperm cells, and not too broadly spread to facilitate rapid scoring.

3. It is important to mix the cell suspension with the liquid agarose when the latter has stabilized at 37°C, to avoid cell damage by heat.

4. The acid unwinding treatment produces DNA denaturation that is only evident in those sperm cells with fragmented DNA. This relative short incubation does not produce detectable denaturation in DNA from spermatozoa without fragmented DNA. It is also true for histone-nucleosome organized chromatin, as in blood leukocytes. It is thought that DNA breaks behave as points of DNA denaturation, single-stranded DNA being produced starting in both sides from the end of the break and proceeding along the DNA helix (10, 17).

5. The lysing step removes proteins, especially the protamines, so the DNA loops that are tightly packed inside the sperm nucleus spread, producing haloes emerging from a central core in those spermatozoa without fragmented DNA (18). If massively broken, i.e., with multiple DNA double-strand breaks, the lysing step allows the DNA fragments to diffuse from the residual core producing a big halo of DNA spots. Nevertheless, this halo may be sometimes confused with the normal halo of spreading of DNA loops from those spermatozoa without fragmented DNA in the case of humans. This lysing-only protocol of SCD is acceptable for the discrimination of spermatozoa with fragmented DNA in those species where the DNA loops are of very small size (19–21). In humans, it is preferable that there is previous incubation in the acid unwinding solution. The production of single-stranded DNA in spermatozoa with fragmented DNA by the treatment with this solution seems to prevent the spreading of DNA fragments, resulting in small halo or no halo at all. This prevention is not total, and when DNA is stained with very high sensitive fluorochromes (GelRed-Biotium; SYBR Green, SYBR Gold-Molecular Probes, etc.), and some residual DNA spots are visualized around the sperm heads from spermatozoa with DNA fragmentation.

6. An important result from using the lysing solution from the kit is that the sperm tails remain preserved. This helps to discriminate mature sperm cells from other possible cell types that could be present in the sample.

7. As an internal control, it is recommended to process a microgel with a control sample of well-known level of DNA fragmentation. This sperm sample should be distributed in aliquots and frozen in liquid nitrogen. An aliquot should be thawed and immediately processed as a microgel close to the sample to be analyzed. A control sample and two different sperm samples can be coprocessed using a single slide. After SCD processing, the control sample should reveal its usual DNA fragmentation yield, thus validating the technique and the results from the other samples.

8. Strong staining with Wright's stain is preferred, to clearly discriminate the peripheral border of the halo (Fig. 1f). If staining is very weak, especially on the region of chromatin dispersion halos, the slide can be restained with Wright's solution. If coloration is too strong, the slide can be discolored by washing gently in tap water, or 10% ethanol if preferred. After air drying, it can be stained again but reducing coloring exposure time.

9. Visualization under fluorescence microscopy is also possible using standard DNA fluorochromes (Fig. 1a–e). After sample dehydration, 20 μL of Counterstaining-Antifading solution (e.g., DAPI in Vectashield) is distributed from a pipette over the slide and covered with a 24 × 60 mm glass cover slip, avoiding trapping air bubbles.

10. Instead of staining, the SCD processed slides may be incubated with a fluorescent-labeled human whole genome probe, which will detect the single-stranded DNA produced from DNA breaks by the unwinding treatment. This is the rationale of the DNA Breakage Detection-Fluorescence *In Situ* Hybridization (DBD-FISH) procedure to detect and quantify DNA breaks (22). The application of this procedure results in only labeling those spermatozoa with small halo, without halo and without halo-degraded, thus demonstrating that these specific types contain fragmented DNA (Fig. 1a'–e'). Accordingly, sequential enzymatic labeling of DNA breaks may be accomplished following the ISNT, or end labeling by the klenow or using the TUNEL assay. This also results in the labeling of those spermatozoa with small halo, without halo and without halo-degraded.

11. The agarose microfilm is very delicate, so the typical DNA denaturation with 70% formamide/2× SSC at 70°C may disrupt it. The halos of the nucleoids in the agarose microgels are very delicate, so that denaturation and washing steps tend

to affect their preservation. Thus, only those slides with well-preserved halos must be analyzed.

12. Incubation in DNA probe washing solutions at high temperature can destroy or damage the microgel or nucleoids. It is recommended not to exceed 44°C in the washing solutions, and not to use long incubation times. When necessary, increasing the formamide concentration and/or decreasing the ionic strength can decrease the washing temperature needed while maintaining high stringency.

13. Overlapping nuclei are not scored. Nuclei showing nullisomy are not directly scored since this may be a consequence of the lack of hybridization. The presence of sperm tails is confirmed under the SpectrumAqua filter set of the microscope. A nucleoid with big or medium halo size (i.e. without DNA fragmentation), is considered disomic when it shows two fluorescent domains of the same chromosome, comparable in size and brightness, and separated by at least one-half diameter of the domain of one signal. Otherwise, a sperm nucleoid with small or without halo (i.e. with DNA fragmentation) is considered disomic when the two signals of similar size, shape, and intensity, are separated by at least one signal diameter. This is a conservative criterion for comparison. Diploidy is established when two distinct chromosome 18 signals and also two-signals for X and/or Y chromosomes are present in the same sperm nucleus. FISH signals from satellite DNA sequences may not be a spot in sperm nucleoids with halos, showing their DNA fibres spreading from a specific point from the core. The origin from which the DNA fibres spread usually has a stronger intensity than that of the diffused fibres (23). This may help to clarify possible questions that may arise in a very few cases. Moreover, DNA probes of smaller size than that of the satellite DNA sequences may be used for aneuploidy assessment.

Acknowledgments

This work was supported by Fondo de Investigaciones Sanitarias (FIS PI070459) and the Consejo de Seguridad Nuclear. Xunta de Galicia INCITE07PXI916201ES.

References

1. World Health Organization. (1999) *WHO laboratory manual for the examination of human semen and sperm-cervical mucus interaction*, 4th edn. Cambridge University Press, Cambridge.

2. Agarwal, A., and Allamaneni, S.S. (2005) Sperm DNA damage assessment: a test whose time has come. *Fertil. Steril.* **84**, 850–853.

3. De Jonge, C. (2002) The clinical value of sperm nuclear DNA assessment. *Hum. Fertil.* **5**, 51–53.

4. Agarwal, A., and Said, T.M. (2003) Role of sperm chromatin abnormalities and DNA damage in male infertility. *Hum. Reprod. Update* **9**, 331–345.
5. Agarwal, A., and Said, T.M. (2004) Sperm chromatin assessment. In: *Textbook of assisted reproductive techniques: laboratory and clinical perspectives.* (Gardner, D.K., ed.). Taylor & Francis, Philadelphia, pp. 93–106.
6. Tesarik, J., Mendoza-Tesarik, R., and Mendoza, C. (2006) Sperm nuclear damage: update on the mechanism, diagnosis and treatment. *Reprod. Biomed. Online* **12**, 715–721.
7. Evenson, D.P., Larson, K.J., and Jost, L.K. (2002) Sperm chromatin structure assay: its clinical use for detecting sperm DNA fragmentation in male infertility and comparisons with other techniques. *J. Androl.* **23**, 25–43.
8. Virro, M.R., Larson-Cook, K.L., and Evenson, D.P. (2004) Sperm chromatin structure assay (SCSA®) parameters are related to fertilization, blastocyst development, and ongoing pregnancy in in vitro fertilization and intracytoplasmic sperm injection cycles. *Fértil. Steril.* **81**, 1289–1295.
9. Muriel, L., Garrido, N., Fernández, J.L., Remohí, J., Pellicer, A., de los Santos, M.J., and Meseguer, M. (2006) Value of the sperm deoxyribonucleic acid fragmentation level, as measured by the sperm chromatin dispersion test, in the outcome of in vitro fertilization and intracytoplasmic sperm injection. *Fertil. Steril.* **85**, 371–383.
10. Evenson, D.P., and Wixon, R. (2006) Clinical aspects of sperm DNA fragmentation detection and male infertility. *Theriogenology* **15**, 979–991.
11. Collins, J.A., Barnhart, K.T., and Schlegel, P.N. (2008) Do sperm DNA fragmentation tests predict pregnancy with in vitro fertilization? *Fertil. Steril.* **89**, 823–831.
12. Enciso, M., Muriel, L., Fernández, J.L., Goyanes, V., Segrelles, E., Marcos, M., Montejo, J.M., Ardoy, M., Pacheco, A., and Gosálvez, J. (2006) Infertile men with varicocele show a high relative proportion of sperm cells with intense nuclear damage level, evidenced by the Sperm Chromatin Dispersion (SCD) test. *J. Androl.* **27**, 106–111.
13. Gallegos, G., Ramos, B., Santiso, R., Goyanes, V., Gosalvez, J., and Fernández, J.L. (2008) Sperm DNA fragmentation in infertile men with genitourinary infection by *Chlamydia trachomatis* and *Mycoplasma*. *Fertil. Steril.* **90**, 328–334.
14. Meseguer, M., Santiso, R., Garrido, N., and Fernández, J.L. (2008) The effect of cancer on sperm DNA fragmentation measured by the sperm chromatin dispersion (SCD) test. *Fertil. Steril.* **90**, 225–227.
15. Fernández, J.L., Muriel, L., Goyanes, V., Segrelles, E., Gosálvez, J., Enciso, M., Laframboise, M., and De Jonge, C. (2005) Simple determination of human sperm DNA fragmentation with an improved sperm chromatin dispersion (SCD) test. *Fertil. Steril.* **84**, 833–842.
16. Muriel, L., Goyanes, V., Segrelles, E., Gosálvez, J., Alvarez, J.G., and Fernández, J.L. (2007) Increased aneuploidy rate in sperm with fragmented DNA as determined by the sperm chromatin dispersion (SCD) test and FISH analysis. *J. Androl.* **28**, 38–49.
17. Ljungman, M. (1999) Repair of radiation-induced DNA strand breaks does not occur preferentially in transcriptionally active DNA. *Radiat. Res.* **152**, 444–449.
18. Tsanev, R., and Avramova, Z. (1981) Nonprotamine nucleoprotein ultrastructures in mature ram sperm nuclei. *Eur. J. Cell Biol.* **24**, 139–145.
19. Enciso, M., López-Fernández, C., Fernández, J.L., García, P., Gosálbez A., and Gosálvez, J. (2006) A new method to analyze boar sperm DNA fragmentation under bright-field or fluorescence microscopy. *Theriogenology* **65**, 308–316.
20. López-Fernández, C., Crespo, F., Arroyo, F., Fernández, J.L., Arana, P., Jonhston, S.D., and Gosálvez, J. (2007) Dynamics of sperm DNA fragmentation in domestic animals II: The Stallion. *Theriogenology* **68**, 1240–1250.
21. Gosálvez, J., Vázquez, J.M., Enciso, M., Fernández, J.L., Gosálbez, A., Bridle, J.R., and López-Fernández, C. (2008) Sperm DNA fragmentation in rams vaccinated with Miloxan. *Open Vet. Sci. J.* **2**, 7–10.
22. Fernández, J.L., and Gosálvez, J. (2002) Application of FISH to detect DNA damage: DNA breakage detection-FISH (DBD-FISH). *Methods Mol. Biol.* **203**, 203–216.
23. Klaus, A.V., McCarrey, J.R., Farkas, A., and Ward, W.S. (2001) Changes in DNA loop domain structure during spermatogeneis and embryogeneis in the Syrian Golden Hamster. *Biol. Reprod.* **64**, 1297–1306.

ERRATUM

Buccal Micronucleus Cytome Assay

Philip Thomas and Michael Fenech

Vladimir V. Didenko (ed.), *DNA Damage Detection In Situ, Ex Vivo, and In Vivo: Methods and Protocols*, Methods in Molecular Biology, vol. 682, DOI 10.1007/978-1-60327-409-8, pp. 235–248, © Springer Science+Business Media, LLC 2011

DOI 10.1007/978-1-60327-409-8_22

The publisher regrets that in chapter 17, page 239, the Buccal Reagent Setup contains an error in the proportions of salts required to prepare the buccal cell buffer. Reagent Setup description:

BUCCAL REAGENT SETUP
Buccal cell buffer To prepare 1 liter of buccal buffer, weigh 1.6 g Tris–HCl, 1.2 g EDTA and 37.2 g of sodium chloride, and dissolve in 600 ml of Milli-Q water. Thoroughly dissolve the salts and adjust the volume to 1,000 ml. Adjust pH to 7.0 and autoclave at 121 1C for 30 min. The buffer will last for up to 3 months when stored at room temperature.

The correct version is:

BUCCAL REAGENT SETUP
Buccal cell buffer To prepare 1 liter of buccal buffer, weigh 1.6 g Tris–HCl, 38.0 g EDTA and 1.2 g of sodium chloride, and dissolve in 600 ml of Milli-Q water. Thoroughly dissolve the salts and adjust the volume to 1,000 ml. Adjust pH to 7.0 and autoclave at 121 1C for 30 min. The buffer will last for up to 3 months when stored at room temperature.

The online version of the original chapter can be found at
http://dx.doi.org/10.1007/978-1-60327-409-8_17

INDEX

A

Abasic sites 134
Absorbance ratios 286
Active caspase-3 .... 40, 41, 57
Adherent cells 7, 107, 179
Aflatoxin B$_1$ 271
Ageing 280
Alexa Fluor 488, 10, 93, 95, 96, 100,
    154, 157, 252–254, 257, 259, 262, 266, 268
Alexa Fluor 647, 33, 95, 98, 100
Alexa Fluor™ series of fluorochromes 11
Alkali-labile sites 134, 142, 144
Alkaline
    comet assay 115
    electrophoresis (*see* Electrophoresis)
    unwinding 134, 135, 137, 143, 145
Alphoid satellite DNA probes 135, 138
AML-5 cells 171, 172
Analysis of
    cells by flow-or image-cytometry 91–100
    ultrasound backscatter signals 172
Aneuploidy 235, 245, 292, 293, 296, 300
Annexin V
    binding 111
    staining 11
Annexin V-Cy5 conjugate 106, 110, 111
Anti-BrdU antibody 10, 11
Anti-digoxigenin antibodies
    alkaline-phosphatase conjugate 68
    gold-conjugated 33
Antifading solution 85, 136, 139, 142, 294, 297, 299
Antioxidant 207. 279
APO-BRDU kit 11, 34
APO-DIRECT kit 34
Apop* ISOL Dual Fluorescence Kit 55, 86
Apop* Peroxidase *In Situ* Oligo Ligation kit ... 51
ApopTag kit 11
Apoptosis
    in AML cells 179
    in cells 165
    detection
        in clinical samples 92
        earlier stages 66
        in situ 45
        by *in situ* ligation 54

in vitro and in vivo 103–113
    in tissue sections 79
    labeling by T7 DNA polymerase 37–46
    or programmed cell death 3
    in tissues 169
Apoptosis-inducing factor (AIF) 56
Apoptotic
    bodies 25, 30
    cell corpse elimination 77
    cell corpses 78, 80–82
    cells criteria for scoring 227–228
    cells detection of (*see* Detection of apoptotic cells)
    DNA fragmentation 16, 39, 79
    DNase I-like cleavage 82
    nucleases cutting properties 39, 59, 77
    thymocyte nuclei 82
Aqua-Poly/Mount mounting medium 6, 9
Archival material 29, 31
Autofluorescence 140, 262
Avidin
    avidin-biotin complex 6
    fluorescein isothiocyanate-conjugated 5
    horseradish peroxidase-conjugated 6

B

Backscatter signal 167, 172
Bacterial artificial chromosomes (BACs) 117
Benzo(a)pyrene
    diol epoxide 272
    DNA adducts 271–277
Binucleate cells
    criteria for selecting 228
    frequency of 230
Biomonitoring 190
Biopsies 166, 182, 185,
    190, 204, 262, 264, 265, 268, 269, 271
Biotin-14-dATP 42, 43, 45
Biotin–dUTP 33, 67, 72
Biotinylated nucleotides 38
Bladder cells 272
Blood
    peripheral blood lymphocytes 209–212, 249–269
    peripheral blood mononuclear cells ... 207–214, 250
    vacutainer blood tubes 220, 222
    white blood cells 189–205

Index

Blunt-ended DNA breaks 39, 40, 77, 81–85
Bone marrow
    cells isolation of 251
    γ-H2AX detection in......................... 249–269
Brain
    aging 29
    Alzheimer's brain tissue sections............................... 21
    apoptosis
        detection in...................................4, 53, 96
        induction by photodynamic
            therapy..................170, 171, 175, 180
    apoptotic rat brain neurons.. 40
    normal rat brain sections nonspecific
        background 86
    photodynamic therapy in rat brain 175
BrdU. *See* 5′-Bromo-2′-deoxyuridine
BrdUTP. *See* 5-Bromo-2′-deoxyuridine-5′-triphosphate
Breaks. *See* DNA breaks
5′-Bromo-2′-deoxyuridine................................92, 192, 200
5-Bromo-2′-deoxyuridine-5′-triphosphate
    (BrdUTP) 10, 11, 92–98
    labeling assay 92
Buccal
    cell
        collection 236, 240
        harvesting 236–238, 240–242
        scoring 239
        staining 238–239, 242, 243
    micronucleus cytome (BMCyt) assay 235
    mucosal tissue 235
Buffered formaldehyde 4, 5

C

Camptothecin...96, 99, 109, 172
Carboxyfluorescein (FAM)... 104
Carcinogen binding, quantitation of............................... 271
Carcinogen-DNA adducts 271
Carcinogenesis... 189, 271
Caspase-3-activated deoxyribonuclease
    (CAD) 39
Caspases.. 92, 103–113
Cell-autonomous and waste-management
    nucleases 78
Cell cycle... 12, 49, 92, 94,
    95, 97, 100, 104, 106, 111, 172–174, 177–178,
    180, 227, 232, 250
Cell cycle inhibitors.. 49
Cell death. *See also* Apoptosis
    non-invasive assessment of 279–287
    other forms of 166, 180
Cell death detection kits.. 10
Cells
    apoptotic................................... 4, 11, 37, 38, 40, 49,
    50, 52–56, 59, 60, 65, 66, 78–82, 85, 91–100,
    103–113, 165, 173, 180, 227–228, 230, 231, 233

bladder. 272
C666-1 .... 171, 175
damaged by mechanical forces.................................... 4
flowchart of the general protocol for
    TUNEL staining of 6
karyolytic...236, 244, 246
karyorrhectic.......................................236, 244–247
non-apoptotic..4, 11, 109,
    110, 112, 113
pyknotic 244, 246
smeared... 273, 275
suspension..6–8, 23, 97,
    107–109, 136–137, 213, 223, 224, 241, 242,
    247, 257, 294, 298
U-932.. 97
undergoing
    active gene transcription 4, 11
    DNA repair... 4
white blood.. 189–204,
    213, 272, 275
Cell shrinkage... 30
Centromere DNA 119
Chemical carcinogens... 271
Chloroform.. 41, 42, 85
Chloromethyl-X-rosamine 105
Chromatin
    condensation and margination.................................. 30
    structure..119–120, 134,
    145, 292
Chromatin-embedded DNA
    photoproducts... 149
Chromatography 195, 297–287
Chromogenic system for TUNEL detection 18
Chromosome-specific satellite DNA probes 135
Chromosome spreads 116
Chronic degenerative diseases 189
Colorimetric staining for light microscopic
    examination 8–9
Comet
    assay......................... 115–131, 134, 250, 292
    tail...115, 117, 119
Confocal microscope 85, 252, 257
Conjugated avidin 5, 6, 10, 97
Conjugates
    streptavidin-Cy2.. 42, 44
    streptavidin-peroxidase....................................42, 43, 46
Controls
    positive tissue control.. 24
    target-specific control 24
Cortical macrophages... 82
Cot-I DNA 123, 126, 128, 131
Counterstain..5, 6, 9, 12, 42,
    45, 112, 120, 136, 138, 139, 142, 144, 151, 152,
    159, 179, 276, 277, 294, 296, 297, 299
CPD photolesion.. 153

Creatinine analysis 282, 283
Criteria for scoring
    apoptotic cells 227–228
    micronuclei 227, 229
    necrotic cells 227, 228
    nuclear buds 226–227
    nucleoplasmic bridges 227, 229
Criteria for selecting binucleate cells 228
Cryostat sections 41–43
CSK buffer 149, 154, 156, 158
Cullin 4A (CUL4A) 150, 151
Cullin 4B (CUL4B) 150, 153, 159
Cultured cells 4–7, 10, 12, 16, 25,
    40, 41, 157, 208, 214, 250
Cyclobutane pyrimidine dimers (CPD) 116, 149
Cy™ series of fluorochromes 11
Cytochrome P450 .... 190, 272
Cytokinesis-block micronucleus cytome
    (CBMN cyt) assay 217–233
Cytome concept 218
Cytometric analysis, markers of
    apoptotic cells 91, 103–113
Cytometry
    flow 6, 8, 10, 12, 92, 94–98, 108–112, 171,
    177–178, 177–179, 250, 252–253, 258, 259
    image 91–100, 105
    laser scanning 92, 94–100, 105
Cytospin preparation of suspension cells 8
Cytostasis 218, 227
Cytostatic and cytotoxic outcomes 230–231
Cytotoxic drugs 92
Cytotoxicity 227

D

DAB. *See* 3,3′Diaminobenzidine
DBD-FISH. *See* DNA breakage detection-fluorescence
    in situ hybridization
DDB2 and *DBB1* genes 150
Deoxyribonucleotides 92, 93, 97
Detection of
    apoptosis and DNA damage in tissue sections 51
    apoptosis by the TUNEL 3–12
    apoptotic cells 4, 53, 96
    cell death 166
    DNA damage 4
    DNA damage effects 165–185
    double-strand DNA breaks 50, 53, 54,
    65–74, 133–146
    tumor regions that respond to therapies 182
3,3′ Diaminobenzidine 5, 19, 22, 24,
    193, 274, 278
Digoxigenin-11-dUTP 31, 33, 67,
    68, 72, 73, 123, 126, 127
Disposal of post-apoptotic corpses 78, 85

DNA
    centromere 116, 119, 229
    digestion 191, 195
    halos 292
    isolation from white blood cells 194–195
    lesions 134, 142, 150, 152, 190
    loops 117, 141, 292, 296, 298
    oxidation 16, 279
    photoproducts 149
    repair 4, 11, 29, 120–121,
    141, 142, 149–159, 229, 245, 250, 280
    satellite 135, 138, 143, 144, 300
    telomeric 199
DNA adducts 189–204, 271–278
DNA-based apoptotic markers 55
DNA breakage detection-fluorescence *in situ*
    hybridization 133–146, 296, 299
DNA breaks
    blunt-ended 39, 40, 81–85
    DNase I-and DNase II-types of 78–80
    double-strand 50, 53, 54, 65–74, 134
    with 3′OH 53, 79
    5′OH DNA breaks 54, 77–86
    with 3′overhangs 67
    single-strand 134
    staggered 59
    at the ultrastructural level 29–34
DNA cleavage. *See* DNA breaks
DNA damage
    comet assay for targeted examination of 115–131
    detection (*see* Detection of, DNA damage)
    free radical-induced 54
    in situ, ex vivo and in vivo,
    in situ labeling of 37–46, 54
    method for the detection of 4, 165, 198, 200
    oxidative 16
    spectroscopic detection of 165–185
    in tissue sections 51
    at the ultrastructural level 29–34
    UV-induced 149–159
DNA degradation 15, 30, 39, 66, 77, 80, 81, 85
DNA fragmentation
    apoptotic 16, 39, 79
    sperm 291–300
DNA polymerase *in situ* nick translation. *See In situ* nick
    translation (ISNT) assay
DNA probes
    alphoid satellite 135, 138
    with blunt ends, or short 3′overhangs 79
    chromosome-specific satellite 135
    classical satellite 135, 138
    for FISH 116–118, 125, 126, 135, 145
    fluorescent 134
    human 135

DNA probes (Continued)
  labeling of 123, 126–127
  telomeric or pancentromeric 136
DNase I....................................... 11, 26, 34, 38, 39,
  54, 78–80, 82, 83, 85, 123, 127
DNase I-and DNase II-types
  of DNA breaks.. 78–80
DNase II... 54, 78–86
DNase II-like breaks in the cytoplasm
  of cortical macrophages 82
DNase II-type cleavage. See DNase I-and DNase II-types
  of DNA breaks
DNase I-like cleavage in apoptotic
  thymocyte nuclei.. 82
DNase I-like nucleases 54
DNase I type enzymes....................................... 39
Dorsal root ganglia (DRG) neurons................. 16
Double-hairpin oligonucleotide probe 81
Double-labeling.. 19–23
Double-strand DNA breaks 50, 53, 54,
  65–74, 133–146
Dried microgel 292, 295, 297
DSB. See Double-strand DNA breaks

E

E. coli DNA polymerase I 37
Electron microscopy 30, 31, 33, 57, 251, 252, 254
Electrophoresis. See also Gel electrophoresis
  alkaline 120, 122, 125
  neutral........................ 120, 122, 125–126, 134
Elimination of apoptotic cell corpses.......... 80–82
  See also Disposal of post-apoptotic corpses
ELISpot... 207–214.
  See also Enzyme-linked immuno spot assay
ELISpot assay kits 211
Embedding
  in an epoxy resin 31
  cells in agarose 124
  of cell suspension 136–137, 294
  microgel.......................................136–137, 294, 297
  mold.. 265
  paraffin 264–265
EM-ISEL assay... 30
End labeling. See In situ (DNA) end labeling
Endogenous
  antioxidant capacity 207
  biotin... 45
  DDB1.. 151, 158
  8-oxodG 287
  peroxidase 9, 10, 15, 201, 202, 276
  phosphatases 68
  toxic agents 29
Endonuclease III 116, 122
Endonucleases, lesion-specific........................ 116

3′ Ends of fragmented DNA 6
Engulfment-mediated DNA degradation 77, 80
Enzyme linked immunosorbent assays
  (ELISA)..........................208, 272, 287
Enzyme-linked immuno spot assay....................... 207–214
Enzymes.................................... 4–6, 11, 20, 22, 26, 30,
  37–40, 44, 45, 51, 54, 56, 65, 73, 74, 78–86,
  105, 107, 112, 116, 119, 122, 125, 131, 207, 272
Enzyme-sensitive sites..................................... 116
Epidermoid carcinoma of the larynx 170
Epithelial cells 190
Epoxide hydrolase enzymes........................... 272
Epoxy resins.. 33
1,N^6-etheno-2′-deoxyadenosine (εdA) 45, 190
3,N^4-etheno-2′-deoxycytidine (εdC) 190
Etheno-2′-deoxyguanosine............................. 189
Etheno-DNA adducts 189–204
Evaluating the specificity of anti-8-OHdG
  antibodies.. 25
Excision repair. See Nucleotide excision repair (NER)
Executioner apoptotic nucleases 39
3′→5′ Exonuclease activity 39, 40, 45
5′→3′ Exonuclease activity 39
Ex vivo
  definition .... v, vi
  labeling of DNA damage........................... 183

F

False positives 11, 56, 247
Feulgen staining 242–243, 246
FISH. See Fluorescence in situ hybridization
FITC-12-dUTP. See Fluorescein-12-dUTP
FITC fluorescence... 86
Fixation
  cell... 92, 256–257
  chemical... 16
  formaldehyde 10
  specimen 264
Fixative
  Diff-Quik 219, 225
  ethanol/glacial acetic acid 238
  formaldehyde 16, 17
FLICA. See Fluorochrome-Labeled Inhibitors
  of Caspases
FLIVO probes 113
FLIVO™.. 113
Flow-and image-cytometry, multiparameter
  analysis of cells by 92
Flowchart of the general protocol for
  TUNEL staining 6
Flow cytometry...............................6, 8, 10, 12,
  92, 94–98, 108–112, 171, 177–178, 177–179,
  250, 252–253, 258, 259
Fluorescein-12-dUTP (FITC-12-dUTP) 33, 72, 73

Fluorescein isothiocyanate-conjugated avidin 5
Fluorescence analysis 96
Fluorescence filter sets 181
Fluorescence in situ hybridization (FISH) 116–120,
    123, 125, 126, 130, 131, 133–146, 292–294,
    296, 297, 299
    comets 116, 118–119
    probes... .... 116, 126
Fluorescence intensity 97, 98, 100,
    141, 144, 258, 268
Fluorescence microscopy 6, 8, 10, 50,
    105, 116, 246, 293, 299
Fluorescent
    antibody enhancer set for digoxigenin
        detection 124, 129, 130
    conjugated avidin 10, 97
        (see also Fluorescein isothiocyanate-conjugated
        avidin)
    detection 10, 12, 18, 54
    domains 300
    human whole genome probe 296, 299
    images 139, 157, 159
    immunostaining 151, 162
    in situ hybridization 115–131, 133, 292, 299
        (see also Fluorescence in situ hybridization)
    label 18, 20, 21, 45, 46, 138, 299
    light source 254
    microscope 65, 84, 85, 257
    probes 18, 235
    staining 9
    tagged avidin 6
        (see also Fluorescein isothiocyanate-conjugated
        avidin)
Fluorochrome-labeled inhibitors of caspases
    (FLICA) 103–113
Fluorochromes 11, 34, 91–93, 97, 99,
    103–113, 135, 137, 139, 143–145, 293, 298, 299
Fluorochrome-tagged deoxyribonucleotides 92, 93
Formaldehyde-fixed, paraffin-embedded
    tissues 43–44
Formamidopyrimidine DNA glycosylase
    (FPG) 116, 122
Formvar 31, 32
Free 3'OH groups 53
Free radical-induced DNA damage 54
Frozen sections 40, 46, 50, 275

G

Gaps in double stranded DNA 38, 40
Gel electrophoresis. See also Electrophoresis
    agarose 127
    pulsed-field 250
    SDS-polyacrylamide 260
    single-cell 292
    (see also Comet, assay)

Gene
    DBB1 150
    DDB2.. 150
    MGMT 119
    TP53 121
Gene amplification 218
General anesthesia 264
Generating 3'PO$_4$/5'OH breaks 54
Generation of free 3'-hydroxy termini 3
Gene-specific
    DNA repair 120–121
    signals 121
Genetic instability 179
Genetic toxicology 218
Gene transcription 4, 11, 34
Genomic DNA 3, 4, 117, 118, 120, 126–128
Genotoxicity
    assay 30
    testing... 115
Gold
    -conjugated anti-digoxigenin antibodies 33
    -conjugated anti-fluorescein immunoglobulins 33
    -conjugated streptavidin 33
    -coupled anti-digoxigenin goat
        immunoglobulins 32
    labeling 34
    quantitative analysis of labeling 34

H

Hairpin
    looped 51, 52, 56
    loopless 51, 52, 56
Hairpin probe 51, 52, 54, 56, 66
Hairpin-shaped oligonucleotides 65, 66
Halosperm® kit 292, 293
γ-H2AX
    detection 249–269
    foci 250, 257
    in human and mouse tissues 250
    levels 250, 259, 262
    in peripheral blood lymphocytes, splenocytes,
        and bone marrow 249–269
    protein. 250
    in xenografts and skin 262–263
Hematoxylin (H&E staining) 5, 9, 45, 57,
    172, 174, 178, 193, 265, 274, 276
Hep-2 cells.. .... 172, 173.
    See also Epidermoid carcinoma of the larynx
H&E staining. See Hematoxylin
Histology analysis 177–179
Histone H2AX 250
Hoechst 33342 .... 5, 6, 9, 12, 112
Horseradish peroxidase 6, 254
HPLC coupled to tandem mass spectrometry
    (LC-MS/MS) 281, 282, 284–287

HPP staining 51
Human biomonitoring 190
Human exposure 271
Human peripheral blood lymphocytes 209, 211–212, 259, 267
Human sperm quality 291
Humidified chamber/Humid chamber 8, 9, 20–22, 25, 26, 32, 43, 44, 70, 71, 84, 100, 139, 275–277
Hybridization mix 128
Hydrogen peroxide 5, 9, 45, 204, 207, 209, 273, 281, 283
Hydrogen peroxide-induced oxidative stress 209
8-Hydroxy-2'-deoxyguanosine (8-OHdG) 16–23, 25–26
4-Hydroxy-2-nonenal (HNE) 190
Hypotonic treatment 233

I

IFN gamma 208, 212
Image analysis
    digital 140–141
    software 135, 140
    system 140
Immunoaffinity
    clean up 191, 196
    column preparation 191, 196
Immunoblotting
    DSB detection limits 250
    γ-H2AX protein 250
Immunoelectron microscopy (IEM) 33
Immunofluorescence detection of BPDE-DNA adducts in
    frozen or paraffin tissue sections 273, 275
    slides containing smeared cells 273
Immunohistochemical
    analysis 249, 271
    detection of
        εdA and εdC 190, 193, 194, 197, 201–204
        DNA adducts 271–278
        8-OHdG 16–20, 22, 23, 25, 26
Immunohistochemistry
    combination with TUNEL assay 15–26
    γ-H2AX in xenografts and skin 262–267
Immunoperoxidase detection of BPDE-DNA adducts in
    frozen or paraffin tissue sections 272–273
    slides containing smeared cells 273
Immunophenotyping 99, 107
In situ
    assays 52–56
    cell death detection kits 10
    definition 50–52
    detection of apoptosis 3–12

detection of DNA strand breaks 50
labeling of DNA damage 39
TUNEL staining of cultured cells and tissue sections 3
In situ (DNA) end labeling .... 29–33, 38, 56, 59. See also In situ end-labeling (ISEL)
In situ end-labeling (ISEL) by
    Klenow polymerase 53–54, 56
    T7 DNA polymerase 37–46
    terminal transferase 49–50
    (see also In situ nick-end labeling; TdT-mediated dUTP nick-end labeling)
In situ hairpin-1 ligation assay .... 51. See also In situ ligation
In situ hybridization .... 115–131, 133, 292, 299. See also Fluorescence in situ hybridization
In situ ligation
    limitations 55–58
    perspectives of methodology 55–58
    review 49
    simplified procedure 65–74
    in tissue sections 65–74
In situ nick-end labeling (ISNEL) .... 37. See also TdT-mediated dUTP nick-end labeling
In situ nick translation (ISNT) assay 30, 38, 79, 292, 299
In situ oligonucleotide ligation technique (ISOL) 51, 52, 55–58, 85, 86
Internal phase of cell disassembly 80
Internucleosomal DNA
    fragmentation 66
    laddering 11
In vitro
    definition v
    markers of apoptotic cells 103–113
In vivo
    definition v
    detection 57
    labeling of DNA damage 39
    markers of apoptotic cells 91, 103–113
Ionizing radiation 121, 134, 142, 145, 231, 250
Ischemic injury 166
ISEL. See In situ end-labeling
ISL. See In situ ligation
ISL-TUNEL co-labeling 53
ISNEL. See In situ nick-end labeling
ISNT. See In situ nick translation
ISOL. See In situ oligonucleotide ligation technique
Isopore polycarbonate filter 149, 152–155, 157

J

Jurkat cells 110

K

Karyolytic cells 236, 244, 246,
Karyorrhectic cells 245–247
Kinase......................53–55, 80, 192, 193, 197, 200
Klenow
    enzyme 53
    fragment 37, 38
    polymerase... 53–54, 56
    polymerase-based labeling 53

L

Labeling
    of DNA breaks for apoptosis detection 38
    of DNA damage 39
    immunohistochemical for 8-OHG................ 26
    in situ....................................77–46, 54, 57, 69
    5'OH blunt ended DNA breaks 84–85
    techniques... 39, 51
    by topoisomerase 77–86
Laser scanning cytometer 92, 106
Laser scanning cytometry............................... 105
Lesion-specific endonucleases 116
Leukocytes 120, 298
Ligation-mediated PCR............................... 50, 59
Ligation probes..............................51–53, 56, 69, 72
Light Green cytoplasmic stain........................ 240
Light microscopic examination 8–9
Light microscopy.................. 6, 9, 10, 30, 179, 180, 243
Light source.. 254
Limitations of TUNEL................................ 55–58
Lipid peroxidation (LPO) 189, 190, 207
Liquid chromatography-tandem mass
    spectrometry 279–288
LMP agarose. See Low melting point (LMP) agarose
Localized UV-irradiation 149
Locally irradiated cells 149–159
Low-dose radiation exposure.......................... 250
Low-frequency ultrasound 183
Lowicryl... 33
Low melting point (LMP) agarose 115, 121, 124, 131, 135,
    136, 293, 294
LR Gold... 33
LR White... 33
Lymphocytes 207–211, 217–233,
    249–269
Lysosomal (phagocytic) nucleases in apoptosis........... 77–78

M

Margination.. 30
Marker for
    the detection of early phase apoptosis...... 103
    oxidative stress 280
β-Mercaptoethanol................ 44, 93, 131, 208, 212, 253

Method for
    DNA adduct detection 273
    identifying apoptotic cells............................ 4
    in situ TUNEL staining of cultured cells
        and tissue sections...3–12
    the localization of DNA strand breaks at the
        ultrastructural level 29–34
    local UV-irradiation through Isopore
        filters.. 157
    measuring chromosome breakage 218
    measuring DNA damage 115–131, 271
    measuring MNi in cultured human and/
        or mammalian cells 217
    TUNEL staining 4, 6, 8, 10–12, 176, 178, 182
Method for detection of
    εdA in human urine.................................... 198
    εdC in human urine.................................... 200
    DNA damage 4, 165–185
    DNA damage using mid-to high-frequency
        ultrasound 165–185
    PAH-DNA adducts................................. 272, 273
Micrococcal nuclease 191
Microgels 141, 299
Micronuclei/micronucleus (MNi) 217–233, 235–247
Microscopy..6, 8–10, 18, 30,
    31, 33, 50, 57, 105, 107–108, 116, 136, 167,
    172, 179, 180, 242, 243, 246, 250–252, 254,
    257, 259, 293–295, 298, 299
Microtomy.. 265
Mitochondrial potential 110
Mitogens 211, 212, 231, 232
Mitotic arrest/catastrophe 165, 166,
    170–172, 174, 180
MitoTracker Red............................105, 106, 110
Modified T7 DNA polymerase ....41, 43–45.
    See also Sequenase™
Monitoring
    cell death... 166
    cell death using high-frequency ultrasound 166
    of chronic degenerative diseases............................... 189
    human exposure 271
    of in vivo exposure to genotoxins............... 217
    levels of DNA damage................................ 271
    patient response to radiotherapy or
        drug treatment 255
    therapy response 185
    using cultured cells as a pollution probe 25
Mononuclear cells 207–214, 231, 250
Mounting medium 5, 6, 8, 9, 18, 20, 23, 154, 219, 239, 252,
    257, 266, 268, 293, 295
Multinucleated cells................................. 226–227
Multiparameter analysis of cells by flow-or image-
    cytometry.. 91
Multisubstrate deoxyribonucleoside kinase 193

N

Native T7 DNA polymerase.................................... 37–46.
  See also Unmodified T7 DNA polymerase
NBT/BCIP 68, 71
Necrosis (oncosis)...................................... 24, 25, 30, 53, 55,
  57–59, 96, 105, 166, 218, 228, 233
Necrotic cells 11, 15, 22–24, 30, 59,
  67, 111, 227, 228, 230, 233, 236
Negative control 10, 11, 25, 26,
  85, 94, 98, 203, 242, 243, 277
Neurons...................................... 16, 17, 19–20, 29, 31, 40, 56
Neutral electrophoresis. *See* Electrophoresis
Neutralization washing.. 137
Nick-end labeling.. 37.
  See also In situ nick-end labeling
Nicks... 34, 39–41, 53, 79
Nick translation .... 30, 38, 79, 123, 126–127, 292.
  See also In situ nick translation
NIH Image Program 34
NMP agarose. *See* Normal melting point agarose
Non-apoptotic cells 4, 11, 109, 110, 112, 113
Non-invasive
  assessment of cell death 182
  assessment of DNA oxidation 279
  detection of cell death.. 166
Non-specific background.........24, 26, 45, 46, 52, 72–74, 86
Normal melting point agarose 121, 135
Nuclear bud 217, 218,
  229–230, 236, 245, 246
Nuclear Division Index.................................... 226, 230, 231
Nuclear DNA..4, 133, 152, 203
Nuclear dyes .... 12.
  See also 4′,6-Diamidino-2-phenylindole
Nuclear pyknosis 30
Nuclear staining... 236, 267
Nucleic acids..37, 118, 280
Nucleoid..................................... 117, 120, 133, 137, 139,
  141, 142, 144, 145, 291, 292, 296, 299, 300
Nucleoplasmic bridge 218, 228, 229
Nucleosome 115, 150, 298
Nucleotide excision repair (NER) 149–150,
  152, 153, 158
Nucleotides............................30, 33, 38, 42, 44, 46, 49, 54,
  58, 65–67, 72, 73, 92, 93, 97, 118, 131, 195, 197

O

5′OH blunt ended DNA breaks 84–85
8-OHdG. *See* 8-Hydroxy-2′-deoxyguanosine
8-OHdG immunohistochemistry 15–26
5′OH DNA breaks 54, 77–86
3′OH DNA end 34, 38, 179
3′OH groups..53, 54, 74, 79
3′OH/5′PO$_4$ breaks 54
3′-OH termini of the DSBs 92, 93

Oligonucleosomal fragments... 30
Oligonucleotide ligation technique. *See In situ*
  oligonucleotide ligation technique (ISOL)
Oligonucleotide probes (oligoprobes)....................... 51, 52,
  65, 81–84, 117, 118, 135
Oligonucleotides 51, 52, 54, 58, 60,
  65, 66, 81, 83, 117, 118, 135
Oligoprobes. *See* Oligonucleotide probes
Oncosis..165, 166.
  See also Necrosis
Oscillating double-hairpin oligonucleotide probe 81
Oscillating nano-size device 80
Other forms of cell death 166, 180
3′-Overhang ligation 52
3′ Overhangs..30, 39, 40, 51–54,
  56, 59, 60, 66, 67, 79. *See also* Overhangs
  at 3′ or 5′ ends
5′ Overhangs...38–40, 45, 53.
  See also Overhangs at 3′ or 5′ ends
Overhangs at 3′ or 5′ ends.............................30, 38–40, 45,
  51–54, 56, 59, 60, 66, 67, 79
Oxidative cell damage 15–26
Oxidative DNA damage... 16
Oxidative stress..................... 16, 17, 19, 189, 207–214, 280
8-Oxo-7,8-dihydro-2′-deoxyguanosine
  (8-OxodG) 279–288

P

Pancentromeric probe... 138, 218
Paraffin-embedded tissue blocks 41, 43, 68, 83
Paraffin-embedded tissue sections................................. 21
Paraffin embedding 264–265
Paraffin sections..8, 262, 263, 275
PARP.. 56.
  See also Poly(ADP-ribose) polymerase
P1-artificial chromosomes (PACs)................................ 117
PBMCs...208–213, 250.
  See also Peripheral blood mononuclear cells
PBS. *See* Phosphate buffered saline
PCR-derived probes..................... 52, 53, 55, 56, 66, 67, 72
PCR fragments..51, 53, 65–74
PCR *in situ* ligation assay 52
PCR primers 67
PCR probe preparation 67–68
PEG-8000.. 68, 70, 73, 74
PEI...197, 201.
  See also Polyethyleneimine
Peptide nucleic acid (PNA) probes................................. 118
Peripheral blood lymphocytes.................209–212, 249–269
Peripheral blood mononuclear cells
  (PBMCs)... 208–213, 250
Peroxidase activity 10, 45, 201, 202, 276
Perspectives of the ISL methodology 58–59
Pfu polymerase 51, 66, 67, 70, 72
Pfu polymerase-synthesized fragment 66

Phagocytic nuclease 77–78
Phagocytic phase of apoptosis 77, 78
Phagocytizing cells 82
Phosphate buffered saline (PBS) 4–9, 17,
    19–23, 25, 41–46, 68, 83, 84, 93–95, 97, 98,
    106, 107, 109, 110, 121, 124, 125, 153–157,
    171, 193, 196, 201–203, 208, 212, 251–253,
    255–259, 265–267, 273–277, 293–295
Phosphatidylserine 105, 111
5' Phosphorylated double-strand DNA breaks 50, 53
(6-4) Photoproducts (6-4PP) 149
Phytohemagglutinin (PHA) 211, 212,
    219, 221–223, 232, 251, 256
5'PO$_4$ breaks .... 54, 80.
    See also 3'OH/5'PO$_4$ breaks
Pol I .... 37, 38.
    See also E. coli DNA polymerase I
Polyacrylamide gel electrophoresis.
    See Gel electrophoresis
Polycarbonate filter 149, 152–155, 157
Polycyclic aromatic hydrocarbon (PAH)
    diol epoxides 272
Polyethyleneimine (PEI) 192, 197, 201
Poly(ADP-ribose) polymerase (PARP) 56, 104
3'PO$_4$/5'OH DNA breaks 54
Positive tissue control 24
^{32}P-postlabeling 189–204
Preparation of
    antibody detection solutions 129
    hybridization mix 128
    immunoaffinity column 191, 196
    PCR-labeled blunt-ended probes for
        in situ ligation 70
    PCR-labeled in situ ligation probes containing
        3'A-overhangs 69
    PEI cellulose 197
    PHA solution
    smeared, frozen or paraffin samples 273
    suspension cells 8
    urine samples 198
Pretreatment 26, 34, 40, 53, 54, 86
Probes for FISH 116–118, 135, 145
Programmed cell death .... 3, 30, 227.
    See also Apoptosis
Proliferation marker Ki-67 56
Properties of
    the major executioner apoptotic nucleases 39
    native (unmodified) T7 DNA polymerase 44
        (see also Native T7 DNA polymerase)
    vaccinia DNA topoisomerase I 81
Propidium iodide (PI) 10–12, 92,
    93, 95–98, 105, 106, 108–111, 130, 179, 252,
    253, 255, 257, 259
Proteinase K 5, 9–11, 21, 25,
    46, 68, 70, 73, 74, 83–86, 191, 193, 194, 201,
    202, 273–275, 277

Protruding 5' DNA ends 38
p21WAF1/CIP1 49
Pyknotic cells 246

Q

QF 195
Quantitation of carcinogen binding 271
Quantitation of staining 273
Quantitative analysis 34, 105, 267
Quantitative ultrasound methods 166–169, 183
Quenching 73, 74

R

Radiation 121, 134,
    142, 150, 171, 173, 176, 184, 231–233, 250
Rat brain 40, 86, 170, 172, 175, 178
Rat dorsal root ganglia (DRG)
    neurons 16, 17, 19–20, 23
Rat heart 85
Rat liver 113, 201, 202, 204
Rat mammary gland 56
Rat pups 16
Rat thymus 46, 50, 82, 83
R&D Systems' Cell and Tissue
    Staining kit 24
Reactive oxygen species (ROS) 16, 29, 207
Reagents for preparation of smeared, frozen
    or paraffin samples 273
Reagents to induce apoptosis 172
Recognition sequence 81, 82
Repair auxiliary complexes 150–153
Repair proteins 149–159
Resin 31, 33, 281
Rhodamine fluorochrome 11

S

Satellite DNA 135, 138, 143, 144, 300
SCD. See Sperm Chromatin Dispersion (SCD) test
SCID mice 171
Selective detection of
    apoptotic cells in paraffin-embedded tissue sections . 65
    both 3'→5' and 5'→3' exonuclease
        activities in situ 53
    the phagocytic phase of apoptosis 77
Self-driven cell disassembly 80, 82
Sequenase™ 45
Simplified in situ ligation 65
Single-base
    3' overhangs 30, 39, 40, 53, 54, 59, 66, 67, 79
    5' overhangs 38–40, 45, 54
Single-cell gel electrophoresis (SCGE) 292
Single nucleotide overhangs .... 39.
    See also Overhangs at 3' or 5' ends
Single-strand and double-strand DNA breaks 53
Single strand breaks 120

Index

Single-stranded DNA regions 133
Single-stranded gaps 38–41
Slide preparation 124, 131,
    135–137, 222, 224–226, 233, 236–238,
    240–242, 247, 265
Small animal imaging 167, 169, 170, 183
Sodium borohydrate (NaBH$_4$) 26
Solid phase extraction 282–284
Specimen collection 264
Spectrum analysis 167–169, 176–178, 181, 182
Sperm
    analysis 295–296
    cell 292, 296–299
    DNA fragmentation 291–300
Sperm chromatin dispersion (SCD) test 291–300
Sperm chromatin structure assay (SCSA) 292
Spleen phosphodiesterase (SPD) 191
Splenocytes 249–269
Standard semen analysis 291
Streptavidin
    Alexa Fluor-488 conjugate
        alkaline phosphatase 68, 211, 213
        Cy3 .... 124, 136, 139
        Cy2 conjugate 42, 44
        fluorescein conjugate 10, 42
    gold-conjugated
        peroxidase conjugate 6, 42
        Texas Red conjugate 42
Suspension cell culture 12, 250
Suspension cells 6–8, 12, 250–252, 255
SV40 transformed human fibroblasts 153
Systemic lupus erythematosus (SLE) 85

T

Taq and Pfu
    labeling techniques 51
    polymerase-based labeling 72
    polymerase *in situ* ligation assay 51
Taq polymerase 60, 66, 67, 69
Target-specific control 24
TBE buffer 125, 135, 137
T4 DNA kinase section pretreatment 53
T4 DNA ligase 40, 50, 51,
    54, 56, 68, 70, 71, 73, 74, 79, 82
T4 DNA polymerase 38, 40
T5 DNA polymerase 38
T7 DNA polymerase
    modified 45
    native 38, 39, 45
T7 DNA polymerase-based *in situ* labeling 38
TdT-mediated dUTP nick-end labeling
    (TUNEL) 3–12, 15–26,
    30, 37, 38, 40, 49, 50, 53–59, 79, 92, 176, 178,
    182, 184, 292, 299
Telomere 116, 119, 120, 142, 218, 229, 235
Telomeric DNA 119
T4 endonuclease V 116, 122
Terminal deoxynucleotidyl transferase (TdT) 4–11,
    15, 18, 20, 22, 24, 30–34, 37, 40, 41, 49, 54, 79,
    92, 93, 95, 98, 99
3' Termini 37, 39
Texas Red-12-dUTP 72
T7 gene 5 protein 38
TH1 cytokines 208
TH2 cytokines 208
Thioredoxin 38
Tissue sections .... 4–9, 12, 17, 18, 20, 21.
    See also Paraffin-embedded tissue sections
T4 kinase. *See* T4-polynucleotide kinase
TLC plates 192, 197, 198, 201
Topoisomerase-based technique .... 55.
    See also Vaccinia topoisomerase I
Topotecan (TPT) 95, 96, 111, 263, 265, 268
T4-polynucleotide kinase 54, 192, 197
Treatment with proteinase K 10
Tumorigenesis 15
TUNEL. *See* TdT-mediated dUTP nick-end labeling

U

Ubiquitin E3 ligase 150–153, 158
U-932 cells 97
Ultrasensitive and specific techniques for detection
    of εdA and εdC 189
Ultrasound backscatter 172, 174, 180, 181, 183
Ultrasound imaging of DNA damage 165–166
Ultrasound imaging of mitotic
    arrest/catastrophe 174
Ultrasound spectrum analysis 167–169, 178, 181, 182
Unfixed cryostat sections 42
Unmodified T7 DNA polymerase .... 41, 43, 44.
    See also Native T7 DNA polymerase
Unwinding solution 134, 135, 137, 143, 298
Unwinding techniques 145
Urine
    collection 282, 283
    extracts 282, 285
    for 8-oxo-7,8-dihydro-2'-deoxyguanosine
        (8-oxodG) analysis 279
    sample 190, 198–200, 283, 285, 287
UV-damaged DNA binding protein 1
    (DDB1) 150, 151, 153, 158
UV-damaged DNA binding protein 2
    (DDB2) 150–153
UV-damaged DNA binding protein complex
    (UV-DDB) 150, 152, 153, 158
UV-induced DNA damage 149–159
UV-induced DNA lesions 150
UV-irradiation 153–155, 157, 158

V

Vaccinia topoisomerase-based labeling............................. 53
Vaccinia topoisomerase I 54, 55, 78, 81, 82, 84–86
Vacutainer blood tubes 220, 222
Vectashield antifade mounting medium 5, 8, 9
Viable cells..226–228, 230–232
V(D)J recombination... 59

W

Waste-control phase 80
Waste-management cells 80
Waste-management nucleases 78
Western blotting......................................253–254, 259–262

White blood cell............................... 189–205, 213, 272, 275
Wright's stain 293, 295, 296, 299

X

Xenografts.. 169, 170, 172, 175, 176, 178, 179, 183, 249–269
Xeroderma pigmentosum (XP) 149, 150
XP cell lines... 150
Xylene...8, 21, 43, 68, 70, 83, 84, 254, 265, 273, 275–278

Y

Yeast artificial chromosomes (YACs) 117, 135